Lecture Notes in Economics and Mathematical Systems

For information about Vols. 1–183, please contact your bookseller or Springer-Verlag

Vol. 184: R. E. Burkard and U. Derigs, Assignment and Matching Problems: Solution Methods with FORTRAN-Programs. VIII, 148 pages. 1980.

Vol. 185: C. C. von Weizsäcker, Barriers to Entry. VI, 220 pages. 1980.

Vol. 186: Ch.-L. Hwang and K. Yoon, Multiple Attribute Decision Making – Methods and Applications. A State-of-the-Art-Survey. XI, 259 pages. 1981.

Vol. 187: W. Hock, K. Schittkowski, Test Examples for Nonlinear Programming Codes. V. 178 pages. 1981.

Vol. 188: D. Bös, Economic Theory of Public Enterprise. VII, 142 pages. 1981.

Vol. 189: A. P. Lüthi, Messung wirtschaftlicher Ungleichheit. IX, 287 pages. 1981.

Vol. 190: J. N. Morse, Organizations: Multiple Agents with Multiple Criteria. Proceedings, 1980. VI, 509 pages. 1981.

Vol. 191: H. R. Sneessens, Theory and Estimation of Macroeconomic Rationing Models. VII, 138 pages. 1981.

Vol. 192: H. J. Bierens: Robust Methods and Asymptotic Theory in Nonlinear Econometrics. IX, 198 pages. 1981.

Vol. 193: J.K. Sengupta, Optimal Decisions under Uncertainty. VII, 156 pages. 1981.

Vol. 194: R. W. Shephard, Cost and Production Functions. XI, 104 pages. 1981.

Vol. 195: H. W. Ursprung, Die elementare Katastrophentheorie. Eine Darstellung aus der Sicht der Ökonomie. VII, 332 pages. 1982.

Vol. 196: M. Nermuth, Information Structures in Economics. VIII, 236 pages. 1982.

Vol. 197: Integer Programming and Related Areas. A Classified Bibliography. 1978 – 1981. Edited by R. von Randow. XIV, 338 pages. 1982.

Vol. 198: P. Zweifel, Ein ökonomisches Modell des Arztverhaltens. XIX, 392 Seiten. 1982.

Vol. 199: Evaluating Mathematical Programming Techniques. Proceedings, 1981. Edited by J.M. Mulvey. XI, 379 pages. 1982.

Vol. 200: The Resource Sector in an Open Economy. Edited by H. Siebert. IX, 161 pages. 1984.

Vol. 201: P. M. C. de Boer, Price Effects in Input-Output-Relations: A Theoretical and Empirical Study for the Netherlands 1949–1967. X, 140 pages. 1982.

Vol. 202: U. Witt, J. Perske, SMS – A Program Package for Simulation and Gaming of Stochastic Market Processes and Learning Behavior. VII, 266 pages. 1982.

Vol. 203: Compilation of Input-Output Tables. Proceedings, 1981. Edited by J. V. Skolka. VII, 307 pages. 1982.

Vol. 204: K.C. Mosler, Entscheidungsregeln bei Risiko: Multivariate stochastische Dominanz. VII, 172 Seiten. 1982.

Vol. 205: R. Ramanathan, Introduction to the Theory of Economic Growth. IX, 347 pages. 1982.

Vol. 206: M.H. Karwan, V. Lotfi, J. Telgen, and S. Zionts, Redundancy in Mathematical Programming. VII, 286 pages. 1983.

Vol. 207: Y. Fujimori, Modern Analysis of Value Theory. X, 165 pages. 1982.

Vol. 208: Econometric Decision Models. Proceedings, 1981. Edited by J. Gruber. VI, 364 pages. 1983.

Vol. 209: Essays and Surveys on Multiple Criteria Decision Making. Proceedings, 1982. Edited by P. Hansen. VII, 441 pages. 1983.

Vol. 210: Technology, Organization and Economic Structure. Edited by R. Sato and M.J. Beckmann. VIII, 195 pages. 1983.

Vol. 211: P. van den Heuvel, The Stability of a Macroeconomic System with Quantity Constraints. VII, 169 pages. 1983.

Vol. 212: R. Sato and T. Nôno, Invariance Principles and the Structure of Technology. V, 94 pages. 1983.

Vol. 213: Aspiration Levels in Bargaining and Economic Decision Making. Proceedings, 1982. Edited by R. Tietz. VIII, 406 pages. 1983.

Vol. 214: M. Faber, H. Niemes und G. Stephan, Entropie, Umweltschutz und Rohstoffverbrauch. IX, 181 Seiten. 1983.

Vol. 215: Semi-Infinite Programming and Applications. Proceedings, 1981. Edited by A.V. Fiacco and K.O. Kortanek. XI, 322 pages. 1983.

Vol. 216: H.H. Müller, Fiscal Policies in a General Equilibrium Model with Persistent Unemployment. VI, 92 pages. 1983.

Vol. 217: Ch. Grootaert, The Relation Between Final Demand and Income Distribution. XIV, 105 pages. 1983.

Vol. 218: P. van Loon, A Dynamic Theory of the Firm: Production, Finance and Investment. VII, 191 pages. 1983.

Vol. 219: E. van Damme, Refinements of the Nash Equilibrium Concept. VI, 151 pages. 1983.

Vol. 220: M. Aoki, Notes on Economic Time Series Analysis: System Theoretic Perspectives. IX, 249 pages. 1983.

Vol. 221: S. Nakamura, An Inter-Industry Translog Model of Prices and Technical Change for the West German Economy. XIV, 290 pages. 1984.

Vol. 222: P. Meier, Energy Systems Analysis for Developing Countries. VI, 344 pages. 1984.

Vol. 223: W. Trockel, Market Demand. VIII, 205 pages. 1984.

Vol. 224: M. Kiy, Ein disaggregiertes Prognosesystem für die Bundesrepublik Deutschland. XVIII, 276 Seiten. 1984.

Vol. 225: T.R. von Ungern-Sternberg, Zur Analyse von Märkten mit unvollständiger Nachfragerinformation. IX, 125 Seiten. 1984

Vol. 226: Selected Topics in Operations Research and Mathematical Economics. Proceedings, 1983. Edited by G. Hammer and D. Pallaschke. IX, 478 pages. 1984.

Vol. 227: Risk and Capital. Proceedings, 1983. Edited by G. Bamberg and K. Spremann. VII, 306 pages. 1984.

Vol. 228: Nonlinear Models of Fluctuating Growth. Proceedings, 1983. Edited by R.M. Goodwin, M. Krüger and A. Vercelli. XVII, 277 pages. 1984.

Vol. 229: Interactive Decision Analysis. Proceedings, 1983. Edited by M. Grauer and A.P. Wierzbicki. VIII, 269 pages. 1984.

Vol. 230: Macro-Economic Planning with Conflicting Goals. Proceedings, 1982. Edited by M. Despontin, P. Nijkamp and J. Spronk. VI, 297 pages. 1984.

Vol. 231: G.F. Newell, The M/M/∞ Service System with Ranked Servers in Heavy Traffic. XI, 126 pages. 1984.

Vol. 232: L. Bauwens, Bayesian Full Information Analysis of Simultaneous Equation Models Using Integration by Monte Carlo. VI, 114 pages. 1984.

Vol. 233: G. Wagenhals, The World Copper Market. XI, 190 pages. 1984.

Vol. 234: B.C. Eaves, A Course in Triangulations for Solving Equations with Deformations. III, 302 pages. 1984.

Vol. 235: Stochastic Models in Reliability Theory. Proceedings, 1984. Edited by S. Osaki and Y. Hatoyama. VII, 212 pages. 1984.

continuation on page 363

Lecture Notes in Economics and Mathematical Systems

Managing Editors: M. Beckmann and W. Krelle

345

A. Cambini E. Castagnoli L. Martein
P. Mazzoleni S. Schaible (Eds.)

Generalized Convexity and Fractional Programming with Economic Applications

Proceedings of the International Workshop
on "Generalized Concavity, Fractional
Programming and Economic Applications"
Held at the University of Pisa, Italy,
May 30 – June 1, 1988

Springer-Verlag
Berlin Heidelberg New York London
Paris Tokyo Hong Kong Barcelona

Managing Editors

Prof. Dr. M. Beckmann
Brown University
Providence, RI 02912, USA

Prof. Dr. W. Krelle
Institut für Gesellschafts- und Wirtschaftswissenschaften
der Universität Bonn
Adenauerallee 24–42, D-5300 Bonn, FRG

Editors

Prof. Dr. Alberto Cambini
Prof. Dr. Laura Martein
Department of Statistics and Applied Mathematics, University of Pisa
Via Ridolfi 10, I-56100 Pisa, Italy

Prof. Dr. Erio Castagnoli
Institute of Quantitative Methods, Bocconi University
Via Sarfatti 25, I-20136 Milano, Italy

Prof. Dr. Piera Mazzoleni
Institute of Mathematics, University of Verona
Via dell'Artigliere 10, I-37129 Verona, Italy

Prof. Dr. Siegfried Schaible
Graduate School of Management, University of California
Riverside, California 92521, USA

ISBN-13: 978-3-540-52673-5 e-ISBN-13: 978-3-642-46709-7
DOI: 10.1007/978-3-642-46709-7

2142/3140-543210 – Printed on acid-free paper

PREFACE

Generalizations of convex functions have been used in a variety of fields such as economics, business administration, engineering, statistics and applied sciences. In 1949 de Finetti introduced one of the fundamental of generalized convex functions characterized by convex level sets which are now known as quasiconvex functions. Since then numerous types of generalized convex functions have been defined in accordance with the need of particular applications. In each case such functions preserve some of the valuable properties of a convex function.

In addition to generalized convex functions this volume deals with fractional programs. These are constrained optimization problems which in the objective function involve one or several ratios. Such functions are often generalized convex. Fractional programs arise in management science, economics and numerical mathematics for example.

In order to promote the circulation and development of research in this field, an international workshop on "Generalized Concavity, Fractional Programming and Economic Applications" was held at the University of Pisa, Italy, May 30 - June 1, 1988. Following conferences on similar topics in Vancouver, Canada in 1980 and in Canton, USA in 1986, it was the first such conference organized in Europe. It brought together 70 scientists from 11 countries.

Organizers were Professor A. Cambini, University of Pisa, Professor E. Castagnoli, Bocconi University, Milano, Professor L. Martein, University of Pisa, Professor P. Mazzoleni, University of Verona and Professor S. Schaible, University of California, Riverside.

This volume constitutes the proceedings of the workshop. All papers at the conference were carefully refereed and a selection of them is published herewith. Thanks are due to all referees for their prompt help to produce a timely volume.

The contributions are organized in four sections: 1. Generalized Convexity, 2. Fractional Programming, 3. Duality and Conjugation and 4. Applications of Generalized Convexity in Management Science and Economics.

Each section contains an introductory article written upon invitation by one of the participants at the conference, which is intended to give deeper insight into the contributions of the section.

We are deeply indebted to CNR (Italian National Research Council) for generous financial support.

Thanks to Mrs. M. Taurisano for her cooperation in preparing the proceedings.

Finally we thank Dr. Müller, Springer Verlag for his outstanding help in the production of this volume.

CONTENTS

CHAPTER I

GENERALIZED CONVEXITY

INTRODUCTION TO GENERALIZED CONVEXITY

S. Schaible

The convexity of sets and the convexity and concavity of functions have been the object of many studies during the past one hundred years.
Early contributions to convex analysis were made by Holder [10], Jensen [11] and Minkowski [16], [17]. During the last forty years various generalizations of convex functions have been introduced [2]. We mention the early work by de Finetti [9], Fenchel [7], Arrow and Enthoven [1], Mangasarian [13], Ponstein [19] and Karamardian [12].
Usually such generalizations were motivated by a particular problem in economics, management science or optimization theory.
Whereas convex analysis has been presented in a number of research monographs and textbooks (see for example [20], [21], [26]), generalized convexity results appeared so far mainly in journals.
Apart from the present volume the proceedings of two earlier conferences on generalized convexity are now available [24], [25]. Also, the first textbook has just been published [2].
The purpose of this first paper at the International Workshop on Generalized Concavity, Fractional Programming and Economic Applications was to provide all participants with the basics of generalized convexity.
It was mainly aimed at those who in the past had not worked in this field. In addition it was meant to serve researchers already working in generalized convexity to gain a better overall perspective of where their particular results fit in.
It is hoped that the readers of this volume like the participants of the conference benefit from this introduction to generalized convexity. Given the purpose of the article, we concentrate on classical generalizations of convexity. For a comprehensive treatment of the subject the reader is referred to [2].

1. Convex functions reviewed

Before we turn to generalized convex functions we briefly review some basics of convex analysis in an n-dimensional Euclidean space $\mathbb{R}^n$.

Definition 1: A set C of IR^n is convex if for every $x^1, x^2 \in C$ and $0 < \lambda < 1$ we have

$$(1) \qquad \lambda x^1 + (1-\lambda) x^2 \in C.$$

Definition 2: A function f defined on a convex set C of IR^n is convex (cx) if for every $x^1, x^2 \in C$ and $0 < \lambda < 1$ we have

$$(2) \qquad f(\lambda x^1 + (1-\lambda) x^2) \le \lambda f(x^1) + (1-\lambda) f(x^2).$$

It is strictly convex (s.cx) if the inequality in (2) is strict for all $x^1 \neq x^2$.

Definition 3: A function f is (strictly) concave on C, if -f is (strictly) convex on C.
In economics and management science we encounter convex and concave functions quite frequently. A cost function is convex if marginal costs are increasing. An example is overtime production cost. Concave cost functions are met as well. They arise for instance if discounts are granted for larger quantities. Turning to functions in more than one variable, let us consider production or utility functions. Of particular interest here are upper level sets $\{x \in C \mid f(x) \ge \alpha\}$ for given levels $\alpha \in IR$. One can show that in case of a concave production or utility upper level sets are convex sets. This is a very desirable feature in economic analysis.
Convex functions also play an important role in optimization theory.

Definition 4: The optimization problem

$$(3) \qquad (P) \qquad \begin{array}{c} f(x) \to \text{Min} \\ \text{subject to} \\ g_i(x) \le b_i \quad i = 1, \dots, m \end{array}$$

is called a convex program if all functions f, g_i are convex on some set C of IR^n.
Convex programs have many useful properties summarized below [14].

Proposition 1: Given a convex program (P), we have:

- P_1: The set of feasible solutions is convex.
- P_2: Any local minimum is a global minimum.
- P_3: The Karush-Kuhn-Tucker optimality conditions are sufficient for a minimum.

P_4: Dual programs (D) of (P) and duality relations between (D) and (P) can be established.

P_5: A minimum of (P) is unique if f is strictly convex.

At this point the question arises whether there are nonconvex functions for which the same properties in optimization theory and in economic analysis are still true. It is not difficult to see that this is so. For instance, one may simply take a monotone transform h(f(x)) of a convex (concave) function f(x) where h is increasing. Then h(f(x)) is often not convex (not concave). However the upper level sets are still convex for concave f and properties P_1 ,........, P_5 still hold. Hence, there are many nonconvex functions which are as useful in economic analysis and optimization as convex functions are. This observation has motivated the study of various types of generalized convex functions.

2. Definitions and properties of generalized convex functions

We mentioned before that function with convex upper level sets are very important in production and utility theory. This property is used to define the first type of a generalized convex function.

Definition 5: A function f defined on a convex set C of IR^n is quasiconcave if its upper level sets

(4) $$\{x \in C \mid f(x) \geq \alpha\}$$

are convex for every $\alpha \in IR$.
f is quasiconvex (qcx) if -f is quasiconcave, i.e, if its lower level sets

(5) $$\{x \in C \mid f(x) \leq \alpha\}$$

are convex for every $\alpha \in IR$.
We can show the following:

Proposition 2: A function f is quasiconvex on a convex set C of IR^n if and only if for every $x^1, x^2 \in C$ and $0 < \lambda < 1$

(6) $$f(\lambda x^1 + (1-\lambda) x^2 \leq \mathrm{Max}\,(f(x^1), f(x^2))$$

If $C \subseteq IR$, then quasiconvex functions are fully characterized as being either monotone increasing, monotone decreasing or first decreasing and then increasing [26]. In view of Prop. 2 a function on a n-dimensional convex set C is quasiconvex if and only if it is quasiconvex on each line segment of C, i.e. if the restriction of f to any line segment of C is either increasing, decreasing or first decreasing and then increasing.
For $C \subseteq IR^2$ we mention the following examples of nonconvex (and nonconcave) functions:

$f(x) = x_1/x_2$, $C = \{x \in IR^2 \mid x_2 > 0\}$ is quasiconvex and quasiconcave on C.
$f(x) = x_1 x_2$, $C = \{x \in IR^2 \mid x_1 \geq 0, x_2 \geq 0\}$ is quasiconcave (but not quasiconvex) on C.

From now on we will restrict ourselves to quasiconvex functions. Similar statements can be proved for quasiconcave functions. In order to obtain some of the properties of convex functions in Section 1 we have to isolate certain subclasses of quasiconvex functions. We introduce

Definition 6: A function f defined on a convex set C of IR^n is strictly quasiconvex (s.qcx) if for every $x^1, x^2 \in C$, $x^1 \neq x^2$ and $0 < \lambda < 1$

(7) $$f(\lambda x^1 + (1-\lambda) x^2) < \mathrm{Max}\,(f(x^1), f(x^2)).$$

f is semistrictly quasiconvex (ss.qcx) if $f(x^1) \neq f(x^2)$ instead of $x^1 \neq x^2$ is assumed above.

A function is quasiconvex, but not semistrictly quasiconvex if there exists a line segment in C over which f is constant, but there is still a point on the extension of that line segment in C with a smaller value of f. Furthermore, a (nonconstant) function is semistrictly quasiconvex, but not strictly quasiconvex, if f is constant only over some line segment at the minimal level of f.
Comparing Definition 2, Proposition 2 and Definition 6 one can show that the following relationships exist for lower semi-continuous functions:

(8)

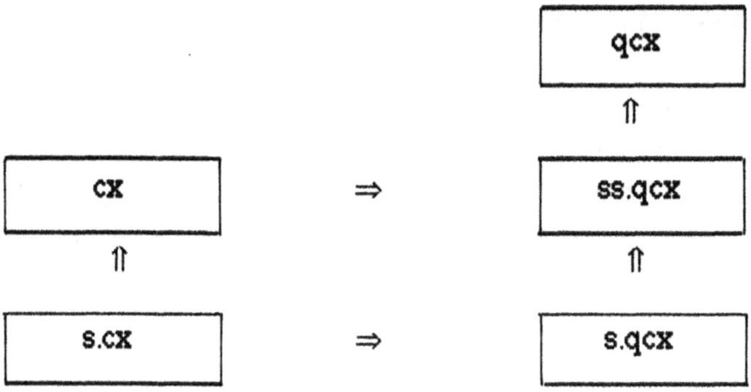

All of these inclusions are strict.

Two more special classes of quasiconvex functions are needed:

Definition 7: A differentiable function f on an open convex set C of $\mathbb{R}^n$ is <u>pseudoconvex</u> (pcx) if for every $x^1, x^2 \in C$

(9) $$(x^1 - x^2)^T \nabla f(x^2) \geq 0 \text{ implies } f(x^1) \geq f(x^2).$$

f is <u>strictly pseudoconvex</u> (s.pcx) if the second inequality in(9) is strict for $x^1 \neq x^2$

For a pseudoconvex function a stationary point ($\nabla f(\bar{x}) = 0$) is a global minimum as (9) shows. A function is pseudoconvex, but not strictly pseudocenvex if f is constant over a line segment at the minimal level of f.

By including pseudoconvex functions we can extend the diagram in (8) (for differentiable functions) as follows:

(10)

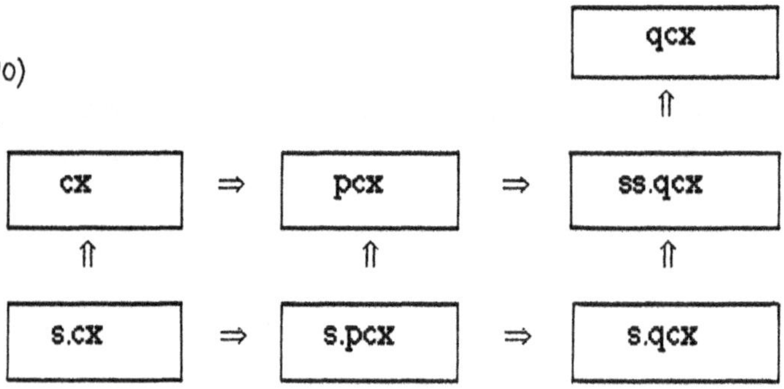

Let us now introduce

Definition 8: The optimization problem

$$(11) \qquad (P) \qquad \begin{array}{c} f(x) \rightarrow \text{Min} \\ \text{subject to} \\ g_i(x) \leq b_i \; i = 1, \ldots, m \end{array}$$

is a quasiconvex program, if all functions f, g_i are quasiconvex on some convex set C of IR^n.
As a generalization of Proposition 1 we obtain [2]:

Proposition 3: Given a quasiconvex program (P), we have:

P_1: The set of feasible solutions is convex.
P_2: Any local minimum is a global minimum, if f is semistrictly quasiconvex.
P_3: The Karush-Kuhn-Tucker optimality conditions are sufficient for a minimum
P_4: Dual programs (D) of (P) and duality relations between (D) and (P) can be established [3], [4], [18].
P_5: A minimum of (P) is unique if f is strictly quasiconvex.

3. Applications of generalized convex programs.

One source of nonconvex, but quasiconvex programs are convex-concave fractional programs [2], [23]. Here the objective function is the ratio of a nonnegative convex and a positive concave function or it is composed of several such ratios.
Single-ratio convex-concave programs arise naturally when rates of economic and/or technical terms are to be optimized. Examples are productivity, return on investment, return/risk, cost/time. Furthermore, single-ratio fractional programs are met as subproblems in large-scale programming or as surrogate problems in stochastic programming. They also occur in game-theory and information theory.
Multi-ratio fractional programs are quasiconvex if the objective function is the maximum of several convex-concave ratios. Such Min-Max fractional programs arise in equilibrium theory in economics, in multi-objective fractional programming, in goal programming with rates as well as in discrete rational approximation in numerical mathematics.
Apart from fractional programs in management science optimization problems in economic analysis are sometimes quasiconvex. A detailed discussion is given in

[2]. Among others the consumer's utility maximization problem is analyzed there. Consider

$$(12) \qquad G(p) = \text{Max}\,\{ F(x) \mid x \geq 0,\ p^T x \leq 1\}$$

where F is the direct utility function, x the consumption vector, p a normalized price vector and $G(p)$ the indirect utility function defined as the optimal value function through (12). The following economic duality theorem can be shown:

Proposition 4. If F is continuous, increasing and quasiconcave in $x \geq 0$, then G is continuous, decreasing and quasiconvex in $p \geq 0$. Furthermore,

$$(13) \qquad F^*(x) = \text{Min}\,\{ G(p) \mid p \geq 0,\ p^T x \leq 1\}$$

coincides with $F(x)$.

Note that both (12) and (13) are quasiconvex programs. A similar relationship can be established between the producer's cost function and production function. Again two quasiconvex programs arise. We mention that additional properties for these pairs of dual functions can be derived if more restrictive types of quasiconvex functions discussed in Section 2 are used [2].

4. Generalized convexity of special families of functions

The definitions of various types of generalized convexity are hard to verify for a given function. Simpler characterizations are needed. These can be obtained by restricting oneself to special families of functions, such as

a) composite functions,

b) products and ratios,

c) quadratic and C^2 - functions.

Many of the earlier criteria in a), b) and c) were derived in [22]. For functions in c) see also [8].

a) Composite functions

We have

Proposition 5: If f is (ss.) qcx on a convex set C of IR^n and G is (strictly) increasing on the range of f, then $G(f)$ is (ss.) qcx on C.

Proposition 6: a) If $f_1, \ldots\ldots, f_m$ are cx on a convex set of IR^n, G is increasing and (ss.) qcx on the convex hull of the range of $(f_1, \ldots, f_m)$, then $G(f_1, \ldots\ldots, f_m)$ is (ss.) qcx on C.
b) If all f_i are affine, then G does not need to be increasing.

Proposition 7: If $f_i, i \in I$, are (ss.) qcx on a convex set C of IR^n, then $\underset{i \in I}{\mathrm{Max}}\, f_i(x)$

is (ss.) qcx on C.

Similar results can be derived for (strictly) pseudoconvex functions. In addition to being useful by themselves, Propositions 5, 6 and 7 help us to derive criteria for other classes of functions, namely products and ratios as well as quadratic functions.

b) <u>Products and ratios</u>
With help of Proposition 6a) we can show

Proposition 8: Given a positive convex function f_o and positive concave functions $f_1, \ldots, f_k$ on a convex set C of IR^n and nonnegative real numbers ai such that

$a_o \geq \sum_{i=1}^{k} a_i$. Then

$$(14) \qquad F(x) = \prod_{i=1}^{k} (f_i(x)^{a_i} / (f_o(x))^{a_o}$$

is semistrictly quasiconcave on C.
Applying Proposition 5 with $G(y) = -y^{-1}$ to $F(x)$ in (14) we obtain

Proposition 9: The function $1/F(x)$ in (14) is ss.qcx on C.

In fact, one can show that $1/F(x)$ is even convex if $a_o \geq 1 + \sum_{i=1}^{k} a_i$.

In the following we mention some special cases of the results above.

$F(x) = \prod_{i=1}^{k} (f_i(x))^{a_i}$ is semistrictly quasiconcave and $1/F(x)$ is convex if f_i are positive concave and $a_i \geq 0$.

$F(x) = f(x)/g(x)$ is semistrictly quasiconcave (semistrictly quasiconvex) if f is nonnegative concave (convex) and g is positive convex (concave). This function arises in single-ratio fractional programming as the objective function.

In multi-ratio fractional programming the objective function $\max_{i \in I} f_i(x)/g_i(x)$ is semistrictly quasiconvex for convex-concave ratios in light of Proposition 7.

$F(x) = f(x)/(x^TAx + \alpha)$ is semistrictly quasiconcave and $1/F(x)$ convex if f is positive concave, A a positive definite matrix and $\alpha > 0$. The convexity of $1/F(x)$ follows from the fact that $(x^TAx + \alpha)^{1/2}$ is still convex.

c) Quadratic functions and C^2 - functions

Quadratic functions $Q(x) = 1/2\ x^TAx + b^Tx$ can be reduced to a normal form without mixed terms of variables with the help of an affine transformation. An analysis of the generalized convexity properties of normal forms together with Proposition 6 b) yelds a full characterization of quadratic functions. In particular, the analysis in [22] shows that all quasiconvex quadratic functions are convex transformable, i.e., there exists an increasing function G such that $G(Q(x))$ is convex. From this we see

Proposition 10: A (nonconvex) quadratic function $Q(x)$ is pseudoconvex on an open convex set C of $\mathbb{R}^n$ if and only if

$$\nabla^2 Q(x) + r(x)\, \nabla Q(x)\, \nabla Q(x)^T$$

is positive semidefinite where

$$r(x) = 1/2\ [\delta - Q(x)]^{-1}$$

for some $\delta > Q(x)$.

In particular, a quadratic form $Q(x) = 1/2\ x^TAx$ is pseudoconvex if and only if either A is positive semidefinite or $A - (x^TAx)^{-1} (Ax)(Ax)^T$ is positive semidefinite.

From Proposition 10 it follows:

Proposition 11: A (nonconvex) quadratic function $Q(x)$ is pseudoconvex on an open convex set C of IR^n if and only if all principal minors of the bordered Hessean

$$(16) \qquad B(x) = \begin{pmatrix} \nabla^2 Q(x) & Q(x) \\ Q(x)^T & 0 \end{pmatrix}$$

are nonpositive, and if a minor is zero, then the associated principal minor of the Hessean $\nabla^2 Q(x) = A$ is nonnegative.

Furthermore, the following two characterizations of pseudoconvex quadratic functions can be derived.

Proposition 12: A nonconvex quadratic function $Q(x)$ is pseudoconvex on an open convex set C of IR^n, if and only if $B(x)$ in (16) has exactly one (simple) negative eigenvalue for all x in C.

Proposition 13: A quadratic function $Q(x)$ is pseudoconvex on an open convex set C of IR^n, if and only if

$$(17) \qquad z \in IR^n \, , \, z^T \nabla Q(x) = 0 \text{ implies } z^T \nabla^2 Q(x) z \geq 0$$

for all x in C.

Similar results as in Propositions 10, 11, 12, and 13 can be obtained for strictly pseudoconvex functions $Q(x)$. They are easier to verify than the above criteria. We mention that a quadratic function $Q(x)$ is quasiconvex on a convex set C of IR^n with nonempty interior C_0 if and only if $Q(x)$ is pseudoconvex on C_0. Therefore, it is sufficient to characterize pseudoconvex functions $Q(x)$ as done in Propositions 10, 11, 12 and 13 above.
In case of quadratic functions in nonnegative variables the criteria above reduce to finite criteria in the parameters A, b of $Q(x) = 1/2\, x^TAx + b^Tx$ [2], [15].
Moving to arbitrary C^2 - functions we find that neither of the conditions in Propositions 10, 11, 12 and 13 are both necessary and sufficient. Whereas, the conditions in Propositions 10 and 11 are sufficient, but not necessary, the reverse is true for the conditions in Propositions 12 and 13.
However, under somewhat more restrictive assumptions the necessary conditions in Proposition 13 become sufficient [5], [6].

Proposition 14: Let f be a C^2 - function on an open convex set C of IR^n such that $\nabla f(x) \neq 0$ for all x in C. Then f is pseudoconvex on C if and only if

(18) $$z \in IR^n, \ z^T \nabla f(x) = 0 \text{ implies } z^T \nabla^2 f(x) z \geq 0$$

for all x in C.

One can also show that f is strictly pseudoconvex on C if the inequality in (18) is strict for $z \neq 0$.

This concludes our introduction to some basic results on generalized convex functions. For a comprehensive treatment of these and other types of generalized convex functions the reader is referred to [2].

References

[1] **Arrow, K.J. and Enthoven A.C.**, Quasi-concave Programming, Econometrica 29,1961, pp. 779-800.

[2] **Avriel, M., Diewert, W.E., Schaible, S. and Zang I.**, Generalized Concavity, Plenum Publishing Corporation, New York and London 1988.

[3] **Crouzeix, J.P.**, Contributions à l'Etude des Fonctions Quasiconvexes, These de Doctorat, Universitè de Clermont-Ferrand II, France 1977.

[4] **Crouzeix, J.P.**, A Duality Framework in Quasiconvex Programming, in: Generalized Concavity in Optimization and Economics, edited by S.Schaible and W.T.Ziemba, Academic Press, New York, 1981.

[5] **Crouzeix, J.P. and Ferland J.A.**, Criteria for Quasi-convexity and Pseudo-convexity: Relationships and Numerical Analysis, Mathematical Programming 23, 1982, pp.193-205.

[6] **Diewert, W.E., Avriel, M. and Zang I.**, Nine Kinds of Quasiconcavity and Concavity, J. of Economic Theory 25, 1981, pp.397-420.

[7] **Fenchel, W.**, Convex Cones, Sets and Functions, Mimeographed Lecture Notes, Princeton, New Jersey, 1951.

[8] **Ferland, J.A.**, Quasi-convex and Pseudo-convex Functions on Solid Convex Sets, Ph.D. Dissertation, Stanford University, Stanford, USA, 1971.

[9] **de Finetti, B.**, Sulle Stratificazioni Convesse, Annali di Matematica Pura ed Applicata (4) 30, 1949, pp.173-183.

[10] **Holder, O.**, Uber einen Mittlwertsatz, Nachr. Ges. Wiss. Goettingen, 1889, pp.38-47.

/11/ **Jensen, J.L.W.V.**, Sur les Fonctions Convexes et les Inègalities Entre les Valeurs Moyeumes, Acta. Math., 30,1906, pp.175-193.

/12/ **Karamardian, S.**, Duality in Mathematical Programming, J. Math. Anal. Appl. 20, 1967, pp.344-358.

/13/ **Mangasarian, O.L.**, Pseudo-convex Functions, J. SIAM Control Ser. A3, 1965, pp.281-290.

/14/ **Mangasarian, O.L.**, Nonlinear Programming, McGraw-Hill, New York, 1969.

/15/ **Martos, B.**, Nonlinear Programming: Theory and Methods, North-Holland, Amsterdam, 1975.

/16/ **Minkowski, H.**, Geometrie der Zahlen, Teubner, Leipzig, 1910.

/17/ **Minkowski, H.**, Theorie der konvexen Korper, insbesondere Begrundung ihres Oberflachenbegriffs, Gesammelte Abhandlungen II, Teubner, Leipzig, 1911.

/18/ **Passy, U. and Prisman E.Z.**, A convex-like Duality Scheme for Quasi-convex Programs, Mathematical Programming 32, 1985, pp.278-300.

/19/ **Ponstein, J.**, Seven Kinds of Convexity, SIAM Review 9, 1967, pp.115-119.

/20/ **Roberts, A.W. and Varberg D.E.**, Convex Functions, Academic Press, Mew York, 1973.

/21/ **Rockafellar, R.T.**, Convex Analysis, Princeton University Press, Princeton, New Jersey, 1970.

/22/ **Schaible S.**, Beitrage zur quasikonvexen Programmierung, Doktorarbeit, Mathematisch-Naturwissenschaftliche Fakultat, Universitat Koln, 1971.

/23/ **Schaible S.**, Analyse and Anwendungen von Quotientenprogrammen, Hain-Verlag, Meisenheim, 1978.

/24/ **Schaible S., and Ziemba W.T.** (eds.), Generalized Concavity in Optimization and Economics, Academic Press, New York, 1981.

/25/ **Singh, C.** (ed.), Continuous-time Programming, Fractional Programming and Multi-objective Programming, Proceedings of the Conference at St.Lawrence University, Canton, New York, July 29-Aug 1,1986, J. of Information and Optimization Sciences 10, No.1, January 1989.

/26/ **Stoer, J. and Witzgall C.**, Convexity and Optimization in Finite Dimensions I, Springer-Verlag, Berlin, 1970.

STRUCTURAL DEVELOPMENTS OF CONCAVITY PROPERTIES

P. Mazzoleni

The papers presented at the Pisa meeting cover a wide range of subjects in optimization and economics, but they put particular emphasis on the structural properties of concavity Besides going into a guided description on the results presented in the three papers of this section, we would like to express the feeling of continued lively interest in our field of research.

Of course our introduction has no pretence of completeness. Its aims is only to encourage an opening of new lines of research.

The literature provides several characterizations and properties for each class of generalized concave function.

Here we confine our attention to the most fruitful ones as regards to the applications.

1. A unifying approach

One of the most interesting aspects of nonlinear programming problems for economic applications is the sensitivity analysis with respect to the parameters and especially to the ones in the constraints.

In this field concavity properties as well as continuity and differentiability conditions have attracted much interest since they form a theoretical basis for sensitivity and stability analysis.

Consider a general parametric optimization problem

(1) $$\mathcal{P}(u) : \max \{ f(x,u) : x \in X(u) \}$$

where $f: IR^n \times IR^k \to IR$ and $X(u)$ is a convex multifunction.

A very synthetic way to study the concavity properties of f is given in Kyparisis and Fiacco (1987).

Definition 1 - A function $\Phi: IR^n \to \overline{IR}$, $\overline{IR} = IR \cup \{+\infty\}$, is called G-concave on a convex set C, if for any $z^0, z^1 \in C$, $z^0 \neq z^1$ and $\alpha \in I$ the following inequality holds:

(2) $$\Phi(\alpha z^0+\bar{\alpha}z^1) \geq G(\Phi(z^0),\ \Phi(z^1),\ \|z^0-z^1\|,\ \alpha)$$

where $G(r^0, r^1, \delta, \alpha)$: $\bar{\mathbb{R}}^2 \times \mathbb{R}^2_+ \to \bar{\mathbb{R}}^2$ is continuous and nondecreasing in (r^0, r^1), $\|\cdot\|$ is an arbitrary norm on $\mathbb{R}^n$, and $I := [0,1]$ the unit interval.
Define the optimal value function

(3) $$f^*(u) := \begin{cases} \sup\limits_x \{f(x,u): x \in X(u)\} & \text{if } X(u) \neq \emptyset \\ -\infty & \text{if } X(u) = \emptyset \end{cases}$$

Then f^* is G-concave:

Theorem 1 - If f is G-concave on $\mathbb{R}^n \times S$ with $\|\cdot\|$ restricted to u and $X(u)$ is a convex multifunction on a convex set $S \subset \mathbb{R}^k$, then f^* is G-concave on S.
Definition 1 certainly includes a great variety of concavity properties which have been proposed in the literature as the following table 1 shows, but it does not contain any indicator linked with the different functions.

tab. 1

concavity	$\alpha r^0+\bar{\alpha}r^1$
uniform concavity	$\alpha r^0+\bar{\alpha}r^1+\alpha\bar{\alpha}d(\delta)$
quasiconcavity	$\min\{r^0,r^1\}$
uniform quasiconcavity	$\min\{r^0,r^1\}+\alpha\bar{\alpha}d(\delta)$
paraconcavity	$\alpha r^0+\bar{\alpha}r^1-\alpha\bar{\alpha}d(\delta)$
paraquasiconcavity	$\min\{r^0,r^1\}-\alpha\bar{\alpha}d(\delta)$

where d: $\mathbb{R}_+ \to \mathbb{R}_+$ is such that $d(t)>0$ for $t>0$, $d(0)=0$ and d is non decreasing: for instance $d(\delta)=a\delta^2$ gives strong concavity.
If the inequality 2 strictly holds, pseudoconcavity properties can also be included; moreover, quasimonotonicity of the optimal value function can be proved.
In the first meeting held in Vancouver (Canada) in 1980, Oettli studied the parametric problem under the assumption that X(u) is quasiconvex, but confined himself to the objective function f(x).
Kyparisis and Fiacco generalized the result to f(x,u).
If we refer to the (α,λ)-concavity introduced by Castagnoli and Mazzoleni (1986), we can link the degree of concavity of the optimal value function to the corresponding degree of the objective function and of the constraints.

Definition 2 (Castagnoli and Mazzoleni (1986)) A function f: $C \to \mathbb{R}$, defined on a convex set $C \subset \mathbb{R}^n$, is said to be (α,λ)-concave, or simply λ-concave, if

(4) $$f(x^0)\le f(x^1) \Rightarrow f(\alpha x^0+\bar{\alpha}x^1) \ge \lambda f(x^0)+\bar{\lambda}f(x^1)\,,\ \forall\, x^0, x^1 \in C,\ \forall \alpha \in I$$

where $\lambda=\lambda(\alpha; x^0, x^1) \in I$.

(α,λ)-concave functions are partially ordered by their λ : concavity weakens as λ increases. For instance classical concavity corresponds to $\lambda=\alpha$, pseudoconcavity to $\bar{\lambda}=\bar{\alpha}\, b(x^0,x^1)$, with $0<b(x^0,x^1)<1$, quasiconcavity to $\lambda=1$.
When f is a single-valued function, an interval of values $[u^0,u^1]$, corresponding to any pair x^0, x^1 in the domain, is described by the point $u^\alpha=\alpha u^0+\bar{\alpha}u^1$ for α increasing from 0 to 1.
Consider the multifunction $\Gamma: C\to \mathcal{P}(\mathrm{IR})$ defined on a convex set $C \subset \mathrm{IR}^n$ and taking values on the power set $\mathcal{P}$ of IR and define $\Gamma_\alpha(x)= \cup\, \{\, \Gamma(x^\beta): \beta \in [0,\alpha]\, \}$ for any $x^\beta = \beta x^0+ \bar{\beta}x^1$.
Then we can weaken the convexity of Γ.

Definition 3 - A multifunction $\Gamma:\ C\to \mathcal{P}(\mathrm{IR})$ is called $(\alpha;\lambda,\mu)$-convex if for any $x^0, x^1 \in C$ and $\alpha \in I$ there exist functions $\lambda=\lambda(\alpha)$, $\mu=\mu(\alpha)$, $0\le\lambda, \mu\le 1$ such that

(5) $$\Gamma_\alpha^{\lambda,\mu}(x) = \cup\ \{\, \Gamma(x^\beta) : \beta \in [\lambda(\alpha), \mu(\alpha)]\, \}$$

We obtain convex multifunctions for $\lambda(\alpha)=\mu(\alpha)=\alpha$; quasiconvex multifunctions for $\lambda(\alpha)=0$, $\mu(\alpha)=1$ and a family of intermediate cases.
If we consider the perturbed optimization problem

$$\pi(u) := \sup\, \{\, f(x) : x \in C,\ u \in \Gamma(x)\, \}$$

we get the optimality conditions:

Theorem 2 - Let $f : C\to \mathrm{IR}$ be an (α,μ)-concave function and $\Gamma : C\to \mathcal{P}(\mathrm{IR})$ an $(\alpha;\lambda,\mu)$-convex multifunction.
Then $\pi(\cdot)$ is $(\alpha;\lambda,\mu)$-concave.

The same results hold if we combine the (α,λ)-convexity and the (G,h)-convexity introduced by Ben-Tal (1977).
It is now easy to apply the previous result to the parametric problem 1 if we merge the vector (x,u) into a new vector y by suitably modifying the multifunction Γ so that the parameter appears only in the constraint set

$$\pi(y) := \sup_{(x,u)} \{\, f(x,y) : (x,u) \in \bar{\Gamma}(y)\, \}$$

$$\bar{\Gamma}(y) := \{\, (x,u) : x \in \Gamma(u),\ y^T u\ge 0\, \}$$

2. Monotone and concave functions

The interest in monotone functions in economics and in mathematical programming is emphasized by the large number of applications.

Consider for instance the monotonicity properties of cost and production functions, of utility functions and the increase in the perturbation function

$$h(u) = \sup \{ f(x) : x \in C, g_i(x) - u_i \geq 0, i = 1, \ldots\ldots, m \}.$$

It is therefore worth recognizing the degree of monotonicity for concave functions.

The class of quasiconcave functions in only one variable consists of monotone functions and functions which are first nondecreasing, then nonincreasing.

The definition of a corresponding line function

$$\varphi_{x,y}(t) = f(x+ty) \qquad t \in IR$$

for any pair $x,y \in C$ allows us to state the corresponding property for functions of n variables, $f: C \rightarrow IR$, defined on a convex subset $C \subset IR^n$ (Martos (1975)).

Indeed the property can be reread as a local directional monotonicity according to the sign of $z-x$ in $\alpha x + \overline{\alpha} z = x + \overline{\alpha}(z-x) = x + ty$.

Chabrillac and Crouzeix (1987) express $z-x$ as a linear combination of a basis $v_1, \ldots\ldots\ldots, v_n$, $t(z-x) = \sum r_i v_i$, and inequality

$$f(x) \leq f(x+ty)$$

becomes

(6) $$f(x) \leq f(x + \sum t_i v_i) :$$

they assume $t_i \geq 0$ for all i, because they consider only one kind of monotonicity.

Therefore the monotonicity property of the line function $\varphi_{x,y}(t)$ is easily generalized to a local monotonicity at each point $a \in C$ such that $\overline{f}(a) < \overline{k}$ where $\overline{f}$ is defined by

$$\overline{f} = \sup \{\lambda / x : x \in \mathrm{cl}\ \mathrm{lev}_{\geq k} f\}$$

and $\overline{k}$ is the (possibly infinite) value such that

$$\dim(\mathrm{lev}_{\geq k} f) \begin{cases} = n & \text{if } k < \overline{k} \\ < n & \text{if } k > \overline{k} \end{cases}$$

3. Differentiability properties

As pointed out in the proceedings of the first meeting held in Vancouver, 1980, one-sided derivatives do not necessarily exist for the various types of generalized concave functions. Diewert (1981) provides several characterizations for subclasses of nondifferentiable generalized concave functions in terms of Dini derivates.

A set of necessary and sufficient conditions is developed for a nonlinear program in which the objective function and the constraints are, respectively, nondifferentiable pseudo-concave and quasiconcave.

It is worth strengthening the role of generalized concavity in order to prove the sufficiency of generalized Kuhn-Tucker conditions for a nonsmooth nonlinear programming problem, even if this is not the place to give further details.

Indeed two different fields of research are now particularly active. The first is trying to develop more and more general definitions of a derivative still preserving the convexity property as a function of the direction, in order to obtain meaningful results for nonsmooth programming. Two different lines of research can be recognized in the attempt of generalizing these properties.

A weaker set of optimality conditions is presented by Ward by requiring only that the directional derivative involved is sublinear and majorizes either the upper or the lower Dini derivate.

Therefore a procedure analogous to the passage from the directional derivative to the subdifferential is developed for the Dini derivate.

Moreover the definition of an upper concave approximate is also generalized from a continuous superlinear function to the upper semicontinuous case.

The second line of research studies the differentiability properties of generalized concave functions and first and second order characterizations of the various extensions.

Going back to Martos (1975), quasiconcave functions are almost everywhere continuous.

More recently, Chabrillac and Crouzeix (1987) prove that they are also differentiable almost everywhere.

A particular definition of derivative following the degree of concavity was analyzed by Castagnoli and Mazzoleni at the Canton meeting, 1986.

Indeed if we pass from the classical concavity to the (α,λ)-concavity the usual inequality

$$f(y)-f(x) \leq \nabla f(x)(y-x)$$

for concave functions holds in a modified form for (α,λ)-concave functions as well.

Alternatively, Castagnoli and Mazzoleni show in the paper presented at this meeting that it is possible to preserve the usual form of the inequality by modifying the definition of a derivative.

Moreover they try to unify the great variety of second order conditions given in the literature for pseudo and quasiconcavity in order to state a flexible condition giving a suitable degree of curvature to the several classes of generalized concave functions.

4. Further weakening of concavity and quasiconcavity

There has been a trend to generalize and refine convexity of functions and sets and to apply them in mathematical programming and in economic problems.
A first order weakening of convexity, the so-called invexity firstly introduced by Hanson (1981) and subsequently extended by Craven (1981) and Martin (1985), allows one to state sufficiency for the Kuhn-Tucker necessary conditions for some nonconvex constrained optimization problems.
Mangasarian (1975) first formulated a second order dual to a nonlinear problem, thus giving the idea of a second order concavity property.
Indeed Bector and Bector (1987) analyze so-called generalized second order pseudo and quasiconcave functions which are based on a perturbation of the first order classical characterization in terms of the Hessian matrix and develop a suitable theory of duality.
But what is even more interesting is the attempt to generalize quasiconcavity itself, and here we recognize three different lines of research.
The first deals with local notions, the second treats equality constraints and the third considers classes of non concave functions and looks for the level of properties still holding from the classic theory.
Let us briefly review the first two aspects.
Starting from the local definition of star-shaped set Kaul and Kaur (1982) define semilocal concavity.

Definition 4 - A numerical function f defined on a set $X \subset IR^n$ is said to be semilocally concave at $x^* \in X$ if X is locally star-shaped at x^* and if, corresponding to x^* and each $x \in X$, there exists a positive number $\beta(x^*,x) \leq a(x^*,x) \leq 1$ such that

(7) $$f(\alpha x+\bar{\alpha}x^*) \geq \alpha f(x)+\bar{\alpha}f(x^*) \qquad 0 < \alpha < \beta(x^*,x)$$

Clearly, if $\beta(x^*,x)=a(x^*,x)=1$ we turn back to the classical concavity.
An analogous generalization can be given for quasiconcavity.

Definition 5 - A numerical function f defined on a set $X \subset IR^n$ is said to be semilocally quasiconcave at $x^* \in X$ if X is locally star-shaped at x^* and, corresponding to x^* and each $x \in X$, there exists a positive number $\beta(x^*,x) \leq a(x^*,x) \leq 1$ such that

$$\left.\begin{array}{l} f(x) \geq f(x^*) \\ 0<\alpha<\beta(x^*,x) \end{array}\right\} \Rightarrow f(\bar{\alpha}x+\alpha x^*) \geq f(x^*)$$

There are examples of functions which are semilocally quasiconcave, but not quasiconcave:

$$f(x) = \begin{cases} 1 & \text{if } x=0 \\ 0 & \text{if } x\neq 0 \end{cases} \quad ,x \in IR^n$$

A number of necessary and sufficient optimality conditions is derived for a class of nonlinear programming problems. More recently Rapcsac (1987) has pointed out that concave and generalized concave functions as well as convex programming problems are defined on convex sets and in general this property is not satisfied by nonlinear programming. For instance the feasible set of nonlinear programming problems constrained by nonlinear equality conditions is a differentiable manifold which is not convex, but arcwise connected through suitable geodesics.

Definition 6 - It is said that a set $X \subset M$, M being a connected Riemaniann $\mathcal{C}^2$ k-manifold, is geodesically convex, if any two points of X are joined by a geodesic belonging to X.

From a coordinate point of view, if a geodesic in M is parametrized by the arc length on the interval [0,b], then it can be considered as a continuous map from [0,b] into M which is of class $\mathcal{C}^2$ on each of a finite number of subintervals $[0,b_1]$, $[b_1,b_2]$,......,$[b_{i-1},b_i]$.....
with $0<b_1<b_2......<b$.

Definition 7 - Let $X \subset M$ be a geodesically convex set. Then it is said that a function $f: X \to IR$ is geodesically concave if its restrictions to all geodesic arcs beloging to X are concave, that is, for every g(s), $s \in [0,b]$,

$$f(g(\alpha b)) \geq \alpha f(g(0)) + \bar{\alpha} f(g(b)) \qquad 0 \leq \alpha \leq 1$$

Second order necessary and sufficient conditions which characterize arcwise-concave functions on surfaces are given in Rapcsac (1987).
Following the procedure developed by Avriel and Zang (1980) for arcwise connected functions from Definition 7 pseudo and quasiconcave functions can also be easily given in the geodesical framework.
For the third aspect let us refer for instance to the issue of the Mathematical Programming Journal entirely devoted to the difference of convex functions, the so called d.c. functions, and to the decomposition of a function in terms of suitable transforms of convex functions (Thon and Thorlund-Petersen (1986)).
In this respect we remark, for instance, as pointed out in Castagnoli and Mazzoleni's paper, that there are functions which are simultaneously λ_1-convex and λ_2-concave. Therefore the difference between a convex function f_1 and λ_1-convex, λ_2-concave function is λ_2-convex and the degree of convexity is still retained.
The nonconcavity of functions which result from suitable transformations of concave functions is better justified if we analyze the quasiconcave case.
Even the simple addition of two quasiconcave functions leads to a function which is no longer quasiconcave, and suitable conditions for the quasiconcavity of $f(x) = \sum_{i=1}^{n} f_i(x)$ have to be added (see for instance Crouzeix and Lindberg (1986)).

Finally, we would like to illustrate briefly the kind of generalization for quasiconcavity developed in the paper presented by Hackman and Passy at the Pisa meeting.
It is well known (Singer (1981, 1982)) that the existence of a separating hyperplane is a necessary and sufficient condition for the absence of a duality gap. This result holds for general abstract spaces and nonconvex functionals.
In the theory of convex duality one usually assumes that the hypograph of a given concave function is a closed convex set and the function is assumed to be lower semicontinuous. In quasiconvex duality theory Passy and Prisman (1985) only require that any level set of the function is equal to the intersection of the open half spaces that contain it.
Correspondingly, a weaker type of quasiconcavity can be defined.
Recall that a function f: $IR^n \to IR$ is upper semicontinuous and quasiconcave if all its upper level sets

$$L_{\geq u}(f) := \{x: f(x) \geq u\},\ u \in IR$$

are closed and convex.
This property can suitably be weakened if for any pair of points $x^o, x^1 \in C \subseteq IR^n$ we no longer consider the line segment $[x^o, x^1]$ defined as $\{ x(\alpha) : x(\alpha) = (\alpha x^o_1 + \bar{\alpha} x^1_1, \ldots, \alpha x^o_n + \bar{\alpha} x^1_n), \alpha \in I \}$, but rather the rectangle $R[x^o, x^1] := [x^o_1, x^1_1] \times [x^o_2, x^1_2] \times \ldots \times [x^o_n, x^1_n]$.

Definition 8 - A subset $C \subseteq IR^n$ is said to be semi-convex if for each two points $x^o, x^1 \in C$ and coordinate i and for all $\alpha \in I$, there exist $\alpha_1, \alpha_2, \ldots, \alpha_{i+1}, \ldots, \alpha_n \in I$ such that the point $(\alpha_1 x^o_1 + \bar{\alpha}_1 x^1_1, \ldots, \alpha_i x^o_i + \bar{\alpha}_i x^1_i, \ldots, \alpha_n x^o_n + \bar{\alpha}_n x^i_n)$ is in C.

It is worth noticing that for each two points x^o, x^1, in C and coordinate i, there exists a path χ, not necessarily continuous, such that the image of χ lies in $R[x^o, x^1] \cap C$ and the image of χ_i is $[x^o_i, x^1_i]$.
Such a new property, which is essentially based on the projection of set C on the coordinate axes, is analyzed both for its mathematical programming properties and for its economic applications.

References

[1] **Avriel M., Zang I.**, Generalized Arcwise-connected Functions and Characterization by Local-Global Minimum Properties, J.O.T.A. 1980, vol. 32.
[2] **Bector C.R., Bector M.K.**, Second Order Duality for a Quasi Convex Programming Problem, Cahier du C.E.R.O. 1987, vol. 21, pp.61-68.
[3] **Ben-Tal A.**, On Generalized Means and Generalized Convex Functions, J.O.T.A., 1977, vol. 21, pp.1-13.

[4] **Castagnoli E., Mazzoleni P.**, Generalized Convexity for Functions and Multifunctions and Optimality Conditions. Tech. Rep. Dept. Oper. Res. Univ. Pisa, 1986, n.134.

[5] **Castagnoli E., Mazzoleni P.**, Verso un unico tipo di concavità, Rivista AMASES, anno 9, 1986, fasc.1, pp.15-31.

[6] **Castagnoli E., Mazzoleni P.**, About Derivatives of some Gemeralized Concave Functions. J. of Information and Optimization Sciences, 1989, vol.10, pp.53-64.

[7] **Castagnoli E., Mazzoleni P.**, Differentiable $(\alpha\lambda)$-Concave Functions, This volume 1988.

[8] **Craven B.D.**, Invex Functions and Constrained Local Minima, Bull. Austral. Math. Soc. 1981, vol.24, pp.357-366.

[9] **Chabrillac Y., Crouzeix J.P.**, Continuity and Differentiability Properties of Monotone Real Funtions of Several Real Variables, Math. Progr. Study. vol.3, 1987, pp. 1-16.

[10] **Crouzeix J.P., Lindberg P.O.**, Additively Decomposed Quasiconvex Functions, Math. Progr., 1986 vol. 35, pp.42-57.

[11] **Hackman S.T., Passy U.**, Semiconvexity, This volume 1988.

[12] **Kaul R.N., Kaur S.**, Sufficient Optimality Conditions using Generalized Convex Functions, Opsearch vol. 19, 1982, pp.212-224.

[13] **Kyparisis J., Fiacco A.V.**, Generalized Convexity and Concavity of the Optimal Value Function in Nonlinear Programming, Math. Progr. 1987, vol. 39, pp.285-304.

[14] **Martin D.H.**, The essence of Invexity, J.O.T.A., 1985, vol. 47, pp.65-76.

[15] **Martos B.**, Nonlinear Programming Theory and Methods, North Holland 1975.

[16] **Ortega I.H., Rheinboldt W.C.**, Iterative Solutions of Nonlinear Equations in Several Variables, Academic Press. 1970.

[17] **Passy U., Prisman E.Z.**, A Convex-like Duality Scheme for Quasiconvex Programs, Math. Progr. 1985, vol. 32, pp. 278-300.

[18] **Rapcsak T.**, Arcwise-convex Functions on Surfaces, Publ. Math. 1987, vol. 24, pp.35-41.

[19] **Schaible S., Ziemba W.T.**, Generalized Concavity in Optimization and Ecomonics, Acad. Press. 1981.

[20] **Thon D., Thorlund-Peterson L.**, Sums of Increasing Convex and Increasing Concave Functions, Oper. Res. Letters, 1986, vol. 5, pp.313-316.

[21] **Ward D.**, Convex Directional Derivatives in Optimization, This volume, 1988.

This research was partially supported by the National Research Council and the Italian-Ministery of Education.

PROJECTIVELY-CONVEX MODELS IN ECONOMICS

S.T. Hackman- U. Passy

Abstract

Convexity (quasi-convexity) provides the traditional framework to describe optimization models in economics. Hackman and Passy (1988) introduced a generalization od convexity called Projective-convexity (or P-convexity). In this paper, we show that P-convexity provides a natural framework for describing a variety of models (e.g., consumer preference, consumer budgeting, resource allocation, multi-output) which do not adhere to the assumption of convexity.

1. Introduction

A function is quasi-convex (or quasi-concave) if its level sets are convex. The well-known separating hyperplane theorem, which asserts that a point not in a convex set may be separated by a hyperplane, may and often is invoked when analysing such functions. It is therefore not surprising that functions modeling consumer and producer behavior such as the direct (indirect) utility (production) function, the cost and profit functions, the conditional indirect (direct) utility (production) function are traditionally assumed to be quasi-convex (or quasi-concave). Indeed, at the heart of the duality approaches in microeconomic theory, which establish dual relationships between the aforementioned economic functions, lies the separation theorem (see Diewert [1982] for an excellent survey, and for applications to the theory of separability, see Blackorby et al. [1978]). But more often that not "the importance of the assumption of convexity in traditional production and economic analysis lies in its analytic convenience rather that its economic realism"

Efforts to generalize the concept of convexity, i.e., convex sets and quasi-concave functions can be categorized into two approaches. The first approach allows a separation between a point and a set by something other than a hyperplane (see, for example, Bellman and Beckenbach [1961], Peixoto [1948], Pollatchek [1973], Ben-Tal and Ben-Israel [1981]). The second approach allows two points in the set to be connected by a more general path than a line segment (see, for example, Martos [1975], Avriel and Zang [1980], Martin [1981]).

The main objectives of research on generalized convexity are to prove convergence of algorithms to find the global minimum of non-convex problems, and to characterize functions which possess only global minimums. However, from an economic perspective, a useful generalization of convexity must fulfill two requirements. First, and foremost, it must admit a rich class of economic models. Second, it must be tractable for analysis, i.e., it must have some useful mathematical structure.

Recently, Hackman and Passy [1988] introduced a generalization of convexity called Projective-convexity or (P-convexity) according to the second approach. It was shown that P-convexity has the requisite mathematical structure. For example, P-convex sets possess a separation property (as in the first approach to generalizing convexity): a point not in a closed P-convex set can be separated by a quadrant, which is an intersection of closed halfspaces generated by orthogonal hyperplanes. As another example, the gradient of a differentiable P-convex function (if non-zero) generates a supporting quadrant to boundary points of the level sets. In essence, a quadrant is to P-convexity what a halfspace is to convexity.

The main purpose of this paper is to show that P-convexity models is a richer class of economic phenomena than quasi-convexity. Models ranging from separable utility (production) functions, consumer budgeting, resource allocation in a multi-technology setting, and multi-product technologies are described in which quasi-convexity can not provide the necessary framework, but P-convexity can.

The outline of this paper is as follow. Section 2 defines P-convex sets, P-concave and P-convex functions. Section 3 describes economic models in which P-convexity naturally arises. Section 4 applies the separation theorem to show that the "non-convex" portion of a P-convex technology is observable.

2. Definition and Motivation

We first define our notation and terminology. Throughout this paper X, X_i, and Y_{ij} refer to finite-dimensional Euclidean spaces, dim (X) denotes the dimension of X, and R denotes the real line. A path in X is a map $\pi : [0,1] \to X$. A continuous path is a path whose coordinate functions are continuous. For two points x and y in X, [x,y] denotes the closed line segment joining x and y. For two points x and y in a product space $\prod_{i=1}^{m} X_i$ the rectangle joining x and y, denoted by R[x,y], is defined to be $[x_1, y_1] \times [x_2,y_2] \times \dots\dots [x_m,y_m]$. All product spaces are endowed with the product topology.

A subset $C \subset X$ is convex if for each two points x,y in C, the line segment [x,y] lies in C, i.e., for all $\lambda \in [0,1]$ the point $((1-\lambda)x_1 + \lambda y_1, \dots\dots ,(1-\lambda)x_n + \lambda y_n)$ is in C. We may reformulate the defining property of convexity in terms of paths: for each two points x,y in C, there exists a continuous path π whose initial point $\pi(0)$ is x, terminal point $\pi(1)$ is y, and whose coordinate functions π_i are given by $\pi_i(\lambda) = (1-\lambda)x_i + \lambda y_i$. An immediate (natural) generalization of convexity is to allow the parameter λ to be different for each coordinate i. This is the essential idea behind P-convexity.

Definition (Hackman and Passy [1988]): A subset $C \subset \prod_{i=1}^{m} X_i$ is Projectively-convex (or P-convex) if for each two points x,y in C and coordinate i, and for all $\lambda \in [0,1]$, there exist $\lambda_1, \lambda_2, \ldots, \lambda_{i-1}, \lambda_{i+1}, \ldots, \lambda_n \in [0,1]$ such that the point $((1-\lambda_1)x_1 + \lambda_1 y_1, \ldots, (1-\lambda_n)x_n + \lambda_n y_n)$ is in C.

It is important to note that P-convexity of a set C is relative to a product space $\prod_{i=1}^{m} Xi$. If $X_i = \prod_{j=1}^{n_i} Y_{ij}$, then the P-convexity of C viewed as a subset of $\prod_i X_i$ implies the P-convexity of C when viewed as a subset of $\prod_{i,j} Y_{ij}$, but not conversely. Thus, every P-convex set C, when viewed as a subset of X, $\dim(X) = \sum_{i=1}^{m} n_i$, is also P-convex.

As in the case of convex sets we may reformulate the defining property of P-convexity in terms of (more general) paths: for each two points x,y in C and coordinate i, there exists a path π (not necessarily continuous) such that the image of π lies in $R[x,y] \cap C$, and the image of π_i is $[x_i,y_i]$. An example of a set which is P-convex in shown in Figures 1(A). An example of a set which is not P-convex is shown in Figures 1(B).
We now turn to defining P-concave (P-convex) functions. Recall that a function $f : R^n \to R$ is upper semicontinuous and quasi-concave if its upper level sets $L_f^{\geq}(u) = \{x : f(x) \geq u\}$, $u \in R$, are each closed and convex. Similarly, a function $f : R^n \to R$ is lower semicontinuous and quasi-convex if its lower level sets $L_f^{\leq}(u) = \{x : f(x) \leq u\}$, $u \in R$, are each closed and convex.

Definition: A function $f : \prod_{i=1}^{m} X_i \to R$ is P-<u>convex</u> (<u>P-concave</u>) if each of its lower (upper) level sets are P-convex.
Note that:
-In the conjugacy duality framework of Passy and Prisman [1985], P-convexity (there called semi-convexity) of the primal bi-function implies that the associated Lagrangian is of a saddle function type. The concept of P-convexity, as an object itself, was not investigated.
- every quasi-convex (quasi-concave) function is P-convex (P-concave),
- every P-convex (P-concave) function is quasi-convex (quasi-concave) in each of its arguments.

The P-convexity (P-concavity) of a function is also relative to its domain. Again, if $X_i = \prod_j Y_{ij}$ then the P-convexity of f defined on $\prod_i X_i$ implies the P-convexity of f when viewed as a function defined on $\prod_{ij} Y_{ij}$, but not conversely. In practice, the domain of f is determined by the variables of a particular problem. It will be convenient in the sequel to refer to a P-convex function f as "P-convex with respect to the partition $(x_1, \ldots\ldots, x_m)$" where x_i is a variable denoting a generic element of X_i.

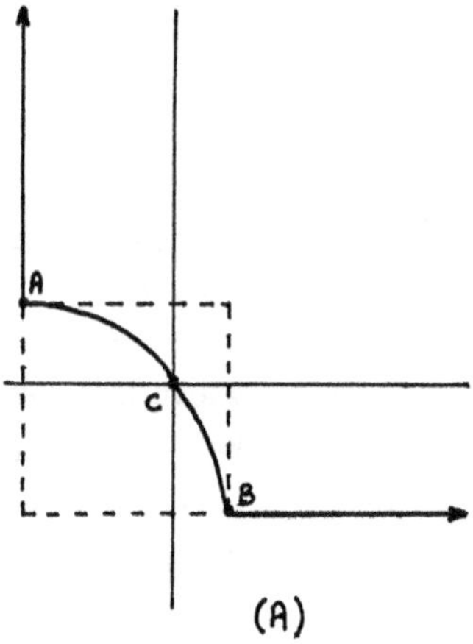

(A)

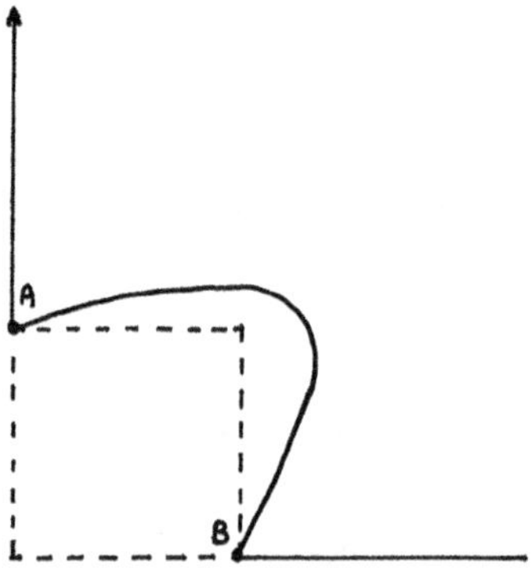

(B)

Figures 1(A) and 1(B) [17]

(A) A P-Convex set ARC ACB lies completely in R[A, B]

(B) A non P-Convex set No path connects A to B which lies in R[A, B]

3 . P-convex Models

3.1 Separable Utility (Production) Functions

Suppose the set C of the commodities of an economy naturally partition into m subsets $C_1, \dots\dots, C_m$. As discussed in Debreu and Koopmans [1982] the commodities in the set C_i could represent (1) the goods and services available in the ith period, (2) those goods and services whose availability is contingent on the ith state of the world or (3) goods and services with related physical characteristics. Let the vector $x = (x_1, \dots\dots, x_m) \in \prod_{i=1}^{m} X_i$ denote the quantities of the commodities consumed, where X_i represents the commodity space associated with the set C_i. If the preference relation of the consumer is represented mathematically by a continuous, complete preorder, then by a Theorem due to Debreu [1954] it may also be represented by a continuous real-valued utility function $\Phi : \prod_{i=1}^{m} X_i \to R$.

If the preference relation is also completely strictly separable (Blackorby et al. [1978, p.126]), then by a representation theorem due to Gorman [1968, p.388-389] (see also Debreu[1959], Koopmans [1972], Blackorby et al. [1978]) m real-valued continuous functions f_i on X_i exist such that $\Phi(x) = \sum_{i=1}^{m} f_i(x_i)$. The model $\Phi(x) = \sum_{i=1}^{m} f_i(x_i)$ also arises naturally in production theory where x_i denotes the vector of quantities of the factors of production associated with the ith elemental production process, and f_i the corresponding production function.

From here, the standard assumption of convexity of preferences is usually invoked to make analyses tractable. The implication is that Φ becomes quasi-concave. There is no a priori reason why the assumption of convexity, which implies a quasi-concave Φ, could not be made or is not valid for each sector's utility function. If the convexity assumption is deemed to be valid, then each f_i becomes a quasi-concave and possible not concave. Under different regularity assumptions, Arrow and Enthoven [1961], Yaari [1977], and Debreu and Koopmans [1982] (and references therein) have shown that if two or more of the f_i's are not concave, then Φ can not be quasi-concave. But, Φ will always be P-concave. More generally, any monotone function of any finite collection of quasi-concave (quasi-convex) functions is necessarily P-concave (P-convex). That is, if $\varphi : R^m \to R$ is non-decreasing or non-increasing in each of its arguments and each f_i, $i = 1,2, \dots\dots, m$, is quasi-concave (quasi-convex) on its domain X_i, then the function $\Phi : X \to R$, $\dim(X) = \sum_i \dim(X_i)$ defined by

$$(3.1) \qquad \Phi(x) = \varphi\,(f_1(x_1), \ldots\ldots\ldots, f_m(x_m))$$

is always P-concave (P-convex). (Here, x_i refers to the appropriate sub-vector of x.) Examples of (3.1) include $\max\,\{f_1, \ldots\ldots, f_m\}$ and $\min\,\{f_1, \ldots\ldots, f_m\}$. If each of the f_i's is non-negative (as is usually the case in production theory), then the product $f_1 \cdot f_2 \cdot \cdot f_m$ is also included in (3.1).

3.2 Consumer Budgeting

Consumer budgeting (Strotz [1957, 1959], Gorman [1959], Blackorby et al. [1978]), or more commonly referred to as "two-stage optimization", is a familiar budgeting procedure: the consumer in the first stage allocates total expenditure to budget categories, and in the second stage allocates expenditures to the commodities of each category. Using the same notation and terminology of Model I, let $p_i \in X_i$ denote the vector of prices of the commodities in set C_i, and let q_i denote the allocation of the overall budget to the ith budget category (modeled by the set C_i). As a function of prices and budget allocations the overall consumer problem may be modeled as follows:

$$(3.2) \qquad \Gamma\,((p_1,q_1), \ldots, (p_m,q_m)) = \max\,\{\, \Phi(x) \mid p_i \cdot x_i \le q_i,\ i = 1, 2, \ldots, m\,\}.$$

We shall refer to the function $\Gamma : \prod_{i=1}^{m} (X_i \times R) \to R$ as the <u>multi-dimensional indirect utility (production) function.</u>

If the budget allocations q_i, $i = 1,2, \ldots\ldots, m$ are optimal, then $\Gamma\,((p_1,q_1), \ldots\ldots, (p_m,q_m))$ as a function of prices equals the indirect utility (production) function $\Gamma_\Phi((p_1, \ldots\ldots, p_m), \sum_{i=1}^{m} q_i)$, which is always quasi-convex (Shephard [1970]). In addition, fixing the prices and viewing Γ solely as a function of the budget allocations, Γ is strictly quasi-convex if, and only if, Φ is quasi-concave (Diewert [1978], Epstein [1975]). If, however, the budget allocations are not optimal, then viewed jointly as a function of prices and budget allocations $\Gamma((p_1,q_1), \ldots\ldots, (p_m,q_m))$ is in general not quasi-convex. If, for example, the utility $\Phi(x) = \sum_{i=1}^{m} f_i(x_i)$ (as in Model I), then $\Gamma((p_1,q_1), \ldots\ldots, (p_m,q_m))$ equals $\sum_{i=1}^{m} \Gamma_{f_i}(p_i,q_i)$, where Γ_{f_i} denotes the indirect utility function associated with f_i. As discussed in Model I, the sum of quasi-convex functions in in general not quasi-convex, but will always be P-convex. As another example suppose $\Phi(x) = \prod_{i=1}^{m} f_i(x_i)$; then $\Gamma((p_1,q_1),\ldots\ldots,(p_m,q_m)) = \prod_{i=1}^{m} \Gamma_{f_i}(p_i,q_i)$.

For a concrete example, let $\Phi : R^2_+ \to R_+$ be a "restricted" Cobb-Douglas production function given by $\Phi(x_1,x_2) = \sqrt{g(x_1) \cdot x_2}$ where $g(x) = \min(x,1)$. The lower level sets of $\Gamma((p_1,1),(p_2,1))$ are depicted in Figure 2. The function Φ models a technology for which there is a limit on the substitution between the factors of production.

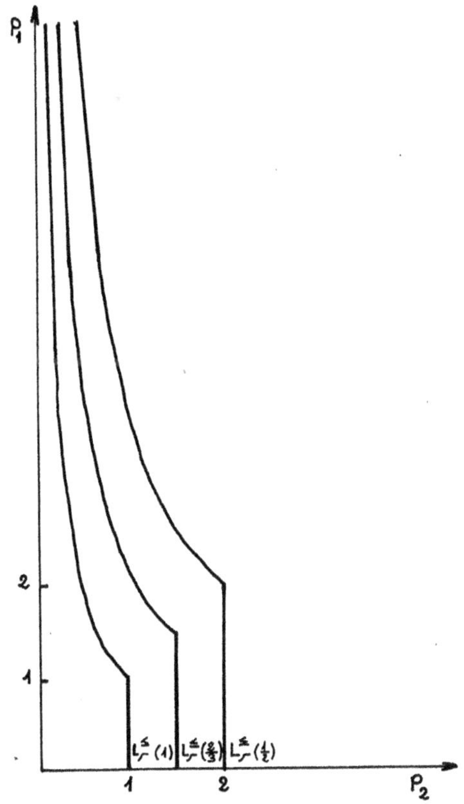

Figure 2 [17]
Level sets of the multi-dimensional indirect utility function obtained from a restricted Cobb-Douglas utility function

The budgeting problem described above is common to many institutional structures. For example, it may model an organizational structure of a firm. In another example, when modeling the dynamics of production or consumer choices the p_i's and x_i's may correspond to the prices and inputs in different time periods. Alternatively, p_i need not represent a price vector. The constraint $p_i \cdot x_i \leq q_i$ could reflect restrictions such as pollution control, or trade restrictions as in the theory of international trade.
Finally, it was shown in Hackman and Passy [1988] that the multi-dimensional indirect production (utility) function is an appropriate dual to a P-concave technology. Namely, the following dual minimization problem recovers the original technology:

$$\Phi(x_1,x_2) = \inf_{(p_1,p_2)} \{ \Gamma((p_1,1),(p_2,1)) : p_1 \cdot x_1 \leq 1,\ p_2 \cdot x_1 \leq 1 \}$$

The duality between $\Phi(x_1,x_2)$ and $\Gamma((p_1,1),(p_2,1))$ expressed above is exactly analogous to the well-known dual relationship between the direct and indirect production (utility) functions. (For a general framework of duality in economics, see Epstein [1981].)

3.3 Resource Allocation in a Multi-Technology Setting

Mundlak [1984] addresses the question of how to represent and measure technology of a sector (such as agriculture) when output is produced by more than one technique. As in Mundlak [1984], let us simplify the discussion by considering only two techniques whose technologies are modeled by production functions $f_1(K_1,L_1)$ and $f_2(K_2,L_2)$, where the factors of production are capital (K) and labor (L). Let us assume that simultaneous production by both techniques is possible. Then one mathematical formulation of the above problem is as follows:

$$(3.3) \qquad \Phi(K,L)=\max \{ f_1(K_1,L_1) + f_2(K_2,L_2) \mid K_1+ K_2= K\ ,\ L_1+L_2= L,\ K_1,K_2,L_1,L_2 \geq 0 \}.$$

If both production functions exhibit constant returns to scale, then so will Φ. Moreover, the unit output level set $L_\Phi(1)$ may be generated from $L_{f_1}(1)$ and $L_{f_2}(1)$ as depicted in Figure 3(A). Thus, under constant returns to scale Φ is quasi-concave.
But what if both production functions exhibit increasing returns to scale? Consider the following simple example: both production functions are Cobb-Douglas, homogeneous of degree, say 3/2, and "symmetric", i.e., $f_1(K_1,L_1) = \sqrt{K_1}L_1$ and $f_2(K_2,L_2) = K_2\sqrt{L_2}$. If both techniques are employed at the optimum, then the Kuhn-Tucker necessary conditions for an optimum imply that $K_2 = L_1$. Consider first the region where $K \geq L$. Write $K = \alpha_0 L$, $\alpha_0 \geq 1$, and $K_2 = \alpha L$, $0 < \alpha \leq 1$ (α is bounded by 1 since $K_2 = L_1 \leq L$). The objective function value for the Kuhn-Tucker solution is $(\alpha\sqrt{\alpha_0 - \alpha} + \alpha\sqrt{1-\alpha})\, L^{3/2}$. On the other hand, if only the first technique is employed, the objective function value is $\alpha_0 L^{3/2}$. Regardless of α,

$\alpha_0 > \alpha\sqrt{\alpha_0 - \alpha} + \alpha\sqrt{1 - \alpha}$. Hence, the true optimum of (3.3) in the region $K \geq L$ is to employ only technique 1. Similarly, due to the symmetry between f_1 and f_2 in the region where
$L \geq K$ the true optimum of (3.3) is to employ only technique 2. Therefore, $\Phi = \max\{f_1, f_2\}$. The unit output level set $L_\Phi(1)$ is depicted in Figure 3(B). Since f_1 and f_2 are monotone and quasi-concave, $L_\Phi(1)$ is P-convex, but, as clearly shown in Figure 3(B), is not convex. Since Φ is homogeneous of degree 3/2, Φ is therefore P-concave, but not quasi-concave. Moreover, the objective function of (3.3) is itself P-concave in the partition $\{(K_1, L_1), (K_2, L_2)\}$. In general, under various combinations of increasing and decreasing returns to scale Φ will be P-concave. We remark that $\Phi(K,L)$ may be viewed as a multi-dimensional indirect production function of Model II where all of the prices are unity.

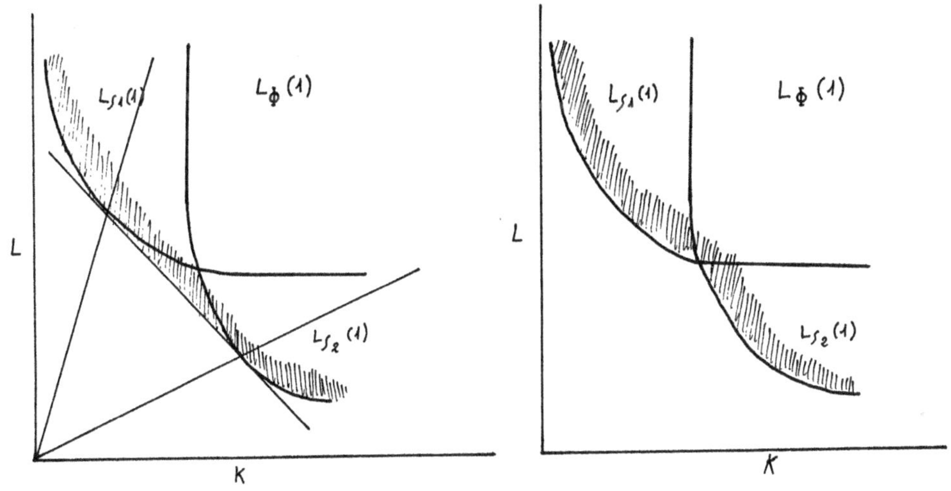

Figure 3 Level sets for a multi-technology form. Shaded area represents $L_\Phi(1)$.

(A) Constant returns to scale case

(B) Increasing returns to scale case

3.4 Multi-Product Firms

In the previous three models only single-product firms were considered. Consider now a multi-product firm. Let $x \in R^n_+$ denote a vector of inputs, and $v \in R^m_+$ denote a vector of outputs. A multi-output technology T is a subset of $R^n_+ \times R^m_+$. If $(x,v) \in T$, then the input vector x is interpreted as being capable of producing the output vector v.

As discussed in Diewert and Parkan [1983], it is possible to model the technology T by a production function. Define $v_1 = F_1(-v_2,......,-v_m,x)$ to be the maximal amount of the first output given that the firm uses x as input and produces at least v_i of product i, $i = 1,2,......, m$. Under Shephard's [1970] axioms for a production technology, Φ_1 is necessarily P-concave in the partition $(v_2,......,v_m,x)$. Similarly,

$$v_j = \Phi_j(-v_1,......,-v_{j-1},-v_{j+1},......,-v_m,x) \text{ is also P-concave for each } j = 2,......,m.$$

In fact, if T is assumed to be P-convex (in the partition (v,x)), then each Φ_j is P-concave in the partition $((v_1,......,v_{j-1},v_{j+1},......,v_m),x)$. Assuming that T is P-convex is less restrictive than the common approach of assuming that T is convex (see, for example, Banker et al. [1984]).

4. Application of the Separation Theorem

An interesting theoretical and practical question is whether or not one would ever observe the "P-convex portion" of the technology, i.e., those boundary points of the level set lying in the interior of the convex hull of the level set (points A, B and C in Figure 4). If the production planner were cost minimizing (one possible behavioral assumption), then points A, B, and C would never be observed. This property of cost minimization is often used to justify that the assumption of convexity of the production possibility sets is not as restrictive as it might appear. In this section we show that if the production planner were output maximizing subject to separate constraints (Model II), then points A, B and C could be observed (see Figure 4).

For each $a \in R^N$ let $H(a)$ denote the closed halfspace $\{x \in R^N : a \cdot x \geq 0\}$. The quadrant generated by vectors $a \in R^n$, $b \in R^m$ is the closed cone $\{(x,y) \in R^n \times R^m: a \cdot x \geq 0, b \cdot y \geq 0\}$. It will be denoted by the symbol $Q[a,b]$. The symbol $(x,y) + Q[a,b]$ denotes the translation of quadrant $Q[a,b]$ to the point (x,y). For a set S the symbols int (S) and ∂S denote the interior of S and boundary of S, respectively.

Theorem 1: (Hackman and Passy [1988]): Let $S \subset R^n \times R^m$ be a closed, P-convex set. If $(0,0) \notin S$, then vectors $a \in R^n$, $b \in R^m$ exist such that $Q[a,b] \cap S$ is empty.

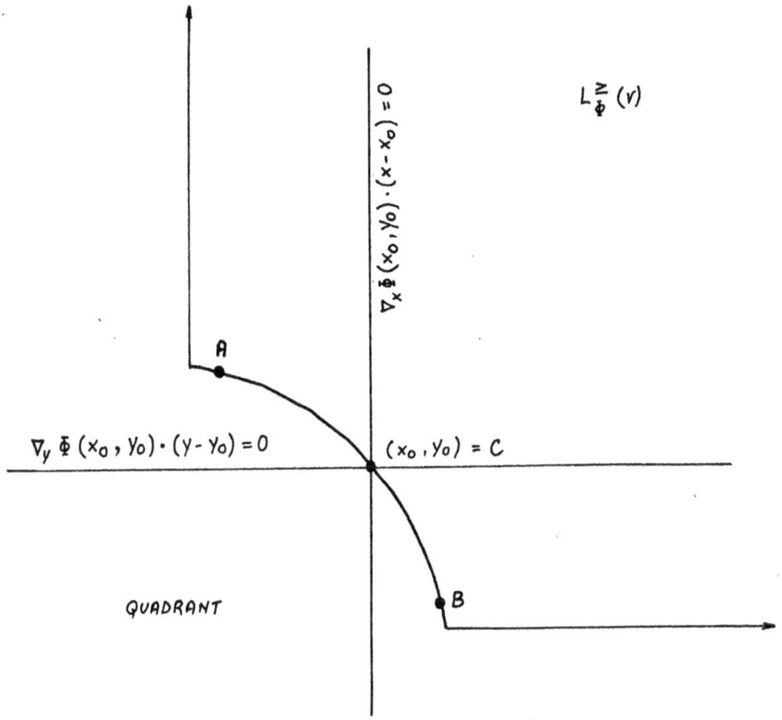

Figure 4 [17]   A P-Convex set with supporting quadrant at point C

Definition: Quadrant (x,y) + Q[a,b] is said to support a set $S \subset R^n \times R^m$ at a point $(x,y) \in S$ if int ((x,y) + Q[a,b]) $\cap$ S is empty.

Corollary : (Hackman and Passy [1988]): Let $S \subset R^n \times R^m$ be a closed P-convex set. If (0,0) $\in \partial S$, then vectors $a \in R^n$, $b \in R^m$ exist such that Q[a,b] supports S at (0,0).

It is an immediate consequence of the above corollary that every boundary point of an upper semicontinuous P-concave technology (like A, B and C of Figure 4) has a set of prices and budgets so that it would be a solution to the optimization problem (3.2) described in Model II. As depicted in Figure 4, when Φ is differentiable, the supporting quadrant is determined by the gradient (Hackman and Passy [1988]).

References

[1] **Arrow K.j. and Enthoven A.C.**, Quasi-concave Programming, Econometrica 21 (1961), 779-800.

[2] **Avriel M. and Zang I.**, Generalized Arcwise-Connected Functions and Characterization of Local-Global Minimum Properties, J. of Optimization Theory and Applications 32 (1980), 407-425.

[3] **Banker R.D., Charnes A. and Cooper W.W.**, Some Models for Estimating Technical and Scale Inefficiencies in Data Envelope Analysis, Management Science 30 (1984), 1078-1092.

[4] **Beckenbach E.F. and Bellman R.**, Inequalities, Spriger, Berlin, 1961.

[5] **Ben-Tal A. and Ben-Israel A.**, F-Convex Functions: Properties and Applications, in Generalized Concavity in Optimization and Economics, S.Schaible and W.T.Ziemba eds., Academic Press, New York, 1981.

[6] **Blackorby C., Primont D. and Russell R.R.**, Duality, Separability, and Functional Structure: Theory and Economic Applications, North-Holland, New York, 1978.

[7] **Debreu G.**, Representation of a Preference Ordering by a Numerical Function, in Decision Processes, R.Thrall, C.Coombs and R.Davis eds., Wiley, New York, 1954.

[8] **Debreu G.**, Theory of Value, Wiley, New York, 1959.

[9] **Debreu G. and Koopmans T.C.**, Additively Decomposed Quasi-convex Functions, Mathematical Programming 21 (1982), 1-38.

[10] **Diewert W.E.**, Hicks Aggregation Theorem and the Existence of a Real Value Added Function, in Production Economics: A Dual Approach to Theory and Applications, Vol. II, M.Fuss and D.McFadden eds., North-Holland, Amsterdan, 1978.

[11] **Diewert W.E.**, Duality Approaches in Microeconomic Theory, in Handbook of Mathematical Economics, Vol. II, K.J. Arrow and M.D. Intrilligator eds., North-Holland, New York, 1982.

[12] **Diewert W.E. and Parkan C.**, Linear Programming Tests of Regularity Conditions for Production Functions, in Quatitative Studies on Production and Prices, W.Eichhorn ed., Wurxburg-Wien, 1983.

[13] **Epstein L.**, A Disaggregate Analysis of Consumer Choice Under Uncertainty, Econometrica 43 (1975), 877-892.

[14] **Epstein L.**, Generalized Duality and Integrability, Econometrica 49 (1981), 855-878.

[15] **Gorman W.M.**, Separable Utility and Aggregation, Econometrica 27 (1959), 469-481.

[16] **Gorman W.M.**, The Structure of Utility Functions, Review of Economic Studies 35 (1968), 369-390.

[17] **Hackman S.T. and U.Passy**, Projectively-convex Sets and Functions, Journal of Mathematical Economics 17 (1988), 55-68.

[18] **Mangasarian O.L.**, Non-linear Programming, McGraw-Hill, New York, 1969.

[19] **Martos B.**, Nonlinear Programming Theory and Methods, North-Holland, Amsterdam, 1975.

[20] **McFadden D.**, Cost, Revenue, and Profit Functions, in Production Economics: A Dual Approach to Theory and Applications, Vol. II, M.Fuss and D.McFadden eds., North-Holland, Amsterdam, 1978.

[21] **Mundlak Y.**, Capital Accumulation: The Choice of Techniques and Agricultural Output, Working Paper 6504 (1984), The Center for Agricultural Economic Research, Rehovoth, Israel.

[22] **Passy U. and Prisman E.Z.**, A Duality Approach to Minimax Results for Quasi-Saddle Functions in Finite Dimensions, Technical Report J-84-5 (1985), Industrial and System Engineering, Georgia Institute of Technology, revised.

[23] **Peixoto M.M.**, Convexity of Cones, Notas Mat. 6 (1948), Livaria Boffini, Rio de Janeiro, in Portuguese.

[24] **Pollatchek M.A.**, Generalized Duality Theory in Nonlinear Programming, Operations Research, Statistics and Economics Mimeograph Series 122 (1973), Faculty of Industrial and Management Engineering, Technion-Israel Institute of Technology.

[25] **Shepard R.W.**, Theory of Cost and Production Functions, Princeton University Press, Princeton, 1970.

[26] **Strotz R.H.**, The Empirical Implications of a Utility Tree, Econometrica 25 (1957), 269-290.

[27] **Strotz R.H.**, The Utility Tree - a Correction and Further Approach, Econometrica 27 (1959), 482-488.

[28] **Yaari M.E.**, A Note on Separability and Quasi-Concavity, Econometrica 45 (1977), 1183-1186.

CONVEX DIRECTIONAL DERIVATIVES IN OPTIMIZATION

D. Ward

Abstract

Broadly speaking, a generalized convex function is one which has some property of convex functions that is essential in a particular application. Two such properties are convexity of lower level sets (in the case of quasiconvex functions) and convexity of the ordinary directional derivative as a function of direction (in the case of Pshenichnyi's quasidifferentiable functions). In recent years, several directional derivatives have been defined that, remarkably, are always convex as a function of direction.
This means that all functions are "generalized convex" in the sense that they have certain convex directional derivatives. As a result, it has become worthwhile to develop generalizations of the Fritz John and Kuhn-Tucker optimality conditions in terms of the subgradients of convex directional derivatives. In this paper, we derive some general versions of these conditions for an inequality-constrained, nondifferentiable, nonconvex mathematical program.

1. Introduction

We consider the mathematical program

(P) $\min \{ f(x) \mid g_i(x) \leq 0, i \in J \}$,

where $f: E \to IR$, each $g_i: E \to IR$, E is a real, locally convex, Hausdorff topological vector space, and J is a finite index set. Necessary optimality conditions for (P) of Fritz John or Kuhn-Tucker type are well known in the case in which f and g_i are differentiable or convex. Such conditions can also be stated for much larger classes of functions through the use of convex directional derivatives (see for example [6, 7, 9, 17, 19, 22-24]). A number of these directional derivatives have been defined, including the following examples.

Definition 1.1. Let $f: E \to IR$ and $x \in E$. Denote by $N(x)$ the class of neighborhoods of x, and by epi f the epigraph of f.
a) The generalized Clarke directional derivative of f at x in the direction y [4, 6, 7, 19, 20] is defined by

$$f^{o}(x;y) := \inf_{\substack{X \in N((x,f(x)))\\ \lambda > 0}} \sup_{\substack{(u,d) \in \operatorname{epi} f \cap X\\ t \in (0,\lambda)}} (f(u+ty)-d)/t.$$

b) The Michel-Penot directional derivative of f at x in the direction y [3, 5, 14-16, 22] is defined by

$$f^{P}(x;y) := \sup_{z \in E} \limsup_{t \downarrow 0} (f(x+tz+ty)-f(x+tz))/t.$$

c) The asymptotic upper Dini derivative of f at x in the direction y is defined by

$$f^{U\infty}(x;y) := \sup_{z \in E} \{ f^{U}(x;z+y) - s \mid f^{U}(x;z) \leq s \},$$

where

$$f^{U}(x;y) := \limsup_{t \downarrow 0} (f(x+ty)-f(x))/t$$

is the upper Dini derivative of f at x in the direction y.

All three of the directional derivatives $f^{o}(x;\circ)$, $f^{P}(x;\circ)$, and $f^{U\infty}x;\circ)$ can be readily demonstrated to be convex and positively homogeneous, although possibly infinite-valued. If E is a normed space and f is locally Lipschitzian near x, then all three are finite-valued and continuous as well.
A number of calculus rules can be derived for $f^{o}(x;\circ)$ and $f^{P}(x;\circ)$ [5, 7, 10, 15, 19, 22], but they will not concern us here. In fact, all that we will require of a directional derivative $f^{A}(x;\circ)$ in the optimality conditions in §2 of this paper is that it be sublinear (i.e., convex and positively homogeneous) and majorize either $f^{U}(x;\circ)$ or the lower Dini derivative

$$f^{L}(x;y) := \liminf_{t \downarrow 0} (f(x+ty)-f(x))/t.$$

This observation leads to the following definition:

Definition 1.2. Let $f : E \to \mathbb{R}$ and $x \in E$, and let f^{A} be some positively homogeneous directional derivative. We say that a continuous sublinear function $h : E \to \mathbb{R}$ is an A upper convex approximate of f at x if $h(y) \geq f^{A}(x;y)$ for all $y \in E$. We will denote the class of all such upper convex approximates by UC(A,f,x).

Remark 1.3. (a) It is possible for UC(A,f,x) to be empty. For example, if $f : \mathbb{R} \to \mathbb{R}$ is defined by $f(x) = |x|^{1/2}$, then UC(L,f,0) and UC(U,f,0) are empty. If E is a normed space, one condition sufficient to guarantee that $f^{o}(x\,;\circ)$, $f^{P}(x\,;\circ)$, and $f^{U\infty}(x\,;\circ)$ are in UC(U,f,x) is that f be locally Lipschitzian near x. Notice that if $h = f^{o}(x\,;\circ)$, then $\partial h(0)$ is nothing other than the Clarke generalized gradient of f at x [6,7].
(b) In a number of references (e.g. [2, 5, 11, 12, 18, 21]), UC(U,f,x) (or some similarly-defined class of functions) has been used in mean-value theorems or in optimality conditions in mathematical programming.

In this paper, we develop general optimality conditions for (P) based on upper convex approximates. We begin in §2 by proving simple necessary conditions involving UC(U,f,x) and UC(L,f,x). The derivation of these conditions requires only a basic calculus rule for f^U and f^L and a little convex analysis. We then examine the special case of a fractional objective function in (P).
In §3, we extend the results of §2 in two directions. First, we allow our classes of upper convex approximates to include proper sublinear functions that are possibly infinite-valued. Second, we prove our optimality conditions not just for L and U upper convex approximates, but for any classes of upper convex approximates whose underlying directional derivatives satisfy a certain calculus rule. We also show that, in optimality conditions for A upper convex approximates, the asymptotic directional derivative

$$f^{A\infty}(x\,;y) := \sup_{z \in E} \{\, f^{A}(x\,;y+z) - s \mid f^{A}(x\,;z) \le s \,\}$$

plays a special role in constraint qualifications.
We will have use for the following notation and terminology in the sequel. For an extended-real-valued function $f : E \to [-\infty,\infty]$, the effective domain of f is the set

$$\operatorname{dom} f := \{\, x \in E \mid f(x) < \infty \,\}.$$

We say that f is proper if dom f is nonempty and f never takes on the value $-\infty$, and that f is lower semicontinuous (l.s.c.) if the epigraph of f (epi f) is closed. If f is finite at x, we say that f is l.s.c. at x if for each $\varepsilon > 0$, there exists $X \in N(x)$ such that $f(y) \ge f(x) - \varepsilon$ for all $y \in X$.
If f is a proper convex function and $x \in \operatorname{dom} f$, the subgradient of f is the subset of the dual space E' defined by

$$\partial f(x) := \{\, x' \in E' \mid \langle y - x, x' \rangle \le f(y) - f(x) \,\}.$$

If λ is a positive real number, it is clear that $\partial(\lambda f)(x) = \lambda\, \partial f(x)$. If $\lambda = 0$, we interpret $(\lambda f)(x)$ as the indicator function of dom f; that is, the function

$$i_{\mathrm{dom}\, f}(x) := \begin{cases} 0 & \text{if } x \in \mathrm{dom}\ f \\ +\infty & \text{otherwise} \end{cases}$$

Then if $\partial f(x)$ is nonempty, we have

$$\begin{aligned} \partial(0f)(x) &= \{ x' \mid \langle y - x, x' \rangle \leq 0 \text{ for all } y \in \mathrm{dom}\ f \} \\ &= \{ x' \mid x' + \partial f(x) \subset \partial f(x) \} \\ &:= 0^{+}\partial f(x), \text{ the recession cone of } \partial f(x). \end{aligned}$$

We will need to keep these facts in mind when we formulate our general optimality conditions in §3.
By a directional derivative, we will mean a mapping that accociates to every $f : E \to IR$ and $x \in E$ a positively homogeneous function $f^{A}(x\,;\circ) : E \to [-\infty, +\infty]$.

2. Dini Derivatives and upper convex approximates

In this section, we develop simple optimality conditions for (P) that rely only upon the calculus of Dini derivatives and the properties of convex sets and functions. The relevant facts about Dini derivatives are contained in the following propositions. We provide a short proof of the inequality in Proposition 2.2.

Proposition 2.1. If $f : E \to IR$ has a local minimum at x, then $f^{L}(x\,; y) \geq 0$ for all $y \in E$.

Proposition 2.2. Let $f_i : E \to IR$, $i = 1, \ldots, n$, and define $f(x) := \max_i f_i(x)$ and $I(x) := \{ i \mid f_i(x) = f(x) \}$. Suppose that f is l.s.c. at x and f_i is continuous at x for each $i \in \{1, \ldots, n\} \setminus I(x)$. Then for each $j \in I(x)$ and $y \in E$,

$$f^{L}(x\,; y) \leq \max\{ f_j^{L}(x\,; y), \max_{I(x)\setminus\{j\}} f_i^{U}(x\,; y) \} \tag{1}$$

Proof. Let $\varepsilon > 0$ and $\lambda > 0$ be given, and denote the function on the right-hand side of (1) by F. Let $y \in E$, and suppose $F(y) \leq r$. Since f is l.s.c. at x and f_i is continuous at x for each $i \in \{1, \ldots, n\} \setminus I(x)$, there exists $\lambda_1 \in (0, \lambda)$ such that $f_i(x + ty) < f(x + ty)$ for all $t \in (0, \lambda_1)$ and all $i \in \{1, \ldots, n\} \setminus I(x)$. Moreover, there exists $\lambda_2 \in (0, \lambda_1)$ such that

$$(f_i(x + ty) - f_i(x))/t \leq r + \varepsilon$$

for all $t \in (0, \lambda_2)$ and all $i \in I(x)\setminus\{j\}$. Finally, there exists t_0 in $(0, \lambda_2)$ such that

$$(f_j(x + t_o y) - f_j(x))/t_o \le r + \varepsilon.$$

Putting these facts together, we see that

$$(f(x + t_o y) - f(x))/ t_o = (\max_{I(x)} f_i(x + t_o y) - f(x))/t_o \le r + \varepsilon.$$

Since ε and λ were arbitrarily chosen, it follws that $f^L(x\,;\,y) \le r$, and the proof of (1) is complete.

We now apply Propositions 2.1 and 2.2 in the proof of necessary optimality conditions for (P).

Theorem 2.3 Let $f(x_o)$ be a local minimum for (P), and define $I(x) := \{ i \in J \mid g_i(x) = 0 \}$. Assume that g_i is continuous at x_o for each $i \in J\setminus I(x_o)$, and that there exist $h \in UC(L,f,x_o)$ and $h_i \in UC(U,g_i,x_o)$ for $i \in I(x_o)$. Then there exist $\lambda_1 \ge 0$, $i \in I(xo) \cup \{0\}$, at least one of which is nonzero, such that

(2) $$0 \in \lambda_o \partial h(0) + \sum_{I(x_o)} \lambda_i \partial h_i(0).$$

If in addition

(3) $$\{y \mid h_i(y) < 0 \text{ for all } i \in I(x_o)\} \neq \emptyset,$$

then $\lambda_o > 0$ in (2).

Proof. Define $F(x) := \max \{ f(x) - f(x_o), \max_J g_i(x) \}$. Then F has a local minimum at x_o, and it follows that F is l.s.c. at x_o. By Propositions 2.1 and 2.2 and the definition of upper convex approximate, we conclude that for all $y \in E$,

$$0 \le F^L(x_o\,;\,y) \le \max \{ f^L(x_o\,;\,y), \max_{I(x_o)} g_i^U(x_o\,;\,y) \}$$

$$\le \max \{ h(y), \max_{I(x_o)} h_i(y) \}.$$

Now call $p := | I(x_o) | + 1$, and define $S_1 := \{ z \in IR^p \mid z \le 0 \}$ and $S_2 := \{ z \mid (h(y),\ldots\ldots h_i(y)\ldots\ldots) \le z$ for some $y \in E \}$, where "$\le$" denotes the usual coordinate-wise ordering in IR^p. Observe that S_1 and S_2 are convex cones with $\text{int } S_1 \cap S_2 = \emptyset$, so there exists a hyperplane separating them. Specifically, there exist λ_i, $i \in \{0\} \cup I(x_o)$, at least one of which is nonzero, such that

$$(4)\qquad \sum_{\{0\}\cup I(x_o)} \lambda_i z_i \le 0 \le \lambda_o h(y) + \sum_{I(x_o)} \lambda_i h_i(y)$$

for all $y \in E$ and $z = (z_i) \in S_1$. It follows from (4) that each $\lambda_i \ge 0$ and that

$$0 \in \partial(\lambda_o h + \sum_{I(x_o)} \lambda_i h_i)(0).$$

Since h and each h_i are continuous and sublinear, we may apply the subgradient sum formula [8, p. 26] to obtain (2).
Finally, suppose that (3) is satisfied and $\lambda_o = 0$. Then (4) implies that $\lambda_i = 0$ for each $i \in I(x_o)$, a contradiction of the fact that at least one $\lambda_i > 0$. We conclude that $\lambda_o > 0$.

Remark 2.4. a) Notice that we have made no explicit continuity assumptions about f and g_i, $i \in I(x_o)$, in Theorem 2.3. We have, however, assumed that f and g_i have Dini derivatives that are majorized by continuous convex functions. We will relax this requirement in §3.
b) The constraint qualification in (3) depends upon the choice of h_i. We will see in §3 that (3) could be replaced by the single constraint qualification

$$(5)\qquad \{ y \in E \mid g_i^{U\infty}(x_o; y) < 0 \text{ for all } i \in I(x_o) \} \ne \emptyset$$

that is independent of the choice of h_i.
c) Notice that in the proof of Theorem 2.3, one could apply a formula for the subgradient of the pointwise maximum of a finite number of convex functions instead of using a separation argument and the subgradient sum formula.
Various sorts of objective functions in (P) can be dealt with in Theorem 2.3 with the help of further calculus rules for Dini derivatives. One example is the fractional program

$$(\text{FP})\qquad \min \{ F(x)/G(x) \mid g_i(x) \le 0,\ i \in J \}.$$

We next derive optimality conditions for (FP) by combining Theorem 2.3 and the following quotient rule. The proof of this quotient rule is straightforward, and we omit it here.

Proposition 2.5 Let $G : E \to (0, \infty)$ be continuous along lines at $x \in E$, and let $F : E \to (0, \infty)$. Call $f := F/G$. Then for all $y \in E$,

$$(6)\qquad f^L(x\,;y) \le (F^L(x\,;y)\, G(x) + (-G)^U(x\,;y)\, F(x))/(G(x))^2.$$

Theorem 2.6 Let $F : E \to (0, \infty)$, and let $G : E \to (0,\infty)$ be continuous along lines at x_0, a local minimizer for (FP). Call $f := F/G$. Assume that g_i is continuous at x_0 for each $i \in J\backslash I(x_0)$, and that there exist $h \in UC(L,F,x_0)$, $H \in UC(U,-G,x_0)$, and $h_i \in UC(U,g_i,x_0)$ for $i \in I(x_0)$. Then there exist $\lambda_i \geq 0$, $i \in I(xo) \cup \{0\}$, at least one of which is nonzero, such that

(7) $$0 \in \lambda_0(\partial h(0) + f(x_0)\partial H(0)) + \sum_{I(x_0)} \lambda_i \partial h_i(0).$$

If in addition (3) holds, then $\lambda_0 > 0$ in (7).

Proof. If $f(x_0) =)$, then x_0 is a local minimizer of F and $0 \in \delta(0)$ for every $h \in UC(L,F,x_0)$ by Proposition 2.1. In this case, (7) holds with $\lambda_i = 0$. We may assume, then, that F is positive-valued. By Proposition 2.5, the function

$$S(x):=h(x)/G(x_0) + H(x)F(x_0)/(G(x_0))2$$

is an elemnt of $UC(L,f,x_0)$ for each $h \in UC(L,F,x_0)$ and $H \in UC(U,-G,x_0)$. By Theorem 2.3, there exists $\mu_i \geq 0$, $i \in I(x_0) \cup \{0\}$, at least one of which is nonzero, such that

(8) $$0 \in \mu_0 \delta S(0) + \sum_{I(x_0)} \mu_i \delta h_i(0)$$

Now by the subgradient sum formula,

$$\delta S(0) = \delta h(0)/G(x_0) + \delta H(0)F(x_0)/(G(x_0))^2.$$

We may then set $\lambda_0 := \mu_0/G(x_0)$, $\lambda_i := \mu_i$ in (8) to obtain (7).

Remark 2.7. If $F(x) := \prod_{i=1}^{n} Fi(x)$, $G(x) := \prod_{i=1}^{n} G_i(x)$ as in [1], on can use Theorem 2.6 and a product rule for Dini derivatives to extend the necessary conditions of [1, Theorem 1] to nonconvex functions.

3. General Optimality Conditions

In this section, we formulate two generalizations of Theorem 2.3 Both involve a broader concept of upper convex approximate than the one we have defined so far.

Definition 3.1. Let $f : E \to IR$ and $x \in E$, and let f^A be some directional derivative. We say that a l.s.c., proper sublinear function $h : E \to (-\infty,\infty]$ is a generalized A upper convex approximate of f at x if $h(y) \geq f^A(x;y)$ for all $y \in E$. We will denote the class of all such upper convex approximates by $GUC(A,f,x)$.

We observe that if $f^A(x;0) \leq 0$, as is the case for f^L and f^U, then $GUC(A,f,x)$ is always nonempty, since it will at least contain the indicator function of the set $\{0\}$. We note also that $\partial h(0)$ is nonempty for every $h \in GUC(A,f,x)$.
For the remainder of this section, we will let f^B and f^D be two directional derivatives that have the following properties:

(a) For all $x \in E$, $f^B(x;0) \leq 0$ and $f^D(x;0) \leq 0$.
(b) For all constant functions c, $(f + c)^B(x;\circ) = f^B(x;\circ)$.
(c) If x is a local minimum of f, then $f^B(x;y) \geq 0$ for all $y \in E$.
(d) If $f(x) := \max_{1 \leq i \leq n} f_i(x)$, f_i is continuos at x for each $i \in \{1,\ldots,n\} \setminus I(x)$, and some hypothesis $H(f,B,D,x)$ is satisfied, then for $j \in I(x)$ and all $y \in E$,

$$(9) \qquad f^B(x;y) \leq \max \{ f_j^B(x;y), \max_{I(x)\setminus\{j\}} f_i^D(x;y) \}.$$

We have seen in §2 that one possible choice for B and D is $B := L$, $D := U$. A suitable hypothesis $H(f,L,U,x)$ would be the assumption that f is l.s.c. at x. Another possibility is $B := K$, $D := k$, where

$$f^K(x;y) := \liminf_{\substack{t \downarrow 0 \\ v \to y}} (f(x + tv) - f(x))/t$$

and

$$f^k(x;y) := \sup_{Y \in N(y)} \limsup_{t \downarrow 0} \inf_{v \in Y} (f(x + tv) - f(x))/t.$$

Hypotheses $H(f, K, k, x)$ that will guarantee (9) are derived in [22-24].
We now present the first of our generalizations of Theorem 2.3. In the statement of this result, we denote

$$F(x) := \max \{ f(x) - f(x_0), \max_J g_i(x) \}$$

as in the proof of Theorem 2.3.

Theorem 3.2. Let $f(x_0)$ be a local minimum for (P), and let B and D be defined as above. Assume that g_i is continuous at x_0 for each i in $J\backslash I(x_0)$; that a hypothesis H(F, B, D, x_0) holds; and that there exist $h \in GUC(B, f, x_0)$ and $h_i \in GUC(D, g_i, x_0)$, $i \in I(x_0)$, such that each h_i is continuous on the interior of its domain and

(10) $$\operatorname{dom} h \cap \bigcap_{I(x_0)} \operatorname{int} \operatorname{dom} h_i \neq \emptyset$$

Then there exist $\lambda_i \geq 0$, $i \in I(x_0) \cup \{0\}$, at least one of which is nonzero, such that

(11) $$0 \in \lambda_0 \partial h(0) + \sum_{I(x_0)} \lambda_i \partial h_i(0),$$

where $\lambda_i \partial h_i(0)$ denotes $0^+ \partial h_i(0)$ if $\lambda_i = 0$. If in addition

(12) $$\operatorname{dom} h \cap \bigcap_{I(x_0)} \{ y \mid h_i(y) < 0 \} \neq \emptyset$$

then $\lambda_0 > 0$ in (11).

The proof of Theorem 3.2 is completely analogous to that of Theorem 2.3. We have made assumption (10) to ensure that the subgradient sum formula may be applied to $\lambda_0 h + \sum_{I(x_0)} \lambda_i h_i$.

It is of course true that the larger the classes of functions GUC(B, f, x_0) and GUC(D, g_i, x_0), the stronger the result obtained in Theorem 3.2. Since $f^K(x; \circ) \leq f^L(x; \circ)$ and $f^k(x; \circ)$, $f^U(x; \circ)$the choice B: = K, D: = k can be used to strengthen Theorem 2.3.

Our second generalization of Theorem 2.3 relies upon a special property of the asymptotic directional derivative f^{A^∞} that we defined in §1. To explain this property, we make the following definition.

Definition 3.3. Suppose $f^A(x; \circ)$ is proper. The function $h \in GUC(A, f, x)$ is said to be a maximal element of GUC(A, f, x), and epi h is said to be a maximal closed convex subcone of epi $f^A(x; \circ)$, if whenever S is a closed convex cone with epi $h \subset S \subset$ epi $f^A(x; \circ)$, then either S = epi h or S = epi $f^A(x; \circ)$.

If $f^A(x; \circ)$ is l.s.c., the existence of a maximal element of GUC(A, f, x) that minorizes a given element of GUC(A, f, x) can be demonstrated with a standard Zorn's Lemma argument. There is also an interesting relationship between maximal elements of GUC(A, f, x) and $f^{A^\infty}(x; \circ)$.

Proposition 3.4. If $f^A(x;\circ)$ is l.s.c. and proper with $f^A(x;0) = 0$, and h is a maximal element of GUC(A, f, x), then $h(\circ) \le f^{A^\infty}(x;\circ)$.

Proof. If epi $f^A(x;\circ)$ is convex, then the only maximal element of GUC(A, f, x) is $f^A(x;\circ)$, and the inequality holds. Suppose epi $f^A(x;\circ)$ is not convex, and let h be a maximal element of GUC(A, f, x) and $(y, r) \in$ epi $f^{A^\infty}(x;\circ)$. Then for all $(z, s) \in$ epi h, $f^A(x; y + z) - s \le r$, so that $(y + z, r + s) \in$ epi $f^A(x;\circ)$. Since $f^{A^\infty}(x;\circ)$ is positively homogeneous, it follows that the set

$$C := \mathrm{cl}(\mathrm{epi}\ h + \bigcup_{\lambda \ge 0} \lambda(y, r))$$

is a closed convex subcone of epi $f^A(x;\circ)$. Now by the maximality of h, we must have C = epi h, and since $(0,0) \in$ epi $f^A(x;\circ)$, it follows that $(0,0) \in$ epi h. Hence $(y, r) \in C$, and so $(y, r) \in$ epi h and the proof is complete.

We can now use Proposition 3.4 to establish an alternate version of Theorem 3.2 in which assumptions (10) and (12) are replaced by conditions involving $f^{B^\infty}(x_0;\circ)$ and $g_i^{D^\infty}(x_0;\circ)$.

Theorem 3.5. Let $f(x_0)$ be a local minimum for (P), and let B and D be defined as above. Assume that g_i is continuous at x_0 for each i in $J\backslash I(x_0)$, that a hypothesis H(F, B, D, x_0) holds, and that $f^B(x_0;\circ)$ and $g_i^D(x_0;\circ)$ are l.s.c. and proper for each $i \in I(x_0)$. Further, suppose that each $g_i^{D^\infty}(x_0;\circ)$ is continuous on the interior of its domain and

(13) $$\mathrm{dom}\ f^{B^\infty}(x_0;\circ) \cap \bigcap_{I(x_0)} \mathrm{int\ dom}\ g_i^{D^\infty}(x_0;\circ) \ne \emptyset$$

Then for each $h \in$ GUC(B, f, x_0) and $h_i \in$ GUC(D, g_i, x_0), there exist $\lambda_i \ge 0$, $i \in I(x_0) \cup \{0\}$, at least one of which is nonzero, such that

(14) $$0 \in \lambda_0 \partial h(0) + \sum_{I(x_0)} \lambda_i \partial h_i(0),$$

where $\lambda_i \partial h_i(0)$ denotes $0^+\partial h_i(0)$ if $\lambda_i = 0$. If in addition

(15) $$\mathrm{dom}\ f^{B^\infty}(x_0;\circ) \cap \bigcap_{I(x_0)} \{\, y \mid g_i^{D^\infty}(x_0; y) < 0 \,\} \ne \emptyset,$$

then $\lambda_0 > 0$ in (11).

Proof. Define F as in Theorem 2.3, and let $h \in GUC(B, f, x_0)$ and $h_i \in GUC(D, g_i, x_0)$, $i \in I(x_0)$. By Zorn's Lemma and Proposition 3.4, there exist $p \in GUC(B, f, x_0)$, $p_i \in GUC(D, g_i, x_0)$ such that $p(\circ) \le \min \{ h(\circ), f^{B^\infty}(x_0; \circ) \}$ and $p_i(\circ) \le \min \{h_i(\circ), g_i^{D^\infty}(x_0; \circ)\}$. As in the proof of Theorem 2.3, we have for all $y \in E$ that

$$0 \le F^B(x; y) \le \max \{ f^B(x_0; y), \max_{I(x_0)} g_i^D(x_0; y) \} \le \max_{I(x_0)} \{ p(y), \max p_i(y) \}.$$

We may then carry out another separation argument to deduce the existence of $\lambda_i \ge 0$, $i \in I(x_0) \cup \{0\}$, at least one of which is nonzero, such that for all $y \in E$,

$$(16) \qquad 0 \le \lambda_0 p(y) + \sum_{I(x_0)} \lambda_i p_i(y).$$

By (13), we may apply the subgradient sum formula to conclude that

$$0 \in \partial(\lambda_0 p)(0) + \sum_{I(x_0)} \partial(\lambda_i p_i)(0) \subset \partial(\lambda_0 h)(0) + \sum_{I(x_0)} \partial(\lambda_i h_i)(0),$$

and (14) follows. Finally, suppose (15) holds, and let y be an element of the set in (15). If $\lambda_0 = 0$, if follows from (16) that

$$0 \le \sum_{I(x_0)} \lambda_i g_i^{D^\infty}(x_0; y),$$

from which we conclude that each $\lambda_i = 0$, a contradiction. Therefore $\lambda_0 > 0$.

The hypotheses of Theorem 3.5 include the assumption that $f^B(x_0; \circ)$ and $g_i^D(x_0; \circ)$ are l.s.c. and proper, an assumption not required in Theorem 3.2. On the other hand, Theorem 3.5 has the advantage of replacing (10) and (12) with (13) and (15), which do not depend on h and h_i.

We conclude this section with a discussion of the special case $B := L$, $D := U$ in Theorem 3.5. We begin by compiling some informations about the directional derivatives f^{L^∞} and f^{U^∞}.

Definition 3.6. [19] The function $f: E \to IR$ is said to be directionally Lipschitzian at $x \in E$ if there exists some $y \in E$ such that

$$\inf_{\substack{Y \in N(y) \\ X \in N((x, f(x))) \\ \lambda > 0}} \quad \sup_{\substack{v \in Y \\ (u, r) \in \text{epi } f \cap X \\ t \in (0, \lambda)}} (f(u + tv) - r)/t < +\infty.$$

The class of directionally Lipschitzian functions is investigated extensively in [4, 20]. In particular, it is shown in [20] that if f is directionally Lipschitzian at x, then $f^{O}(x;\circ)$ is continuous on the interior of its domain.

The following facts about $f^{L^{\infty}}$ and $f^{U^{\infty}}$ are direct consequences of properties of f^{L} and f^{O}.

Proposition 3.7. Let $f: E \to \mathbb{R}$ and $x \in E$.

a) If f is directionally Lipschitzian at x and $f^{L}(x;\circ)$ (resp., $f^{U}(x;\circ)$) is proper, then $f^{L^{\infty}}(x;\circ)$ ($f^{U^{\infty}}(x;\circ)$) is continuous on the interior of its domain, which is nonempty.

b) If E is a normal space and f is locally Lipschitzian near x, then $f^{L^{\infty}}(x;\circ)$ and $f^{U^{\infty}}(x;\circ)$ are finite and continuous on E.

Proof. We first note that the inequalities

(17) $$f^{L}(x;\circ) \le f^{L^{\infty}}(x;\circ) \le f^{O}(x;\circ)$$

and

(18) $$f^{U}(x;\circ) \le f^{U^{\infty}}(x;\circ) \le f^{O}(x;\circ)$$

are true in general. To see that $f^{U^{\infty}}(x;\circ) \le f^{O}(x;\circ)$, let $\varepsilon > 0$ be given and suppose $f^{O}(x;y) \le r$ and $f^{U}(x;y) \le s$. Then there exist $X \in N((x, f(x)))$ and $\lambda_1 > 0$ such that for all $(u, d) \in \operatorname{epi} f \cap X$ and $t \in (0, \lambda_1)$,

(19) $$(f(u + ty) - d)/t \le r + \varepsilon/2.$$

Moreover, there exists $\lambda_2 \in (0, \lambda_1)$ such that for all $t \in (0, \lambda_2)$,

(20) $$(f(x + tz) - f(x))/t \le s + \varepsilon/2.$$

Now by (20), we may choose $\lambda \in (0, \lambda_2)$ such that for all $t \in (0, \lambda)$, $(x + tx, f(x) + ts + t\varepsilon/2) \in X \cap \operatorname{epi} f$. We deduce from (19) that

$$(f(x + tz + ty) - f(x) - ts - t\varepsilon/2)/t \le r + \varepsilon/2.$$

or

$$(f(x + tz + ty) - f(x))/t - s \le r + \varepsilon.$$

Since ε and (z, s) were arbitrarily chosen, we have $f^{U^{\infty}}(x;y) \le r$, and so the second inequality in (18) is valid. The other inequalities can be established similarly.

Statement (b) now follows directly from (17) and (18), since if f is locally Lipschitzian near x, $f^L(x;\circ)$ and $f^o(x;\circ)$ are finite and continuous on E. If f is directionally Lipschitzian at x, there exist $y \in E$ and $Y \in N(y)$ such that $f^o(x;\circ)$ is bounded above on Y (see [20]). If in addition $f^L(x;\circ)$ (respectively, $f^U(x;\circ)$) is proper, then (17) (resp., (18)) implies that $f^{L^\infty}(x;\circ)$ ($f^{U^\infty}(x;\circ)$) is bounded above and $> -\infty$ on Y. Therefore assertion (a) is valid.

We now give two special cases of Theorem 3.5 for B: = L, D: = U.

Corollary 3.8. Let $f(x_0)$ be a local minimum for (P), where g_i is directionally Lipschitzian at x_0 for each $i \in I(x_0)$ and continuous at x_0 for each i in $J \setminus I(x_0)$. Assume that $f^L(x_0;\circ)$ and $g_i^U(x_0;\circ)$ are proper and l.s.c. for each $i \in I(x_0)$, and that

(21) $$\operatorname{dom} f^{L^\infty}(x_0;\circ) \cap \bigcap_{I(x_0)} \operatorname{int} \operatorname{dom} g_i^{U^\infty}(x_0;\circ) \neq \emptyset.$$

Then for each $h \in GUC(L, f, x_0)$ and $h_i \in GUC(U, g_i, x_0)$, there exist $\lambda_i \geq 0$, $i \in I(x_0) \cup \{0\}$, at least one of which is nonzero, such that

(22) $$0 \in \lambda_0 \partial h(0) + \sum_{I(x_0)} \lambda_i \partial h_i(0)$$

where $\lambda_i \partial h_i(0)$ denotes $0^+\partial h_i(0)$ if $\lambda_i = 0$. If in addition

(23) $$\operatorname{dom} f^{L^\infty}(x_0;\circ) \cap \bigcap_{I(x_0)} \{\, y \mid g_i^{U^\infty}(x_0; y) < 0 \,\} \neq \emptyset,$$

then $\lambda_0 > 0$ in (22).

Proof. Let B: = L, D: = U in Theorem 3.5. By Proposition 3.7, each $g_i^{U^\infty}(x_0;\circ)$, $i \in I(x_0)$, is continuous on the interior of its domain. The assertions then follow from Theorem 3.5.

Corollary 3.9. Let $f(x_0)$ be a local minimum for (P), where f and g_i are locally Lipschitzian near x_0 for each $i \in I(x_0)$ and continuous at x_0 for each i in $J \setminus I(xo)$. Then for each $h \in GUC(L, f, x_0)$ and $h_i \in GUC(U, g_i, x_0)$, there exist $\lambda_i \geq 0$, $i \in I(x_0) \cup \{0\}$, at least one of which is nonzero, such that

(24) $$0 \in \lambda_0 \partial h(0) + \sum_{I(x_0)} \lambda_i \partial h_i(0),$$

where $\lambda_i \partial h_i(0)$ denotes $0^+\partial h_i(0)$ if $\lambda_i = 0$. If in addition

(25) $$\bigcap_{I(x_0)} \{ y \mid g_i^{U^\infty}(x_0; y) < 0 \} \neq \emptyset,$$

then $\lambda_0 > 0$ in (24).

Proof. Since f and g_i, $i \in I(x_0)$, are locally Lipschitzian, $f^L(x_0; \circ)$ and $f^U(x; \circ)$ are finite and continuous, and Proposition 3.7 implies that (21) is satisfied. The assertions then follow from Corollary 3.8.

Remark 3.10. (a) Among the possible choices for h and h_i in Corollary 3.9 are $f^O(x_0; \circ)$, $f^P(x_0; \circ)$, and $f^{U^\infty}(x_0; \circ)$.

(b) In Corollary 3.9, if h and h_i are majorized by $f^O(x_0; \circ)$, then $0^+\partial h(0) = 0^+\partial h_i(0) = \{0\}$, since $\partial f^O(x_0; \circ)(0)$ is bounded whenever f is locally Lipschitzian near x_0 [7].

(c) In the special case $B: = K$, $D: = k$ in Theorem 3.5, one hypothesis $H(F, K, k, x_0)$ that will guarantee (9) is the assumption that each g_i is directionally Lipschitzian at x_0 and (13) holds. Assumption (13) thus serves to "kill two birds with one stone" in this case. It is also true under this hypothesis that each $g_i^{k^\infty}(xo; \circ)$ is continuous on the interior of its domain. The details of the case $B: = K$, $D: = k$ are carried out in [13].

(d) One result from the literature that may be deduced from Corollary 3.9 is [9, Theorem 6]. In fact, Corollary 3.9 can be used to derive the conclusion of [9, Theorem 6] under a weaker constraint qualification than the one given in [9].

4. Conclusion

We have seen that necessary optimality conditions of Fritz John and Kuhn-Tucker type for (P) can be based upon any l.s.c. sublinear majorants of the Dini directional derivatives, or indeed upon any l.s.c. sublinear majorants of directional derivatives f^B and f^D that satisfy the calculus rule (9). In these optimality conditions, the asymptotic directional derivatives f^{B^∞} and f^{D^∞} play a special role.

References

[1] **Borwein J.M.**, Fractional programming without differentiability. Mathematical Programming 11 (1976) 283-290.

[2] **Borwein J.M.**, Stability and regular points of inequality systems. Journal of Optimization Theory and Applications 48 (1986) 9-52.

[3] **Borwein J.M., Fitzpatrick S.P., and Giles J.R.**, The differentiability of real functions on normed linear space using generalized subgradients, Journal of Mathematical Analysis and Applications 128 (1987) 512-534.

[4] **Borwein J.M. and Strojwas H.M.**, Directionally Lipschitzian mappings on Baire spaces, Canadiana Journal of Mathematics 36 (1984) 95-130.

[5] **Caligaris O. and Oliva P.**, Necessary conditions for local Pareto minima, Bollettino della Unione Matematica Italiana 5-B (1986) 721-750.

[6] **Clarke F.H.**, A new approach to Lagrange multipliers, Mathematics of Operations Research 1 (1976) 165-174.

[7] **Clarke F.H.**, Optimization and nonsmooth analysis -John Wiley and Sons, New York, 1983.

[8] **Ekeland I. and Temam R.**, Convex analysis and variational problems - North-Holland, Amsterdam, 1976.

[9] **Hiriart-Urruty J.-B.**, On optimality conditions in nondifferentiable programming, Mathematical Programming 14 (1978) 73-86.

[10] **Hiriart-Urruty J.-B.**, New concepts in nondifferentiable programming, Bull. Soc. Math. France, Memoire 60 (1979) 57-85.

[11] **Ioffe A.D.**, Necessary conditions for a local minimum. 1. A reduction theorem and first-order conditions, STAM J.Cont. Opt. 17 (1979) 245-250.

[12] **Jeyakumar V.**, On optimality conditions in nonsmooth inequality-constrained minimization, Numerical Functional Analysis and Optimization 9 (1987) 535-546.

[13] **Merkovsky R.R. and Ward D.E.**, Upper D.S.L. approximates and nonsmooth optimization, preprint, 1988.

[14] **Michel P. and Penot J.-P.**, Calcul sous-differentiel pour des fonctions lipschitziennes et non lipschitziennes, C.R. Acad. Sci. Paris 298 (1984) 269-272.

[15] **Penot J.-P.**, Variations on the theme of nonsmooth analysis: another subdifferential, in: Demyanov V.F. and Pallaschke D., eds., Nondifferentiable optimization: motivations and applications, Springer-Verlag, Berlin, 1985, pp. 41-54.

[16] **Penot J.-P. and Terpolilli P.**, Cones tangents et singularites, C.R. Acad. Sci. Paris 296 (1983) 721-724.

[17] **Pschenichnyi B.N.**, Necessary Conditions for an extremum, Marcel Dekker, New York, 1971.

[18] **Pschenichnyi B.N. and Khachatryan R.A.**, Constraints of equality type in nonsmooth optimization problems, Soviet Mathematics Doklady 26, (1982) 659-662.

[19] **Rockafellar R.T.**, Directionally Lipschitzian functions and subdifferential calculus, Proceedings of the London Mathematical Society 39 (1979) 331-355.

[20] **Rockafellar R.T.**, Generalized directional derivatives and subgradients of nonconvex functions, Canadian Journal of Mathematics 32 (1980) 257-280.

[21] **Studniarski M.**, Mean value theorems and sufficient optimality conditions for nonsmooth functions, Journal of Mathematical Analysis and Applications 111(1985)313-326.

[22] **Ward D.E.**, Convex subcones of the contingent cone in nonsmooth calculus and optimization, Transactions of the Americam Mathematical Society 302 (1987) 661-682.
[23] **Ward D.E.**, Directional derivative calculus and optimality conditions in nonsmooth mathematical programming, to appear in Journal of Information and Optimization Sciences.
[24] **Ward D.E. and Borwein J.M.**, Nonsmooth calculus in finite dimensions, SIAM J. Cont. Opt. 25 (1987) 1312-1340.

DIFFERENTIABLE (α, λ) - CONCAVE FUNCTIONS

E. Castagnoli - P. Mazzoleni

Abstract

We analyze some properties of a recently introduced concept, the (α, λ)-concavity, useful to state a unified theory both in applications and optimization problems.
Two structural properties, monotonicity and concavifiability, are studied; but what is more interesting, once and twice differentiable (α, λ)-concave functions are characterized in order to find a continuous link between concavity and quasiconcavity.

1. Basic Properties

Recently, we have introduced a definition of concavity for a function that allows us to unify several kinds of concavity already known in the literature.
Throughout this paper, we will use the following notation: we denote the closed interval [0, 1], by I and the open one by I°. A bar over a variable stands for its complement to one: e.g., $\bar{\alpha}$ is $1-\alpha$. All vectors are to be intended as column vectors; therefore, the gradient of a function is to be read as a row vector. Finally, C denotes any convex subset of R^n.

Definition 1. [Castagnoli and Mazzoleni (1989a). A function $f: C \rightarrow R$ is said to be (α, λ)-concave, or simply λ-concave, if:

$$f(x) \leq f(y) \Rightarrow f(\alpha x + \bar{\alpha} y) \geq \lambda f(x) + \bar{\lambda} f(y) \qquad \forall\ x, y \in C\ \forall\ \alpha \in I$$

where:

$$\lambda = \lambda(\alpha; x, y) \in I$$

Moreover, the function is said to be explicitly $(\alpha\lambda)$-concave whenever the second inequality holds only if $f(x) < f(y)$; and it is said to be (α, λ)-concave in the strict sense whenever the second inequality is strict, for $\alpha \in I°$. Finally, a function $f: C \rightarrow R$ is said to be (α, λ)-convex if $-f$ is (α, λ)-concave; similar definitions apply to the other cases.
Well known examples of (α, λ)-concave functions are the following. When $\lambda = 1$, we obtain quasiconcave functions. For $\lambda = \alpha$, instead, we find classical concave functions. When $\bar{\lambda} = \bar{\alpha}\, b(x, y)$, with $0 < b(x, y) \leq 1$, explicitly (α, λ)-concave functions are nothing but pseudoconcave functions.
Finally, explicitly $(\alpha, 1)$-concave functions in the strict sense, that is functions such that:

$$f(x) < f(y) \Rightarrow f(\alpha x + \bar{\alpha} y) > f(x) \qquad \forall\ x, y \in C\ \forall\ \alpha \in I°$$

are the explicitly quasiconcave functions of Martos (1975).
We have shown in Castagnoli and Mazzoleni (1989b) that Definition 1 can be restated as follows by making symmetric the function λ:

Definition 1'. A function $f : C \to R$ is said to be (α, λ)-concave, or simply λ-concave, if:

$$f(\alpha x + \bar{\alpha} y) \geq \lambda f(x) + \bar{\lambda} f(y) \quad \forall\, x, y \in C\, \forall\, \alpha \in I$$

where:

$$\lambda = \lambda(\alpha; x, y) \in I$$

is such that:

$$\lambda(\alpha; x, y) = \bar{\lambda}(\bar{\alpha}; y, x)$$

Remarks 1. Is is immediate to show that:

$$\lambda(0; x, y) \geq 0$$

and:

$$\lambda(1; x, y) = 1 \quad \forall\, x, y \in C$$

2. The function λ describes the degree of concavity of an (α, λ)-concave function. When $\alpha \leq \lambda \leq 1$, all intermediate cases between classical concavity and quasiconcavity are generated; moreover, $0 < \lambda \leq \alpha$ characterizes a higher degree of concavity than the classical case.
3. In other words, we can say that (α, λ)-concavity is based on differently weighed arithmetic means in the domain (with weights $\alpha, \bar{\alpha}$) and in the codomain (with weights $\lambda, \bar{\lambda}$) of the function. This suggests that the most interesting case is obtained when λ depends on x and y through $f(x)$ and $f(y)$. In this case, λ will be denoted as $\Lambda = \Lambda[\alpha; f(x), f(y)]$. In a more general setting, λ could be assumed dependent on $x, y, f(x), f(y)$:

$$\lambda = \lambda[\alpha; x, y, f(x), f(y)]$$

However, this possibility will not be explored here.
It is always possible to refer to Λ : it suffices to define $\Lambda[\alpha; f(x), f(y)]$ as the supremum of $\lambda(\alpha; x, y)$ when x and y vary so that $f(x)$ and $f(y)$ are kept constant.

Theorem 2. [Castagnoli and Mazzoleni (1989a)]. An (α, λ_1)-concave function is also (α, λ_2)-concave if $\lambda_2 \leq \lambda_1$.
(α, λ)-concave functions are thus partially ordered by their (maximal function) λ : concavity weakens as λ increases. In particular, this implies that every (α, λ)-concave function is $(\alpha, 1)$-concave, that is quasiconcave.
Theorem 3. A function $f : C \to R$ is (α, λ)-concave if and only if the functions $\phi_{x,h}(t) = f(x+th)$, defined for any choice of $t \in R$ such that $x, x+th \in C$ are (α, γ)-concave. Moreover,

$$\gamma = \gamma_{x,y}(\alpha, t, \tau) = \lambda(\alpha ; x + th, x + \tau h)$$

Proof. (a) Let f be an (α, λ)-concave function. It is:

$$\phi(\alpha t + \bar{\alpha}\tau) = f[x + (\alpha t + \bar{\alpha}\tau) h] = f[\alpha(x + th) + \bar{\alpha}(x + \tau h)] \geq$$
$$\lambda f(x + th) + \bar{\lambda} f(x + \tau h) = \gamma \phi(t) + \bar{\gamma} \phi(\tau)$$

(b) Viceversa, let ϕ be an (α, γ)-concave function:

$$f(\alpha x + \bar{\alpha} y) = \phi_{x,y-x}(\bar{\alpha}) \geq \gamma(\alpha ; 0, 1)\,\phi(0) + \bar{\gamma}(\alpha ; 0, 1)\,\phi(1) = \lambda f(x) + \bar{\lambda} f(y)$$

□

Theorem 4. A function f is pseudoconcave if and only if it is explicitly (α, λ)-concave with $\bar{\lambda}$ linear in $\bar{\alpha}$.

Proof. (a) Pseudoconcavity of f requires that, if $f(x) < f(y)$:

$$f(\alpha x + \bar{\alpha} y) \geq f(x) + \bar{\alpha} B(x, y)$$

with $0 < B(x, y) \leq f(y) - f(x)$.
Let us set $B(x, y) = b(x, y)[f(y) - f(x)]$, where $0 < b(x, y) \leq 1$. The previous inequality can be written:

$$f(\alpha x + \bar{\alpha} y) \geq f(x) + \bar{\alpha} b(x, y)[f(y) - f(x)] = [1 - \bar{\alpha} b(x, y)] f(x) + \bar{\alpha} b(x, y) f(y)$$

which proves explicit (α, λ)-concavity of f with $\bar{\lambda} = \bar{\alpha} b(x, y)$.
(b) Viceversa, explicit (α, λ)-concavity of f requires that, if $f(x) < f(y)$:

$$f(\alpha x + \bar{\alpha} y) \geq f(x) + \bar{\lambda}[f(y) - f(x)]$$

Therefore, if $\bar{\lambda}$ is linear in $\bar{\alpha}$, pseudoconcavity is proved by setting:

$$\bar{\lambda}[f(y) - f(x)] = \bar{\alpha} B(x, y).$$

□

Example. Consider the function $f(x) = x^3$. This function is ((α, λ)-concave with:

$$\bar{\lambda} = \frac{3\,\bar{\alpha}^2 (y-x)^2 + 6\,\bar{\alpha} x\,(y-x) + 3\,x^2}{y^2 + xy + x^2}$$

If we restrict attention to functions λ not depending on x and y, the maximum degree of concavity λ to have (α, λ)-concavity is $\lambda = 1 - \bar{\alpha}^3$ (and therefore $\bar{\lambda} = \bar{\alpha}^3$). In fact, by taking $x^3 \leq y^3$ (that is $x \leq y$):

$$f(\alpha x + \bar{\alpha} y) = (\alpha x + \bar{\alpha} y)^3 = \alpha^3 x^3 + 3\alpha\bar{\alpha} x y(\alpha x + \bar{\alpha} y) + \bar{\alpha}^3 y^3$$
$$\lambda f(x) + \bar{\lambda} f(y) = (1 - \bar{\alpha}^3) x^3 + \bar{\alpha}^3 y^3 = \alpha^3 x^3 + 3\alpha\bar{\alpha} x^3 + \bar{\alpha}^3 y^3$$

and we only need to compare:

$$x y(\alpha x + \bar{\alpha} y) \; : \; x^3$$

with the $\geq$ sign holding between the two terms.
The function $f(x) = x^3$ is strictly monotonic but not pseudoconcave; however, it is more than quasiconcave ($\lambda = 1 - \bar{\alpha}^3 < 1$). It is (α, λ)-convex with $\lambda = \alpha^3$ as well. More generally, it is possible to show that the functions defined by $f(x) = x^k$, k odd integer, are (α, λ)-concave with $\lambda = 1 - \bar{\alpha}^k$ and (α, λ)-convex with $\lambda = \alpha^k$.
The function defined by $f(x) = e^x$ is (α, λ)-concave with:

$$\bar{\lambda} = \frac{e^{\bar{\alpha}(y-x)} - 1}{e^{y-x} - 1}$$

2. Monotonicity properties

Let f be a function defined on $C \subseteq R^n$. For every closed segment $[x, y] \subset C$, let us consider the line function defined by $\phi(t) = f(x + th)$, $h = y - x$, $t \in I$.
It is well known [Martos (1975)] that a necessary and sufficient condition for quasiconcavity of f is that every $[x, y] \subset C$ can be partitioned in two subintervals, one of them at least being closed, such that the corresponding line function ϕ is increasing on the first one and decreasing on the second; in particular one of the intervals can be empty or degenerate in a single point. In brief, we can say that ϕ is monotonic or maximal.
For pseudoconcave functions, a similar result holds: a function f is pseudoconcave if and only if the corresponding line functions are strictly monotonic or maximal.
A similar characterization holds also for (α, λ)-concave functions.
We first give the following:

Definition 5. Let ϕ be a real valued function defined on a real interval. We will say that ϕ is k-increasing if:

$$x \geq y \;\Rightarrow\; \phi(x) \geq \phi(y) + k$$

where $k \geq 0$ can depend on x and y.
It is obvious that, if $k = 0$ for each x, y, k-increasingness reduces to simple increasingness.

We will say that ϕ is <u>k-decreasing</u> if $-\phi$ is k-increasing and that ϕ is k-monotonic if it is k-increasing or k-decreasing.
We remark that, in the scalar case, Definition 5 conveniently extends the following definition of uniform increasingness:

Definition 6. [Ortega and Rheinboldt (1970)]. The function $F: D \to R^n$, is said to be <u>increasing</u> if:

$$[F(x) - F(y)]^T (x-y) \geq 0 \qquad \forall x, y \in D$$

F is said to be <u>strictly increasing</u> if the strict inequality holds whenever x is different from y. Finally F is said to be <u>uniformly increasing</u> if a number $\delta > 0$ exists such that:

$$[F(x) - F(y)]^T (x-y) \geq \delta \| x-y \|^2 \quad \forall x, y \in D$$

Theorem 7. A function $f: C \to R$ is (α, λ)-concave if and only if, for every segment $[x, y] \subset C$, the line function $\phi(t)$, $t \in I$, is maximal or k-monotonic with $k = \bar{\lambda}[f(y) - f(x)]$, $f(y) > f(x)$.

Proof. (a) If ϕ is k-increasing on I:

$$\phi(\bar{\alpha}) \geq \phi(0) + k \qquad \forall \bar{\alpha} \in I$$

therefore $(y-x = h)$:

$$f[x + \bar{\alpha}(y-x)] \geq f(x) + k$$
$$f(\alpha x + \bar{\alpha} y) \geq f(x) + k$$

and it is sufficient to take λ such that $k = \bar{\lambda}[f(y) - f(x)]$ to have the (α, λ)-concavity of f.
If ϕ is k-decreasing on I:

$$\phi(\bar{\alpha}) \geq \phi(1) + k$$

that is:

$$f(\alpha x + \bar{\alpha} y) \geq f(y) + k$$

which shows again (α, λ)-concavity of f.
Whenever ϕ is maximal, one can partition I in three subintervals $I^1 = [0, t^1\rangle$, $I^2 = \langle t^1, t^2\rangle$, $I^3 = \langle t^2, 1]$ such that f is k-increasing on I^1, constant (and maximal) on I^2 and k-decreasing on I^3.
We have:

$$\phi(u) \geq \phi(v) \qquad \forall u \in I^2, \ \forall v \in I^1, I^3$$

Letting:

$$u = x + \bar{\alpha}_2(y-x), \ v = x + \bar{\alpha}_1(y-x) \quad \text{or } v = x + \bar{\alpha}_3(y-x)$$

whenever $v \in I^1$ or $v \in I^3$, the preceding inequality becomes:

$$f[x+\bar{\alpha}_2(y-x)] \geq f[x+\bar{\alpha}_1(y-x)] = \phi(\bar{\alpha}_1) \geq \phi(0)+k = f(x)+k$$
$$f[x+\bar{\alpha}_2(y-x)] \geq f[x+\bar{\alpha}_3(y-x)] = \phi(\bar{\alpha}_3) \geq \phi(1)+k = f(y)+k$$

As above, this implies (α, λ)-concavity of f.

(b) Let f be (α, λ)-concave. If : $\max(f(z) : z \in [x, y]) > \max(f(x), f(y))$, ϕ is maximal in β, $\beta \in I^\circ$, such that $z^* = x + \beta(y-x)$ is the maximum point for f on (x, y).
If $\max(f(z) : z \in [x, y]) = f(y)$, we have:

$$f(y) \geq f(\alpha x + \bar{\alpha} y) \geq f(x) + \bar{\lambda}[f(y) - f(x)]$$

that is:

$$\phi(1) \geq \phi(\bar{\alpha}) \geq \phi(0) + k$$

k-increasingness is obtained by noticing that an interval I is associated with every segment $[x, y]$, $x, y \in C$.
The case of $\max(f(z) : z \in [x, y]) = f(x)$ is dealt with in a similar way.

□

If f is quasiconcave, $k = 0$; if f is pseudoconcave:

$$k = \bar{\lambda}[f(y) - f(x)] \geq \bar{\lambda} B(x, y) > 0$$

if f is concave:

$$k = \bar{\alpha}[f(y) - f(x)] \qquad f(y) > f(x)$$

A characterizing property of differentiable pseudoconcave functions consists in the sufficiency of the condition $\nabla f(x) = 0$ to ensure that x is a maximum point.

Theorem 8. [Diewert, Avriel and Zang (1981)]. Let $f: C \to R^n$, C open, f a differentiable function. Then f is pseudoconcave if and only if $x \in C$, $\nabla f(x) h = 0 \Rightarrow \phi(t) = f(x+th)$ reaches a maximum in $t = 0$.
The condition $\nabla f(x) h = \phi'(0) = 0$ is thus sufficient for ϕ to be maximal in $t = 0$.
We can restate this in the following way:
(i) If $\phi'(0) = \nabla f(x) h = 0$, ϕ is maximal in $t = 0$.
(ii) If $\phi'(0) = \nabla f(x) h \neq 0$, ϕ is strictly monotonic in $t = 0$.
A similar property holds for (α, λ)-concave functions with $\lambda < 1$:
(i) If $\phi'(0) = \nabla f(x) h = 0$, ϕ is maximal in $t = 0$.
(ii) If $\phi'(0) = \nabla f(x) h \neq 0$, ϕ is strictly k-decreasing or k-increasing (in a different way, taking $t \in I$, ϕ is k-increasing in $t = 0$ or in $t = 1$)

3. Differentiable (α, λ)-concave functions

The following theorem generalizes well known inequalities for concave, pseudoconcave and quasiconcave functions.

Theorem 9. [Castagnoli and Mazzoleni (1989b)]. Let $f: C \to R$, C open, be a differentiable (α, λ)-concave function. Then there are two non negative numbers μ and μ^*, in general depending on x and y, such that:

$$(1) \qquad \mu(x, y)\,[\,f(x) - f(y)\,] \leq \nabla f(y)(x-y)$$

$$(2) \qquad \mu^*(x, y)\,[\,f(y) - f(x)\,] \leq \nabla f(x)(y-x)$$

Viceversa, if inequalities (1) and (2) hold, it is:

$$(\alpha\mu + \bar{\alpha}\mu^*)\, f(\alpha x + \bar{\alpha} y) \geq \alpha\mu\, f(x) + \bar{\alpha}\mu^*\, f(y)$$

that, whenever μ and μ^* are not simultaneously zero, is equivalent to (α, λ)-concavity of f with:

$$\lambda = \lambda(\alpha\,; x, y) = \alpha\mu \,/\, (\alpha\mu + \bar{\alpha}\mu^*)$$

Remark. 1. We remember that it is always possible to take:

$$\mu = \begin{cases} \limsup\limits_{\alpha \to 0^+} \lambda/\alpha & \text{if } f(x) > f(y) \\ \liminf\limits_{\alpha \to 0^+} \lambda/\alpha & \text{if } f(x) < f(y) \\ \text{any positive number} & \text{if } f(x) = f(y)) \end{cases}$$

$$\mu^* = \begin{cases} \liminf\limits_{\bar{\alpha} \to 0^+} \bar{\lambda}/\bar{\alpha} & \text{if } f(x) > f(y) \\ \limsup\limits_{\bar{\alpha} \to 0^+} \bar{\lambda}/\bar{\alpha} & \text{if } f(x) < f(y) \\ \text{any positive number} & \text{if } f(x) = f(y)) \end{cases}$$

2. In particular, if λ admits partial derivative with respect to α and $\lambda(0) = 0$, one can take:

$$\mu = \lim_{\alpha \to 0^+} \lambda(\alpha\,; x, y) \,/\, \alpha = \lambda'_+(0; x, y)$$

$$\mu^* = \lim_{\bar{\alpha} \to 0^+} \bar{\lambda}(\alpha\,; x, y) \,/\, \bar{\alpha} = \lim_{\alpha \to 1^-} (\,\lambda(1; x, y) - \lambda(\alpha\,; x, y)\,) \,/\, (1 - \alpha) = \lambda'_-(1; x, y)$$

3. Moreover, if λ is symmetric:

$$\lambda(\alpha : x, y) = \overline{\lambda}(\alpha : y, x)$$

it is also:

$$\mu = \lambda'_+(0; x, y) = \lambda'_-(1; y, x)$$
$$\mu^* = \lambda'_-(1; x, y) = \lambda'_+(0; y, x)$$

In the sequel, we will write briefly λ' instead of λ'_+ or λ'_-.
4. Under the assumption of the preceding Remark 2, the two inequalities (1) and (2) can be summarized in one inequality:

$$\lambda'(1; x, y)[f(y) - f(x)] \le \nabla f(x)(y-x)$$

5. Summing together both sides of the inequalities (1) and (2) we find:

$$(3) \qquad (\mu^* - \mu)[f(y) - f(x)] \le [\nabla f(x) - \nabla f(y)](y-x)$$

Example. For the function f: R→R defined by $f(x) = x^3$ one has:

$$\mu = 3y^2/(y^2 + xy + x^2) \qquad \mu^* = 3x^2/(y^2 + xy + x^2)$$

When $0 < \lambda < \alpha$, (α, λ)-concavity of a function can be often characterized in terms of uniform monotonicity of its gradient.
We first give the following definition, which restates Definition 6 for the differentiable case:

Definition 10. A differentiable function f: C→R, C open, is said to be uniformly concave if there exists A > 0 such that:

$$[\nabla f(x) - \nabla f(y)](y-x) \le -A(x-y)^T(x-y) \quad \forall x, y \in C$$

Theorem 11. An (α, λ)-concave function such that:

$$(\mu^* - \mu)[f(y) - f(x)] > 0 \qquad x, y \in C$$

is uniformly concave and viceversa.

Proof. (a) Let f be an (α, λ)-concave function. From inequality (3), it follows that:

$$[\nabla f(x) - \nabla f(y)](x-y) \ge -A(x-y)^T(x-y)$$

whenever:

$$(\mu - \mu^*)[f(x) - f(y)] \ge -A(x-y)^T(x-y)$$

that is:

$$0 < A \leq \frac{(\mu^* - \mu)\,[\,f(y) - f(x)\,]}{(x-y)^T\,(x-y)}$$

(b) If f is uniformly concave, it is certainly concave.

□

We consider now the relationships among pseudoconcave, quasiconcave and (α, λ)-concave functions with reference to first order conditions.
Let us take a pseudoconcave function. We consider only the case where $f(x) < f(y)$. Since $\bar{\lambda} / \bar{\alpha} = b(x, y)$, it is $\mu^* = b(x, y) > 0$ and we can write inequality (2) as follows:

$$\nabla f(x)(y-x) \geq b(x, y)\,[\,f(y) - f(x)\,] = B(x, y) > 0$$

Notice that, since $\lambda/\alpha = (1 - \bar{\alpha} b(x, y)) / \alpha$, μ is not limited from above; thus, inequality (1) is not significant.
The well known implication for pseudoconcave function:

$$f(x) < f(y) \Rightarrow \nabla f(x)(y-x) > 0$$

however, is now stated in a more detailed way which allows one to distinguish, by means of the parameter $b(x, y)$, different levels of pseudoconcavity.
For quasiconcave functions, in the case $f(x) < f(y)$, one has $\mu^* = \bar{\lambda} / \bar{\alpha} = 0$; hence, the usual (and weaker) inequality (2) holds:

$$\nabla f(x)(y-x) \geq 0$$

In brief: for quasiconcave functions, $\mu^* = 0$ and $\mu = +\infty$; for pseudoconcave functions $0 < \mu^* < +\infty$ and $\mu = +\infty$; for concave functions, $\mu = \mu^* = 1$.
Consider now the points $x-v$ and x; inequality (1) becomes:

$$\mu\,[\,f(x-v) - f(x)\,] \leq \nabla f(x)(-v)$$

By taking x and $x+v$, instead, inequality (2) becomes:

$$\mu^*\,[\,f(x+v) - f(x)\,] \leq \nabla f(x)(v)$$

Combining these two inequalities, we get:

$$\mu^*\,[\,f(x+v) - f(x)\,] \leq \nabla f(x)\,v \leq \mu\,[\,f(x) - f(x-v)\,] = -\mu\,[\,f(x-v) - f(x)\,]$$

Therefore, $\nabla f(x)\, v$ belongs to a cone. To represent it geometrically, let us introduce cartesian axes X, Y by setting $X = f(x + v) - f(x)$ and $Y = \nabla f(x)\, v$.

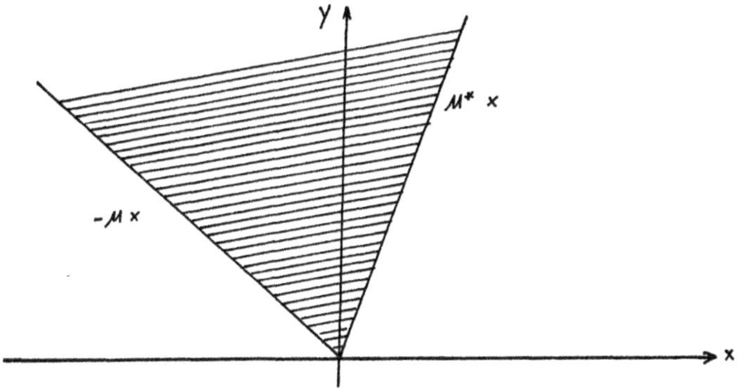

For pseudoconcave functions, only the left-hand side inequality is maintained:

$$b(x, y)\,[\, f(x + v) - f(x)\,] = \mu^*\,[\, f(x + v) - f(x)\,] \leq \nabla f(x)\, v$$

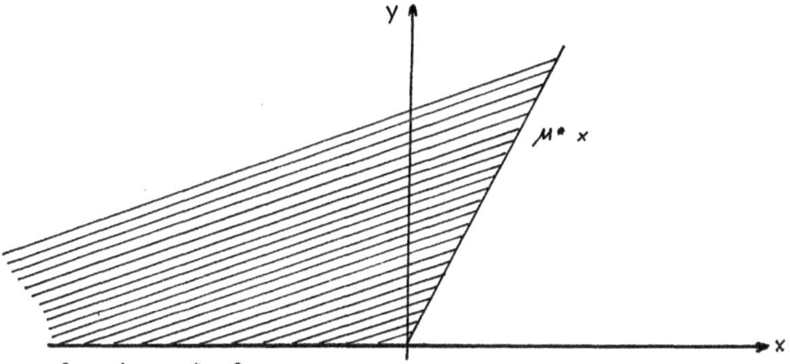

For quasiconcave functions, $\mu^* = 0$:

$$0 \leq \nabla f(x)\, v$$

and the cone degenerates into the nonnegative half-plane.

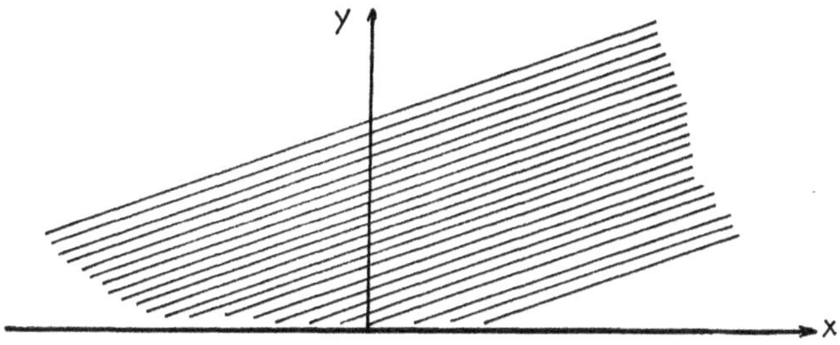

In brief, we have seen that the usual inequality:

$$f(y) - f(x) \le \nabla f(x)(y-x)$$

for concave functions holds in a modified form for (α, λ)-concave functions as well. We will now show that it is possible to preserve the usual form of the inequality by modifying the definition of derivative.

In Castagnoli and Mazzoleni (1989b), the following new definition of derivative was introduced:

Definition 12. Let $f: C \to R$, C open, be an (α, λ)-concave function with $\lambda < 1$. We will call "derivative" evaluated in x in the direction v ($\| v \| = 1$) the number $\mathcal{D}_v f(x)$, whenever existing, such that for $\bar{\alpha} \to 0^+$:

$$f[\alpha x + \bar{\alpha}(x + v)] \sim \lambda f(x) + \bar{\lambda}[f(x) + \mathcal{D}_v f(x)]$$

For quasiconcave functions, this definition leads to define:

$$\mathcal{D}_v f(x) = \lim_{\bar{\alpha} \to 0'} f(x + \bar{\alpha} v) - f(x)$$

if the limit exists and it is non positive; otherwise, we say that $\mathcal{D}_v f(x)$ does not exist.

For these new "derivatives", the following relationships hold:

$$\mathcal{D}_{x-y} f(y)\, \mu = D_{x-y} f(y) = \nabla f(y)(x-y)$$
$$\mathcal{D}_{y-x} f(x)\, \mu^* = D_{y-x} f(x) = \nabla f(x)(y-x)$$

Inequalities in the usual form:

$$f(x) - f(y) \le \mathcal{D}_{x-y} f(y)$$
$$f(y) - f(x) \le \mathcal{D}_{y-x} f(x)$$

follow immediately for (α, λ)-concave functions. By summing them up, we find:

$$\mathcal{D}_{x-y} f(y) - \mathcal{D}_{y-x} f(x) \ge 0$$

which shows a sort of decreasingness for these "derivatives".

4. Concavifiability of (α, λ)-concave functions.

General results concerning composition of functions are known for vector functions [Martos (1975)]. We summarize them below, restricting attention only to real-valued functions.

Theorem 13. Let $f : C \to R$ be a quasiconcave (pseudoconcave, concave) function. Il $g: f(C) \to R$ is increasing (increasing with positive derivative, increasing and concave), then the composite function defined by $g[f(x)]$ is quasiconcave (pseudoconcave, concave) as well.
A similar property holds also for (α, λ)-concave functions:

Theorem 14. Let $f : C \to R$ be an (α, λ)-concave function. If $g: f(C) \to R$ is increasing and (λ, β)-concave, the composite function defined by $g[f(x)]$ is (α, β)-concave through the composite function $\beta(\lambda)$.

Proof.

$$g[f(\alpha x + \bar{\alpha} y)] \geq g(\lambda f(x) + \bar{\lambda} f(y)] \geq \beta(\lambda) g[f(x)] + \bar{\beta}(\lambda) g[f(y)]$$

□

Definition 15. A function $f: C \to R$ is said to be <u>concavifiable</u> or <u>g-concave</u> if a continuous and increasing function $g: f(C) \to R$ exists such that the composite function $g[f(x)]$ is concave.
A necessary condition for concavifiability is pseudoconcavity; a sufficient condition is strong pseudoconcavity [Zang (1981)].
Under the assumption of twice differentiability, consideration of the second derivative of $g[f(x)]$ leads to the negative semidefinitness of the matrix:

$$\nabla^2 g[f(x)] = g'[f(x)] \nabla^2 f(x) + g'[f(x)] \nabla f(x)^T \nabla f(x)$$

as a necessary and sufficient condition for concavity of $g[f(x)]$.
In turn, this leads to the condition that the augmented hessian matrix:

$$H(x, \rho) = \nabla^2 f(x) - \rho \nabla f(x)^T \nabla f(x)$$

be semidefinite negative for each $\rho(x) \geq \rho_0(x)$, where:

$$\rho_0 = \sup_{\substack{\nabla f(x) h = 0 \\ \|h\| = 1}} \frac{h^T \nabla^2 f(x) h}{[\nabla f(x) h]^2} < +\infty$$

However, there are pseudoconcave functions for which it does not exist any finite value ρ such that $H(x, \rho)$ is semidefinite negative. Therefore, they are not concavifiable [Zang (1981) Crouzeix and Lindberg (1986)]: in such cases, $\rho_0(x) = +\infty$.

The value of ρ_0 allows one to recognize the type of concavity: if $\rho_0(x) = 0$ we have the classical concavity; if $\rho_0(x) < 0$ (>0), concavity is stronger (weaker).
The following theorems about the relationships between (α, λ)-concavity and concavifiability also hold: they are nothing but a rephrasing of what has been discussed above.

Theorem 16. Every concavifiable function is (α, λ)-concave with $\lambda<1$.

Proof. By assumption, there exists a continuous and strictly increasing function g such that the composite function defined by $g[f(x)]$ is concave:

$$g\,[\, f(\alpha x + \bar{\alpha} y)\,] \geq \alpha g\,[\, f(x)\,] + \bar{\alpha} g\,[\, f(y)\,]$$

that is:

$$f(\alpha x + \bar{\alpha} y) \geq g^{-1}(\alpha g\,[\, f(x)\,] + \bar{\alpha} g\,[\, f(y)\,])$$

and f is (α, λ)-concave with λ such that:

$$\lambda f(x) + \bar{\lambda} f(y) = f(x) + \bar{\lambda}[\, f(y) - f(x)] = g^{-1}(\alpha g\,[\, f(x)\,] + \bar{\alpha} g\,[\, f(y)\,])$$

If $f(x)=f(y)$, the equality holds trivially as an identity.
If $f(x) < f(y)$, the equality holds for:

$$(4) \qquad \bar{\lambda} = \frac{g^{-1}(\alpha g\,[\, f(x)\,] + \bar{\alpha} g\,[\, f(y)\,]) - f(x)}{f(y) - f(x)}$$

We remark that since the g-mean of $f(x)$ and $f(y)$ lies between these two values, $\bar{\lambda}$ belongs to $I°$. □

In short: the class of concavifiable functions is a proper subclass of the (α, λ)-concave functions with $\lambda < 1$ (obvious, since concavifiable functions are a proper subclass of pseudoconcave functions).
In the next theorem, we will assume λ depending on $f(x)$ and $f(y)$: it will be denoted with $\Lambda[\,\alpha\,; f(x). f(y)\,]$. This simplifies formally the result and makes it more meaningful:

Theorem 17. Let $f\colon C\to R$ be an (α, λ)-concave differentiable function. If $\Lambda'\,[\,1\,; f(x). f(y)\,]$ is positive and finite for each $x, y \in C$, then f is concavifiable.

Proof. We can write:

$$(5) \qquad \Lambda'[\,1\,; f(x). f(y)\,]\,[\, f(y) - f(x)\,] \leq \nabla f(x)\,(y - x)$$

Consider now any function $g: f(C) \to R$ that is strictly increasing, differentiable and (α, v)-concave with $v(\alpha; u, z)$ such that $v'(1; u, z) = [\Lambda'(1; u, z)]^{-1}$; f.i., if $\Lambda'(1; u, z) > 1$ $\mu = \alpha \Lambda'(1; u, z)$.

For such a function it is:

$$v'(1; u, z)\,[\,g(z) - g(u)\,] \le g'(u)(z-u)$$

or equivalently, by setting $u = f(x)$ and $z = f(y)$:

$$\frac{g[\,f(y)\,] - g[\,f(x)\,]}{\Lambda'[1; f(x), f(y)\,]} \le g'[\,f(x)\,][\,f(y) - f(x)\,]$$

Hence:

$$f(y) - f(x) \ge \frac{g[\,f(y)\,] - g[\,f(x)\,]}{\Lambda'[1; f(x), f(y)\,]\, g'[\,f(x)]}$$

Substitution in (5) shows that:

$$\frac{g[\,f(y)\,] - g[\,f(x)\,]}{g'[\,f(x)\,]} \le \nabla f(x)(y - x)$$

and finally

$$g[\,f(y)\,] - g[\,f(x)\,] \le \nabla\, g[\,f(x)\,]\,(y - x)$$

which proves concavity of the function defined by $g[f(x)]$, that is concavifiability of f.

□

It is worth noticing that Theorem 17 is nothing but a corollary of Theorem 14. Since g is (α, v)-concave and f (α, Λ)-concave, the function defined by $g[f(x)]$ is $(\alpha, v(\Lambda))$-concave. In order to obtain its concavity (i.e., the concavifiability of f), it has to be $v[\,\Lambda(\alpha)\,] = \alpha$ for all $x, y \in C$. To this purpose, it suffices that $v' = 1/\Lambda'$.

Theorem 16 provides a necessary condition for concavifiability with no differentiability assumption.

Under this further assumption, the condition expressed by Theorem 17 becomes necessary as well. In fact, the differentiable and strictly increasing g must verify (4):

$$\frac{g^{-1}(\alpha g\,[\,f(x)\,] + \bar{\alpha} g\,[\,f(y)\,]) - f(x)}{f(y) - f(x)} = \bar{\lambda}$$

Letting $f(x) = v$, $f(y) = z$, we get:

$$g\,(\lambda v + \bar{\lambda} z) = \alpha g(v) + \bar{\alpha} g(z)$$

which shows <u>(λ, α)-linearity</u> of g, corresponding to simultaneous (λ, α)-concavity and (λ, α)-convexity.

Noticing that (λ, α)-linearity is nothing but (α, λ^{-1})-linearity, where $\lambda^{-1} = v$ is the inverse function of λ, we find the condition given in Theorem 17:

$$v' = 1 \,/\, \Lambda$$

Finally, it is possible to characterize differentiable (α, v)-linear functions when v is linear in α (or $\bar{v}$ is linear in $\bar{\alpha}$). In such a case, the equation:

$$g\,(\alpha u + \bar{\alpha} z) = v g(u) + \bar{v} g(z)$$

can be rewritten as:

$$\frac{g[\,z + \alpha(u - z)\,] - g(z)}{\alpha} = v/\alpha\,[\,g(u) - g(z)\,]$$

Since $v/\alpha = \mu\,(u, z)$ is not depending on α, we get:

$$g'(z)(u - z) = \mu\,(u, z)\,[\,g(z) - g(u)\,]$$

Assuming (as it is always possible to do) $u = 0$ and $g(0) = 0$, the following simple differential equation is obtained:

$$z g'(z) = \mu\,(0, z)\, g(z)$$

Its general solution:

$$g(z) = k e^{M(z)}$$

where $M(z)$ is a primitive of $\mu(0, z)/z$, characterizes differentiable (α, λ)-linear function whenever v is linear in α [Mereau and Paquet (1974)].

5. Twice differentiable (α, λ)-concave functions

We give here some conditions, different from the one indicated in the preceding section, of (α, λ)-concavity for a twice differentiable function.
In the following theorem, we deal again with an augmented hessian matrix: we set $y = x + th$, $x + th \in C$, $t \in R$ and assume that $f(y) = f(x + th) > f(x)$ and λ is symmetric.

Theorem 18. Let $f: C \to R$ be a twice differentiable (α, λ)-concave function. If $\lambda'(1; x, x+th)$ is increasing in t and if there exists the partial gradient of $\lambda'(1; x, y)$ with respect to y (denoted by $\lambda'_2(1; x, y)$), the matrix:

$$\lambda_2'(1; x, x)\, \nabla f(x) + \nabla^2 f(x)$$

is semidefinite negative for each $x \in C$.
Viceversa, if there exists a vector function η such that the matrix:

$$\eta(x)\, \nabla f(x) + \nabla^2 f(x)$$

is semidefinite negative for each $x \in C$, f is (α, λ)-concave with λ such that:

$$\lambda'(1; x, x + th) = 2/\,(2 - th^T \eta(x))$$

Proof. **(a)** Inequality (2) becomes:

$$\nabla f(x)\, th \geq \lambda'(1; x, x + th)\, [\, f(x + th) - f(x)\,]$$

There exists a number $\tau \in (0, t)$ such that:

$$\nabla f(x)\, th \geq \lambda'(1; x, x + th)\, \nabla f(x + \tau h)\, th$$

By assumption:

$$\lambda'(1; x, x + \tau h) \leq \lambda'(1; x, x + th)$$

so that:

(6) $$\nabla f(x)\, h \geq \lambda'(1; x, x + \tau h)\, \nabla f(x + \tau h)\, h$$

The value of $\lambda'(1; x, x)$ is undefined, but one can always take:

$$\lambda'(1; x, x) = 1 \qquad \forall\ x \in C$$

In fact, since:

$$f(x + th) - f(x) \sim \nabla f(x) th$$

for $t \to 0^+$ and:

$$\nabla f(x)\, th \geq \lambda'(1; x, x+th)\,[\, f(x+th) - f(x)\,]$$

we can always write:

$$\lambda'(1; x, x+th) \sim 1 \qquad\qquad \text{for } t \to 0^+$$

Inequality (6) can therefore be rewritten as follows:

$$[\, \lambda'(1; x, x)\, \nabla f(x) - \lambda'(1; x, x+\tau h)\, \nabla f(x, x+th)\,]\, h \leq 0$$

Alternatively, by letting: $P(x, y) = \lambda'(1; x, y)\, \nabla f(y)$:

$$[\, P(x, x+\tau h) - P(x, x)\,]\, h \leq 0$$

There exists a number $\theta \in (0, \tau)$ such that:

$$h^T P_2(x, x+\theta h)\, h \leq 0$$

where P_2 is the partial jacobian matrix of P with respect to y. We get:

$$P_2(x, y) = \lambda'_2(1; x, y)\, \nabla f(y) + \lambda'(1; x, y)\, \nabla^2 f(y)$$

and the last inequality becomes:

$$h^T[\, \lambda_2'(1; x, x+\theta h)\, \nabla f(x+\theta h) + \lambda'(1; x, x+\theta h)\, \nabla^2 f(x+\theta h)\,]\, h \leq 0$$

Letting t (and therefore τ and θ) tend to zero, we get:

$$h^T[\, \lambda_2'(1; x, x)\, \nabla f(x) + \nabla^2 f(x)\,]\, h \leq 0$$

(b) Viceversa, it is:

$$h^T \eta(x)\;\; \nabla f(x) h + h^T\, \nabla^2 f(x) h \leq 0 \qquad \forall h \neq 0$$

When $t \to 0^+$, we get:

$$f(x+th) - f(x) = t\nabla f(x)h + t^2/2\; h^T\, \nabla^2 f(x)h + o(t^2)$$

$$\leq t\nabla f(x)h - t^2/2\; h^T\, \eta(x)\;\; \nabla f(x)h = t\left[\, 1 - t^2/2\; h^T\, \eta(x)\,\right] \nabla f(x)h$$

and finally:

$$\nabla ft(x)\, th \geq 2/(2 - th^T \eta(x))\,[\, f(x+th) - f(x)\,]$$

which proves (α, λ)-concavity of f, since $2 - th^T \eta(x) > 0$, at least in a neighbourood of $t = 0$.

□

Theorem 18 allows one to rediscover already known results when λ depends on $f(x)$ and $f(y)$:

$$\Lambda = \Lambda\, [\,\alpha\, ; f(x)\, , f(y)\,]$$

In such a case, in fact:

$$\lambda_2'(1; x, y) = \Lambda_2'\,[\,1; f(x)\, , f(y)\,]\, \nabla\, f(y)^T$$

and the condition expressed by it concerns the negative semidefinitness of the matrix:

$$\Lambda_2'\,[\,1; f(x)\, . f(y)\,]\;\; \nabla\, f(x)^T \nabla\, f(x) + \nabla^2\, f(x)$$

In particular, if f is concave, $\Lambda_2' = 0$, and the well known condition on the hessian matrix is reobtained. If f is pseudoconcave, writing $\beta[f(x), f(y)]$ for $b(x, y)$:

$$\Lambda = 1 - \overline{\alpha\beta}\, [\, f(x)\, , f(y)\,]$$

In this case, the condition requires the negative semidefinitness of the matrix:

$$\beta_2\; [\, f(x)\, , f(y)\,]\, \nabla\, f(x)^T\, \nabla\, f(x) + \nabla^2\, f(x)$$

as it is well known [Mereau and Paquet (1974), Crouzeix and Ferland (1982)]: in this case, however, β_2 takes a more precise meaning and allows one to rank types of pseudoconcavity.

Remark. If $\nabla f(x)h = 0$, it has to be $h^T\, \nabla f(x)h \leq 0$. Moreover, if $\nabla f(x) = 0$, it is easy to find [Crouzeix (1980)] that:

$$h^T\, \nabla^2\, f(x)h < 0 \qquad\qquad \forall h \neq 0$$

Under different assumptions, we give a couple of other results:

Theorem 19. Let f be a twice differentiable (α, λ)-concave function. If:

$$\lim_{t \to 0^+}\; [\, \mu\, (x, x + th) - \mu^*(x, x + th)\,]\, /\, t = r(x)$$

is finite, then:

(7) $$h^T \nabla^2 f(x)h \le r(x) \nabla f(x)h \qquad \forall h \neq 0$$

Viceversa, if there exists a function r verifying (7), then f is (α, λ)-concave with λ such that:

$$\mu(x, x + th) - \mu^*(x, x + th) \sim tr(x)$$

for $t \to 0^+$ (if $r(x) \le 0$, f is concave).

Proof. (a) Since:

$$\nabla f(x + th) - \nabla f(x) = th^T \nabla^2 f(x) + o(t), \quad f(x + th) - f(x) = \nabla f(x)th + o(t)$$

by letting $y = x + th$, inequality (3) becomes:

$$[\nabla f(x + th) - \nabla f(x)]\, th \le [\mu(x, x + th) - \mu^*(x, x + th))][f(x + th) - f(x)]$$

For $t \to 0^+$, we get:

$$t^2 h^T \nabla^2 f(x)h \le [(\mu(x, x + th) - \mu^*(x, x + th) / t]\, t^2 \nabla f(x)h$$

and finally:

$$h^T \nabla f(x)h \le r(x) \nabla f(x)h$$

(b) Viceversa, assume that (7) holds. By multiplying by t $r(x)$, we get, for $t \to 0^+$:

(8) $$[f(x + th) - f(x)]\, tr(x) \ge t^2 h^T \nabla^2 f(x)\, h + o(t^2)$$

Since:

$$th^T \nabla^2 f(x) = \nabla f(x + th) - \nabla f(x) + o(t)$$

(8) can be rewritten as follows:

$$[f(x + th) - f(x)]\, tr(x) \ge t\ [\nabla f(x + th) - \nabla f(x)]\, h$$

$$[f(x + th) - f(x)]\, (\mu - \mu^*) \ge [\nabla f(x + th) - \nabla f(x)]\, th$$

showing (α, λ)-concavity of f.

□

Let us consider the class of (α, λ)-concave functions such that only the inequality:

(9) $$\nabla f(x)(y - x) \geq \mu^* [f(y) - f(x)]$$

with $\mu^* > 0$, remains meaningful. Pseudoconcave functions belong to this class: $\mu^* = b(x, y), f(y) > f(x)$.

Theorem 20: Let $f: C \to R$ be a twice differentiable (α, λ)-concave functions. If:

$$\lim_{t \to 0^+} 2 \frac{\mu^*(x, x + th) - 1}{t\, \mu^*(x, x + th)} = r(x)$$

is finite, then:

(10) $$h^T \nabla^2 f(x)h \leq r(x) \nabla f(x)h \qquad \forall h \neq 0$$

viceversa, if (10) holds, f is (α, λ)-concave with λ such that:

$$2(\mu^* - 1) \sim \mu^* t r(x)$$

for $t \to 0^+$.

Proof. (a) Write (9) with $y = x + th$ and subtract $\mu^* \nabla f(x) th$ from both sides:

$$(1 - \mu^*) \nabla f(x)th \geq \mu^* [f(y) - f(x) - \nabla f(x)th]$$

For $t \to 0^+$, it follows:

$$(1 - \mu^*) \nabla f(x)th \geq \mu^*(t^2/2)\, h^T \nabla^2 f(x)h + o(t^2)$$

Dividing by $t^2 \mu^*$, we get:

$$2 \frac{1 - \mu^*}{t\mu^*} \nabla f(x)h \geq h^T \nabla^2 f(x)h$$

which proves the first part of the assertion.

(b) If (10) holds with:

$$r(x) \sim \frac{2(\mu^* - 1)}{t\mu^*}$$

for $t \to 0^+$:

$$\frac{2(1 - \mu^*)}{t\mu^*} \nabla f(x)h \geq h^T \nabla^2 f(x)h$$

we can write:

$$(1 - \mu^*) \nabla f(x)th \geq \mu^*[\, f(x + th) - f(x) - \nabla f(x)th \,]$$

that is:

$$\nabla f(x)th \geq \mu^*[\, f(x + th) - f(x) \,]$$

□

A unifying result can be obtained if we refer to the definition of μ and μ^*:

$$\mu(x, x + th) = \lambda'_+(0; x, x + th)$$

$$\mu^*(x, x + th) = \lambda'_-(1; x, x + th)$$

Therefore we get:

$$\lim_{t \to 0^+} \frac{\mu^*(x, x + th) - \mu(x, x + th)}{t} = \lim_{t \to 0^+} \frac{\lambda_-'(1; x, x + th) - \lambda_+'(0; x, x + th)}{t} = h^T \lambda_2'(1; x, x)$$

and the second order condition is the one obtained in Theorem 18.
A similar result holds for pseudoconcave functions, according to:

$$\lim_{t \to 0^+} \frac{\beta[\, f(x), f(x + th) \,] - \beta[\, f(x), f(x) \,]}{t} = h^T \beta_2 \,[\, f(x), f(x) \,]$$

Consider now the basic definition of the second order derivative for a function f of one real variable:

$$f''(x) = \lim_{h \to 0} \frac{2}{h^2} \left(\frac{f(x+h) - f(x-h)}{2} - f(x) \right) = \lim_{h \to 0} \left(\frac{f(x+h) - f(x)}{h} - \frac{f(x+h) - f(x)}{-h} \right) /h=$$

$$= \lim_{h\to 0} \frac{R^+(x,h) - R^-(x,h)}{h}$$

The perturbed derivatives we introduced in section 3:

$$\frac{f(\alpha x + \bar{\alpha}(x+v)) - f(x)}{\bar{\alpha}} \sim (\bar{\lambda}/\bar{\alpha})\, D_v f(x) \qquad \bar{\alpha}\to 0^+$$

$$\frac{f(\alpha(x-v) + \bar{\alpha}x) - f(x)}{\alpha} \sim (\lambda/\alpha)\, D_{-v} f(x) \qquad \alpha\to 0^+$$

lead to a new second order derivative:

$$D^2_v\, f(x) = \lim_{\bar{\alpha}\to 0^+} \frac{f(\alpha x+\bar{\alpha}(x+v)) - f(x)}{\bar{\alpha}} + \lim_{\alpha\to 0^+} \frac{f(\alpha(x-v)+\bar{\alpha}x)) - f(x)}{\alpha} =$$

$$= \lim_{\bar{\alpha}\to 0^+} (\bar{\lambda}/\bar{\alpha})\, D_v f(x) + \lim_{\alpha\to 0^+} (\lambda/\alpha)\, D_{-v} f(x)$$

provided that the two limits exist.
By summing up the inequalities:

$$\mu\,[\, f(x) - f(y)\,] \le \mu\, D_{x-y} f(y)$$

$$\mu^*[\, f(y) - f(x)\,] \le \mu^*\, D_{y-x} f(x)$$

we get:

$$D^2_v f(x) = D_v f(x) + D_{-v} f(x) \le 0$$

stating the second order condition for (α, λ)-concavity in the classical form.

6. Miscellaneous results

The following theorem about maxima holds:

Theorem 21. For every (α, λ)-concave function, a strict local maximum is global. If $\lambda < 1$, every local maximum is global.

Proof. We only prove the second part. It shows that the local-global property holds not only for pseudoconcave functions but more generally for all functions such that:

$$\lambda(\alpha, x, y) < 1 \qquad \forall \alpha \in I^\circ$$

Let x' be a local maximum and x'' a global one:

$$f(\alpha x' + \bar{\alpha} x'') \geq \lambda f(x') + \bar{\lambda} f(x'') > f(x') \qquad \forall \alpha \in I^\circ$$

which is a contradiction.

□

The next theorem holds only when λ is a symmetric function:

Theorem 22. Every linear combination with positive weights of (α, λ)-concave functions, with λ symmetric, is (α, λ)-concave.

Among many possible ones, we give some other results about reciprocal, ratio and power of (α, λ)-concave functions.

Theorem 23. The reciprocal $G = 1/g$ of positive (α, λ)-convex function g is (α, γ)-concave with:

$$\gamma = \frac{\lambda G(y)}{\lambda G(y) + \bar{\lambda} G(x)} = \frac{\lambda g(x)}{\lambda g(x) + \bar{\lambda} g(y)}$$

The reciprocal of a positive (α, λ)-concave function is (α, γ)-convex.

For negative functions the result holds in the opposite way.

Proof. (a) Let g be an (α, λ)-convex function. Suppose $g(x) \geq g(y)$, that is $G(x) \leq G(y)$:

$$G(\alpha x + \bar{\alpha} y) = \frac{1}{g(\alpha x + \bar{\alpha} y)} \geq \frac{1}{\lambda g(x) + \bar{\lambda} g(y)} =$$

$$= \frac{G(x)\, G(y)}{\lambda G(y) + \bar{\lambda} G(x)} = \frac{\lambda G(y)}{\lambda G(y) + \bar{\lambda} G(x)} G(x) + \frac{\bar{\lambda} G(x)}{\lambda G(y) + \bar{\lambda} G(x)} G(y)$$

(b) Let now g be an (α, λ)-concave function and assume $g(x) \le g(y)$, that is $G(x) \ge G(y)$:

$$G(\alpha x + \bar{\alpha} y) = \frac{1}{g(\alpha x + \bar{\alpha} y)} \le \frac{\lambda G(y)}{\lambda G(y) + \bar{\lambda} G(x)} G(x) + \frac{\bar{\lambda} G(y)}{\lambda G(y) + \bar{\lambda} G(x)} G(y)$$

and since:

$$\frac{\lambda G(y)}{\lambda G(y) + \bar{\lambda} G(x)} \le \lambda$$

we get:

$$G(\alpha x + \bar{\alpha} y) \le \lambda\, G(x) + \bar{\lambda} G(x)$$

□

Example. The function defined by $f(x) = (x^2 + 1)^{-1}$ is (α, γ)-concave with:

$$\gamma = (\,\alpha y^2 + 1\,) \,/\, (\,\alpha y^2 + \bar{\alpha} x^2\,)$$

Theorem 24. Let f and g be respectively an (α, λ_1)-concave function and a positive (α, λ_2)-concave function. The ratio $F = f/g$ is an (α, γ)-concave function with:

$$\gamma = \frac{\lambda_1 g(x)}{\lambda_2 g(x) + \bar{\lambda}_2 g(y)}$$

Proof. Let λ_1 and λ_2 be symmetric functions:

$$F(\alpha x + \bar{\alpha} y) = \frac{f(\alpha x + \bar{\alpha} y)}{g(\alpha x + \bar{\alpha} y)} \ge \frac{\lambda_1 f(x) + \bar{\lambda}_1 f(y)}{\lambda_2 g(x) + \bar{\lambda}_2 g(y)} =$$

$$= \frac{\lambda_1 g(x)\, F(x) + \bar{\lambda}_1 g(y)\, F(y)}{\lambda_2 g(x) + \bar{\lambda}_2 g(y)}$$

□

Theorem 23 and 24 allow one to obtain results for product function as well. We remark a special case: let f be a positive (α, λ)-concave function. The function f^k (k positive integer) is $(\alpha, 1 - \bar{\lambda}^k)$-concave. Whenever k is odd, the result holds also without the positivity assumption.

References

[1] **Ben-Tal A.**, On generalized Means and Generalized Concave Functions, J.O.T.A. (1977), vol.21, pp.1-13.

[2] **Castagnoli E. and Mazzoleni P.**,Towards a unified type of concavity, J.I.O.S. (1989a) vol.10, n.1, pp.225-240.

[3] **Castagnoli E. and Mazzoleni P.**, About Derivatives of some Generalized Concave Functions, J.I.O.S. (1989b) vol.10, n.1, pp.53-64.

[4] **Couzeix J.P.**, On second Order Conditions for Quasiconvexity, Math. Prog. (1980), vol.18, pp.349-352.

[5] **Crouzeix J.P. and Ferland J.A.**, Criteria for Quasiconvexity and Pseudoconvexity: Relationships and Comparisons, Math. Prog. (1982), vol.23, pp.193-205.

[6] **Crouzeix J.P. and Lindberg**. Additively Decomposed Quasiconvex Functions, Math.Progr. (1986), pp.42-57.

[7] **Diewert W.E., Avriel M., and Zang I.**, Nine Kinds of Quasiconcavity and Concavity, J.E.T. (1981), pp.397-420.

[8] **Martos B.**, Nonlinear Programming: Theory and Methods, North Holland (1975).

[9] **Mereau P. and Paquet J.**, Second Order Conditions for Pseudoconvex Functions, SIAM J. of Appl. Math. (1974), vol.27 n.1, pp.131-137.

[10] **Ortega J.M. and Rheinboldt W.C.**, Iterative Solution of Nonlinear Equations in Several Variables, (1970), Academic Press, New York.

[11] **Schaible S. and Ziemba W.T.**, (eds.), Generalized Concavity in Optimization and Economics, Academic Press, New York. (1981).

[12] **Zang I.**, Concavifiability of C^2-functions: An Unified Approach, in Schaible and Ziemba (eds.)(1981), pp.131-152.

This research was partly supported by the National Research Council and by the Italian-Ministery of Education

ON THE BICRITERIA MAXIMIZATION PROBLEM

L. MARTEIN

Abstract

The bicriteria maximization problem has been studied mainly by several authors mainly with the aim of establishing the connectedness of the set E of all efficient points. In this paper we will introduce a parametric real-valued function which allows us either to derive a parametric representation of E in a general form or to find sequential methods for generating E.

1. Introduction

The bicriteria maximization problem P has been studied by several authors [3, 4, 5, 6, 7, 8] mainly with the aim of establishing the connectedness of the set E of all efficient points. In this paper we will introduce a parametric real-valued function $z(\theta)$ which allows us to derive a parametric representation of E in a more general form than the one given in [8]; furthermore we will point out that the properties of the function $z(\theta)$ are strictly related to the connectedness of E and that they can play an important role in finding sequential methods for generating E [7].
The aforesaid approach allows us to obtain new results and known ones.

2. A parametric representation of the set of all efficient points

Let us consider the bicriteria maximization problem

$$P: \max \ (f_1(x), \ f_2(x)), \ x \in R$$

where f_1, f_2 are real-valued continuous functions on the nonempty compact set $R \subset IR^n$.
A point $x^0 \in R$ is said to be efficient for P if there does not exist a point $x \in R$ such that $f_i(x) \geq f_i(x^0)$ $i = 1, 2$ where at least one of these inequalities is strict.
Several authors [3, 4, 5, 8] have pointed out that set E of all efficient points of P is related to the optimal solutions of a suitable parametric problem. Following this idea, we will consider the parametric problem stated in a form which allows us to establish some important relations and to obtain knows results and new ones.
Set:

$$P(\theta): z(\theta) \doteq \max f_2(x),\ x \in R(\theta) \doteq \{ x \in R : f_1(x) \geq M - \theta, \theta \geq 0 \}$$

where $M \doteq \max f_1(x),\ x \in R$.

Let us note that $P(\theta)$ has optimal solutions for any $\theta \geq 0$, since the feasible region $R(\theta)$ is a compact set.

Denote with $S(\theta)$ the set of optimal solutions for $P(\theta)$ and with $S'(\theta)$ the set of optimal solutions to the problem

$$\max f_1(x),\ x \in S(\theta) \tag{2.1}$$

The following theorem gives the relation between E and the optimal solutions to the parametric problem $P(\theta)$.

Theorem 2.1. Consider the bicriteria maximization problem P. Then

$$E = \bigcup_{\theta \in [0, M-m]} S'(\theta) \tag{2.2}$$

where $m \doteq \min f_1(x),\ x \in R$.

Proof. If $\bar{x}$ is an efficient point for problem P then it is also an optimal solution for $P(\bar{\theta})$ with $\bar{\theta} = M - f_1(\bar{x})$ and it is easy to verify that $\bar{x} \in S'(\theta)$; consequently $E \subseteq \bigcup_{\theta \in [0, M-m]} S'(\theta)$.
Now we must show that $E \supseteq \bigcup_{\theta \in [0, M-m]} S'(\theta)$. Let $\bar{x}$ be an optimal solution for problem (2.1) with $\theta = \theta' \in [0, M-m]$; $\bar{x}$ is also an optimal solution for problem $P(\theta'')$ where $\theta'' = M - f_1(\bar{x})$ and $\theta'' \leq \theta'$. Suppose that $\bar{x}$ is not an efficient point for P, that is, there exists $x^0 \in R$ such that the following inequalities:

$$f_1(x^0) \geq f_1(\bar{x}) \tag{2.3a}$$

$$f_2(x^0) \geq f_2(\bar{x}) \tag{2.3b}$$

hold where at least one of these is strict. Since (2.3a) implies $x^0 \in R(\theta'')$, we have $f_2(x^0) \leq f_2(\bar{x})$, that is, from (2.3b), $f_2(x^0) = f_2(\bar{x})$. As a consequence $x^0 \in S(\theta')$, so that x^0 is a feasible point for problem (2.1) with $\theta = \theta'$ and thus (2.3a) can hold only as an equality and this is absurd. This completes the proof.

The previous theorem points out that an optimal solution for $P(\theta)$ is not necessarily an efficient point for P, so that, in general, $E \neq \bigcup_{\theta \in [0, M-m]} S(\theta)$, nevertheless E can be characterized as the union of suitable sets $S(\theta)$.

As we will see, the function $z(\theta)$ plays an important role in finding this characterization. The following theorem states some important properties of the function $z(\theta)$.

Theorem 2.2. Consider the parametric function $z: [0, M-m] \to \mathbb{R}$, $z(\theta) = \max_{x \in R(\theta)} f_2(x)$.

Then i) and ii) hold.

i) $z(\theta)$ is a nondecreasing function;

ii) $z(\theta)$ is an upper semicontinuous function.

Proof. i) It follows by noting that $\theta_1 < \theta_2$ implies $R(\theta_1) \subseteq R(\theta_2)$ so that $z(\theta_1) \le z(\theta_2)$.

ii) The thesis follows immediately from theorem in [13]. For sake of simplicity we will give a direct proof.

Set $\theta_0 \in [0, M-m]$; we must prove that the upperlimit of $z(\theta)$, $\theta \to \theta_0$, is not greater than $z(\theta_0)$.

Let $\{ \theta_n \} \subset [0, M-m]$ be such that $\theta_n \to \theta_0$ and let x_n be an optimal solution for problem $P(\theta_n)$. Since R is a compact set there exists a subsequence $\{ x_s \} \subset \{ x_n \}$ such that $x_s \to x^0 \in R$.

Taking into account that f_1 is continuous and $f_1(x_s) \ge M - \theta_s$ $\forall s$, we have, $f_1(x^0) \ge M - \theta_0$, so that $x^0 \in R(\theta_0)$.

On the other hand $\overline{\lim}_{\theta \to \theta_0} z(\theta) = \overline{\lim}_{x_s \to x^0} f_2(x_s) = (x_0) \le z(\theta_0)$. This completes the proof.

Taking into account i), ii) of theorem (2.2), it is possible to find a suitable set of indices I, $0 \in I$, such that $[0, M-m] = \bigcup_{i \in I} [\theta_i, \theta_{i+1}]$, $\theta_0 = 0$; where the subintervals $]\theta_i, \theta_{i+1}[$, $i \in I$, verify the following properties:

- if $z(\theta)$ is constant in $]\theta_i, \theta_{i+1}[$ then it is either increasing in $]\theta_{i+1}, \theta_{i+2}[$ or constant in $]\theta_{i+1}, \theta_{i+2}[$ with $z(\theta_{i+1}) \neq \lim_{\theta \to \theta_{i+1}} z(\theta)$;
- if $z(\theta)$ is increasing in $]\theta_i, \theta_{i+1}[$ then either $\theta_{i+1} = M-m$ or $z(\theta)$ is constant in $]\theta_{i+1}, \theta_{i+2}[$.

Set $I_1 = \{ i \in I : z$ is increasing in $]\theta_i, \theta_{i+1}[\}$ and set $J = \{ i \in I - I_1 : z(\theta_i) = \lim_{\theta \to \theta_1} z(\theta) \}$.

Now we are able to establish the following results:

Lemma 2.1. Let $\theta' \in]0, M-m[$ be such that

$$z(\theta) < z(\theta'), \ \forall \theta < \theta'. \tag{2.4}$$

Then every optimal solution for $P(\theta')$ is binding at the constraint $f_1(x) \ge M-\theta'$ and, furthermore, it is also an efficient point for problem P.

Proof. Let x' be an optimal solution for $P(\theta')$ and, suppose that $f_1(x') > M-\theta'$. Then x' is optimal for problem $P(\theta_1)$, $\theta_1 = M-f_1(x') < \theta'$ so that $z(\theta_1)=z(\theta')=f_2(x')$ and this contradicts (2.4).

Now we must show that x' is an efficient point for P; to this aim it is sufficient to prove that both a) and b) hold:

a) $f_1(x) \geq f_1(x')$ implies $f_2(x) \leq f_2(x')$, $\forall x \in R$;

b) $f_1(x) > f_1(x')$ implies $f_2(x) < f_2(x')$, $\forall x \in R$.

If $f_1(x) \geq f_1(x')$ we necessarily have $f_2(x) \leq f_2(x')$ since $x \in R(\theta')$ and a) is satisfied. On the other hand $f_1(x) > f_1(x')$ implies that x is an interior point for $R(\theta')$ so that it is not an optimal solution for $P(\theta')$, i.e. $f_2(x) < f_2(x')$. This completes the proof.

Lemma 2.2. Suppose that $z(\theta)$ is constant in $[a, b]$. Then any optimal solution for $P(\theta)$, $\theta \in]a, b[$ binding at the constraint $f_1(x) \geq M-\theta$ is not an efficient point for P.

Proof. Let x' be an optimal solution for $P(\theta)$, with $f_1(x') = M-\theta$ and let x^0 be an optimal solution for $P(a)$. Since $z(a) = z(\theta)$ we have $f_2(x^0) = f_2(x')$ and furthermore $f_1(x^0) \geq M-a > M-\theta = f_1(x')$, so that x' is not an efficient point for P and this completes the proof.

Theorem 2.3. <u>Consider the bicriteria maximization problem P.</u>
Then

$$(2.5) \qquad E = \Big(\bigcup_{\substack{\theta \in]\theta_i, \theta_{i+1}[\\ i \in I_1}} S(\theta) \Big) \cup \Big(\bigcup_{i \in J \cup (0)} S(\theta_i) \Big)$$

Proof. Let us note that $\theta' \in]\theta_i, \theta_{i+1}[$, $i \in I_1$ is such that $z(\theta) < z(\theta')$ $\forall \theta < \theta'$ and that the same property holds for any θ_i, $i \in J$.

From Lemma 2.1 we have $E \supseteq \Big(\bigcup_{\substack{\theta \in]\theta_i, \theta_{i+1}[\\ i \in I_1}} S(\theta) \Big) \cup \Big(\bigcup_{i \in J \cup (0)} S(\theta_i) \Big)$

On the other hand, an efficient point $x' \in E$ is an optimal solution for the problem

$$\begin{cases} \max f_2(x), & x \in R \\ f_1(x) \geq f_1(x') \doteq M - \theta' \end{cases}$$

Since x' is binding at the constraint $f_1(x) \geq M - \theta'$, we have, from Lemma 2.2, that $\theta' \in \Big(\bigcup_{i \in I_1}]\theta_i, \theta_{i+1}[\Big) \cup \Big(\bigcup_{i \in J \cup (0)} \theta_i \Big)$. This completes the proof.

Corrolary 2.1. Suppose that $z(\theta)$ is a semi-strictly quasiconcave function and set $\theta^* = \max H(\theta)$, where $H(\theta) \doteq \{ \theta : z(\theta)$ is increasing in $[0, \theta] \}$.
Then

$$E = \bigcup_{\theta \in [0, \theta^*]} S(\theta) \tag{2.6}$$

Proof. Is is sufficient to note that the semi-strictly quasiconcave function $z(\theta)$ is either a constant function or an increasing function or a function which is increasing in $[0, \theta^*]$ and constant in $]\theta^*, M-m]$, so that the thesis follows from theorem 2.2.
Let us note that if $z(\theta)$ is constant then $E=S(0)$, while if $z(\theta)$ is increasing then

$$E = \bigcup_{\theta \in [0, M-m]} S(\theta).$$

This completes the proof.

The characterization of E as the union of suitable sets $S(\theta)$ plays an important role in finding sequential methods to generate E (see [4,7]) and also in studying the connectedness of E (see [5,8]).
As pointed out in corollary 2.1., E assumes a simple form when $z(\theta)$ is a semi-strictly quasiconcave function.
A class of problem which ensures the semi-strictly quasiconcavity of the function $z(\theta)$ is one where at least one of the objective functions is semi-strictly quasiconcave; this is shown in the following theorem:

Theorem 2.4. Let us consider problem P where the feasible region is a convex set and let us suppose that f_2 is a semi-strictly quasiconcave function. Then $z(\theta)$ turns out to be a semi-strictly quasiconcave function.

Proof. Let us suppose that the nondecreasing function $z(\theta)$ is not semi-strictly quasiconcave, then there exist $\theta_1, \theta^*, \theta_2$ with $\theta_1 < \theta^* < \theta_2$ such that $z(\theta^*) = z(\theta_1) < z(\theta_2)$. Let x_1 and x_2 be optimal solutions for $P(\theta_1)$ and $P(\theta_2)$ respectively; since x_1, x_2 are not necessarily binding at the parametric constraint, we set

$$\theta' \doteq M-f_1(x_1) \le \theta_1 \ , \ \theta'' \doteq M-f_1(x_2) \le \theta_2.$$

Now we will prove that $\theta^* \in]\theta', \theta''[$, that is $\theta^* < \theta''$, since $\theta^* > \theta'$ is obvious.
If $\theta^* \ge \theta''$, we have $f_1(x_2) = M-\theta'' \ge M-\theta^*$, so that $x_2 \in R(\theta^*)$; consequently $f_2(x_2) \doteq z(\theta_2) \le z(\theta^*)$ and this is absurd.
The continuity of the function f_1 implies the existence of $x^* \in]x_1, x_2[$ such that $f_1(x^*) = M-\theta^* \in]M-\theta', M-\theta''[$.
Consider now therestriction of the function f_2 on the line-segment $[x_1, x_2]$. We have $f_2(x^*) \le z(\theta^*) = z_1(\theta_1) \doteq f_2(x_1) < f_2(x_2)$ and this contradicts the semi-strictly quasiconcavity of f_2.
The proof is complete.

For the subclass of bicriteria maximization problems where at least one of the objective functions is semi-strictly quasiconcave, we obtain, as a direct consequence of corollary 2.1 and theorem 2.4, the following result given in [8].

Corollary 2.2. Suppose f_2 is semi-strictly quasiconcave and the feasible region is a convex set. Then (2.6) holds.

3. On the connectedness of E.

In the previous section we outlined the role played by the function $z(\theta)$ in order to achieve a characterization of the set of all efficient points of the bicriteria maximization problem P. Now we will point out that $z(\theta)$ is also related to the connectedness of E, in the sense that the semi-strictly quasiconcavity of z is a necessary condition for E to be connected.
Let us note that connectedness of a set is referred to arcwise-connectedness.
This last property is established in the following theorem:

Theorem 3.1. Consider the bicriteria maximization problem P and assume that E is connected.
Then the function $z(\theta)$ is semi-strictly quasiconcave.

Proof. Let us suppose that the nondecreasing function $z(\theta)$ is not semi-strictly quasiconcave, then there exist $\theta_1, \theta^*, \theta_2$, with $\theta_1 < \theta^* < \theta_1$ such that $z(\theta^*) = zx_1 \; a < z(\theta_2)$.
Let us consider $x_1 \in S'(\theta_1)$ and $x_2 \in S'(\theta_2)$, so that x_1 and x_2 are optimal solutions for $P(\theta_1)$ and $P(\theta_2)$ respectively and also efficient points for P.
Set: $\theta' \triangleq M - f_1(x_1) \le \theta_1$, $\theta'' \triangleq M - f_1(x_2) \le \theta_2$; we have (see proof of theorem 2.4) $\theta' < \theta^* < \theta''$.
Since E is connected, there exists a continuous function $\Phi : [0,1] \to IR^n$ such that $\Phi(0) = x_1$, $\Phi(1) = x_1$ and $\Phi(t)$ is an efficient point for P $\forall t \in]0,1[$. Consider the function $\phi = f_1 \circ \phi : [0,1] \to IR$; since ϕ is a continuous function, we have $\phi([0,1]) \triangleq A \supseteq [f_1(x_2), f_1(x_1)] = [M-\theta'', M-\theta']$, so that there exists $t^* \in]0,1[$ such that $\phi(t^*) = M-\theta^*$. It follows that $x^* = \Phi(t^*)$ is an efficient point which is binding at the constraint $f_1(x) \ge M-\theta^*$ and this is absurd (see lemma 2.2).
This completes the proof.
The following example shows that the semi-strctly quasiconcavity of the function $z(\theta)$ is not, in general, a sufficient condition for E to be connected.

Example 3.1. Consider the following bicriteria maximization problem

$$P: \begin{cases} \max\ (-x_1, x^2_1 + x^2_2) \\ 1 \le x_1 \le 4,\ -1 \le x_2 \le 1 \end{cases}$$

It is easy verify that the set of all efficient points is $E = [a_1, a_2] \cup [b_1, b_2]$, where $a_1 = (1,1)$, $a_2 = (4,1)$, $b_1 = (1,-1)$, $b_2(4,-1)$, so that E is disconnected.
On the other hand, $z(\theta) = (1-\theta)^2 + 1$, $\theta \in [0, 3]$, and thus $z(\theta)$ turns out to be a semi-strictly quasiconcave function because it is increasing in $[0, 3]$.

Remark 3.1. Let us consider the parametric problem

$$P_1(\theta) : z_1(\theta) \triangleq \max\ f_1(x),\ x \in R,\ f_2(x) \ge M_2 - \theta,\ \theta \ge 0$$

where $M_2 \triangleq \max f_2(x),\ x \in R$.
Obviously, we can apply all the previously stated results to problem $P_1(\theta)$. Consequently a necessary condition for E to be connected is that both $z(\theta)$ and $z_1(\theta)$ are semi-strictly quasiconcave functions.
In Example 3.1 the function $z_1(\theta)$ is emi-strictly quasiconcave because $f_1(x) = -x_1$ is also semi-strictly quasiconcave, so that the semi-strictly quasiconcavity of $z_1(\theta)$ and $z(\theta)$ is not sufficient to guarantee the connectedness of E. Taking into account (2.6), the loss of conncetedness is related to the disconnectedness of sets $S(\theta)$. A particular subclass of bicriteria maximization problems for which E turns out to be connected is one where both of the objective functions are semi-strictly quasiconcave (see [8]).

4. Concluding Remarks

In this paper we have pointed out the role played by function $z(\theta)$ in characterizing the set of all efficient points of a bicriteria maximization problem as the union of suitable sets of optimal solutions to the parametric problem $P(\theta)$. Obviously the possibility of finding an efficient sequential method which is able to solve $P(\theta)$ is strictly related to the structure of problem P, that is, to the properties of the objective functions and the constraints.
The aforesaid results and the ones obtained in [7] suggest the possibility of generating E for particular subclasses of bicriteria maximization problems by means of a suitable revision of the parametric methods given in [9, 10, 11, 12].

References

/1/ **Avriel M.**, Non linear programming: Analysis and Methods, Prentice Hall, Englewood Cliffs, N.Y., 1976.

/2/ **Avriel M., Diewert W.E., Schaible S., Zang I.**, Generalized concavity, Plenum Press. New York and London, 1988.

/3/ **Benson H.P.**, Vector maximization with two objective functions, Journal of Optimization Theory and Applications, vol.28, pp.253-257, 1979.

/4/ **Choo E.U., Atkins D.R.**, Bicriteria linear fractional programming, Journal of Optimization Theory and Applications, vol. 36, pp.203-220, 1982.

/5/ **Choo E.U., Atkins D.R.**, Connectedness in multiple linear fractional programming, Management Sciences, vol.29, pp.250-255, 1983.

/6/ **Choo E.U., Schaible S., Chew K.P.**, Connectedness of the efficient set in three criteria quasiconcave programming, Cahiers du C.E.R.O., vol.27, pp.213-220, 1985.

/7/ **Martein L.**, On generating the set of all efficient points of a bicriteria linear fractional problem, This volume.

/8/ **Schaible S.**, Bicriteria Quasiconcave Programs, Cahiers du C.E.R.O., vol. 25, pp.93-101, 1983.

/9/ **Cambini A., Martein L., Pellegrini L.**, A decomposition algorith for a particular class of nonlinear programs, Actes du I° Colloque AFCET-SMF,Tome III, pp.179-189, Palaiseau (France), 1978.

/10/ **Cambini A., Martein L., Sodini C.**, An algorithm for two particular nonlinear fractional programs, Methods of Operations Research, vol. 45, pp.61-70, Athenaum/Hain/Honstein, 1983.

/11/ **Cambini A., Martein L., Schaible S.**, On maximizing a sum of ratios, Journal of Information and Optimization Sciences, vol.10, n.1, 1989.

/12/ **Martein L., Schaible S.**, On solving a linear program with a quadratic constraint, Rivista Matematica per le Scienze Economiche e Sociali, anno 10, n.2.

/13/ **Bank B., Guddat J., Klatte D., Kummer B., Tammer K.**, Non-linear parametric Optimization, Akademie-Verlag-Berlin, 1982.

CHAPTER II

FRACTIONAL PROGRAMMING

FRACTIONAL PROGRAMMING-SOME RECENT RESULTS

C. R. Bector - A. Cambini

Abstract

This paper contains results in fractional programming most of which are published after 1982 by different researchers working in the field. The main focus is placed on mathematical results that lead to or indicate the algorithmic developments of fractional programs containing single or multiple ratios. An attempt is made to identify some research directions that exist in the field.

Introduction

Recently, Schaible (1983) and Schaible and Ibaraki (1983) published survey papers in the field of fractional programming. In those surveys most of the results they have included pertain to fractional programs that contain a single ratio in the obijective, although some results correspond to generalized fractional programs too.

They have barely touched multiobjective fractional programs. However, since then many more interesting results have become available in the literature [for example see Mond and Weir (1982), Datta (1982), Bector and Jolly (1984), Jeyakumar (1985), Ferland and Potvin (1985), Cambini and Martein (1985 a, 1985 b), Weir (1985) Bhatt and Rosenbloom (1985), Nykowski and Zolkiewski (1985), Crouzeix et al. (1985), (1986), Bector and Bector (1986), Bector and Chandra (1986 a, 1986 b, 1986 c), Bector and Bhatia (1986), Bector and Chandra (1987), Vartak and Gupta (1987), Bector et al. (1987), Singh (1988), Bector et al. (1988 a, 1988 b, 1988 c), Sniedovich (1988), Benadada and Ferland (1988), Cambini and Sodini (1988), Cambini et al, (1989) and Bector et al. (1989)] and the intent of the present survey is to cover them.

We shall divide the paper mainly in four sections. In section 1 we introduce some notations definitions and in Section 2 we consider single objective fractional programs. Section 3 will contain the results on 'generalized fractional programs', whereas, in Section 4 we shall deal with optimization of 'multiobjective fractional programs' in Pareto sense.

1. Notations and Definitions

For $x, y \in R^n$, by $x \leqq y$ we mean $x_i \le y_i$ for all i; $x \le y$ means $x_i \le y_i$ for all i and $x_s < y_s$ for at least one s, $1 \le s \le n$. By $x < y$ we mean $x_i < y_i$ for all i. Let h be a numerical function defined on some open set D in R^n containing the convex set $\Gamma \subseteq D$. Let C denote the class of single valued continuous functions and C^p, p = 1, 2, ..., (p being finite) denote the class of all those $h \in C$, every p-th order partial derivative of which exists and is continuous. For $h \in C^1$ let $\nabla_x h$ denote the gradient of h and for $h \in C^2$ let $\nabla^2{}_x h$ denote the Hessian of h. We define a kernel function $K : \Gamma \times \Gamma \times [0,1] \rightarrow R_+$ such that $K = K[x, y, \lambda] > 0$ and is continuos in λ, where $\lambda \in [0,1]$.

Let $k = \lim_{\lambda \to 0} K$, or define $k : \Gamma \times \Gamma \rightarrow R_+$ such that $k = k(x, y) > 0$. We assume that the kernel function is uniformly bounded. Then on Γ the function h is called

1. Beevex [Bector (1968), Bector and Singh (1988)] with respect to

(i) K if for all x, y in Γ and $\lambda \in [0,1]$

$$h(x + (1 - \lambda) y) \leqq K \lambda f(x) + (1 - K \lambda) f(y)$$

(ii) k if for all x, y in Γ, and h differentiable,

$$k [h(y) - h(x)] \geqq (y - x)^T \nabla_x h(x)$$

2. Bonvex [Mond and Weir (1983), Bector and Bector (1986)] if for all x, y in Γ and for all $\eta \in R^n$,

$$[h(y) - h(x)] \geqq (y - x)^T [\nabla_x h(x) + \nabla^2{}_x h(x) \eta] - 1/2\, \eta^T \nabla^2{}_x h(x) \eta$$

3. Pseudobonvex [Mond and Weir (1983), Bector (1986)] if for all x, y in Γ and for all $\eta \in R^n$,

$$(y-x)^T [\nabla_x h(x) + \nabla^2{}_x h(x) \eta] - 1/2\, \eta^T \nabla^2{}_x h(x) \eta \geqq 0 \Rightarrow h(y) \geqq h(x)$$

4. Strictly Pseudobonvex [Bector and Bector (1986)] if for all x, y in Γ and for all $\eta \in R^n$,

$$(y-x)^T [\nabla_x h(x) + \nabla^2{}_x h(x) \eta] - 1/2\, \eta^T \nabla^2{}_x h(x) \eta \geqq 0 \Rightarrow h(y) > h(x)$$

5. Quasibonvex if for all x, y in Γ and for all $\eta \in R^n$,

(i) [Mond and Weir (1983), Bector and Bector (1986)]

$$h(y) \leqq h(x) \Rightarrow (y-x)^T [\nabla_x h(x) + \nabla^2{}_x h(x) \eta] - 1/2\, \eta^T \nabla^2{}_x h(x) \eta \leqq 0, \qquad \text{or}$$

(ii) [Bector and Bector (1986)]

$$h(y) < h(x) \Rightarrow (y-x)^T [\nabla_x h(x) + \nabla^2{}_x h(x) \eta] - 1/2\, \eta^T \nabla^2{}_x h(x) \eta \leqq 0,$$

h is called strictly beevex (strctly bonvex) if in the definition of a beevex (bonvex) function strict inequality holds.

h is called beecave (strictly beecave) on Γ with recpect to K (or k respectively, when h is differentiable) iff -h is beevex (strictly beevex) on Γ with recpect to K (or k respectively, when h is differentiable). A function h is pseudilinear on Γ when it is both beevex and beecave on Γ . We similarly define boncave and generalized pseudoboncave functions and assume that the reader is familiar with the definitions of other generalized convex functions [Mangasarian (1969) and Craven (1982)].

2. Single Ratio Fractional Programming

As in Schaible (1982) suppose f, g, f_i, g_i, $i \in P = \{ 1, 2, ..., p \}$ and h_j, $j \in M = \{ 1, 2, ... m \}$ are real valued functions defined on the set Γ where Γ is as defined above in Section 1. Let $h \in R^n$ whose components are h_j, $j \in M = \{ 1, 2, ... m \}$. We now consider the following single ratio fractional program (FP):

(1) (FP) $\phi(x) = f(x)/ g(x)$

over the set

(2) $S = \{ x \in \Gamma : h(x) \geq 0 \}$

We assume that $f(x) \geq 0$ and $g > 0$ on S and if g is affine function then f need not be restricted to nonnegativity condition on S. In (1) and (2), if at least one of the functions f, g, and h_i, $i = 1, 2, ..., m$ be nonlinear then the (FP) is called nonlinear fractional program (NLFP) and if all of f, g, and h_i, $i = 1, 2, ..., m$ be linear then the (FP) is called a linear fractional programming problem (LFP). In (1) when both f and g are linear on S then ϕ is a pseudolinear function on S and when f is convex and g is concave on S then ϕ is a beevex function on S, with respect to an appropriate kernel function. We now discuss some of the recent results discovered by different researchers in single ratio fractional programs.

(i) Linear Fractional Programming Problem (LFP): Bector and Jolly (1984) considered a number of LFP's including integer LFP, 0-1 LFP and linear fractional transportation problem and using the linearization technique of Bector and Bhatt (1975) and Bhatt (1981) suggested a finite iteration technique to solve such problems. Recently, Bhatt (1989) compared this linearization technique with some of the already existing techniques to solve a LFP and showed that under certain assumptions the methods proposed by Isbell and Marlow (1956), Mangasarian (1969); Bitran and Novaes (1976) and Bhatt (1981) are algorithmically equivalent in the sense that they all generate the same (finite) sequence of points leading to an optimal solution of LFP. Martos (1964) suggested an algorithm to solve a LFP with bounded feasible region and experiments performed by Bitran (1979) showed the superiority of Martos' procedure over the others. Unfortunately Martos' algorithm does not process the problem when the feasible region is unbounded in the sense that, in this case, the algorithm is not able to find a solution or show that no solution exists. This is the reason why Cambini and Martein (1985 b) put forward the so-called "modified version of Martos' algorithm" based on the concept of optimal level solution; technically, this differs from Martos' method on account of a particular choice of pivot, but it is capable of solving a LFP whatever the fasible region. Recently, Cambini and Sodini analyzed the fractional program $\sup f(x) = [(c^T x + c_0)/(d^T x + d_0)]$, subject to $Ax = b$, $x \geq 0$ with respect to the variations in the vector (c, c_0) and suggested an algorithm to solve a parametric LFP in which in place of (c, c_0) the vector $(c + \theta u, c_0 + \theta u_0)$ is taken in the numerator and the set of feasible solutions is bounded or unbounded.

(ii) Nonlinear Fractional Programming Problem: Chandra, Craven and Mond (1983) considered the following pair of fractional programs and, under appropriate assumptions, established the symmetric duality relations between them.

(PS) Minimize $[f(x, y)/ g(x, y)]$

subject to

$$g(x, y)\,\alpha - f(x, y)\,\beta \leq o$$
$$y^T [g(x, y)\,\alpha - f(x, y)\,\beta] \geq o$$
$$\alpha \in \partial_2 f(x, y)\,, \beta \in \partial_2 f(x, y)\,, x \geq o$$

(DS) Minimize $[f(u, v)/ g(u, v)]$

subject to

$$g(u, v)\,\xi - f(u, v)\,\eta \leq o$$
$$y^T [g(u, v)\,\xi - f(u, v)\,\eta] \geq o$$
$$\xi \in \partial_1 f(u, v)\,, \eta \in \partial_1 g(u, v)\,, v \geq o$$

where f and g are functions from $R^n \times R^m$ to R, $f(.\,, y)$ and $g(x\,, .)$ are convex, $f(u\,, .)$ and $g(.\,, v)$ are concave, $\partial_1 f(u, v)$ denotes the subdifferential of $f(.\,, v)$ and $\partial_2 f(x, y)$ denotes the subdifferential of $f(x\,, .)$. $\partial_1 g(u, v)$ and $\partial_2 g(x, y)$ are defined similarly.

Under appropriate restrictions, Cambini and Martein (1985 a) extended Fenchel and Lagrange duals to fractional programs in such a way that the Fenchel and Lagrange duals for nonlinear programs can be obtained as special cases by setting the denominator function equal to 1. Furthermore, they established the equivalence of Fenchel and Lagrange duals in fractional programs. Singh (1982) and (1984) incorporated respectively set inclusive constraints and minmax objective function in the study of the fractional programs.

Other worth mentioning papers that extend FP duality to generalized convex functions are Mond and Weir (1982) and Weir (1985).

Another direction in which work in duality for FP has been done is second order duality [Jeyakumar (1985), Bector and Chandra (1986), (1987) and Vartak and Gupta (1987)]. Jeyakumar (1985) discussed Schaible (1976) and Bector (1973) type second order duality of FP's under ρ-convexity conditions on the functions involved in the problem. Bector and Chandra (1986) extended Jeyakumar's problem by including a nondifferentiable term in each of the numerator and the denominator and provided slightly simpler proofs of the duality theorems. Later motivated by Mond (1974) and Mangasarian (1975), Bector and Chandra (1987) introduced Schaible (1976) type duality for FP's and proved the duality theorems under the conditions of (generalized)- bonvexity [Bector and Chandra (1985) and Mond Weir (1981)] on f, g and h_i $i = 1, 2, \ldots\ldots\ldots m$. Vartak and Gupta (1987) considered Schaible (1976) and Bector (1973) type second order duality for FP's under η-convexity [Kaul and Kaur (1985)] assumptions. Very recently Sniedovich (1988) introduced a new format for fractional programming problems. More specifically, he reassesed the theoretic base underlying fractional programming, justified that fractional programming problems be conceived as pseudolinear [Bector (1969) and Bector and Singh (1988)] problems leaning on first order necessary and sufficient optimality condition and called for the recasting of them in the form of the parametric approach [Ibaraki (1983), Schaible (1976)].

3. Generalized Fractional Programming (GFP)

It has been found that in some nonlinear fractional programming problems more than one ratios occurs in the objective function. Such problems can be stated as

(3) (GFP_1) $$\inf \{ \max_{1 \le i \le p} \phi_i(x) : x \in S \}$$

(4) (GFP_2) $$\sup \{ \sum_{i=1}^{p} \phi_i(x) : x \in S \}$$

where, $\phi_i(x) = f_i(x)/ g_i(x)$, $i = 1, 2, \ldots\ldots, p$ are real valued functions on S, $g_i(x) > o$ and when $g_i(x)$ is not affine on S, then, $f_i(x) \geq 0$. GFP_1 is referred to as generalized fractional program [Crouzeix, Ferland and Schaible (1983), Jagannathan and Schaible (1983)]. Literature available on GFP_2 is very scanty [Cambini, Martein and Schaible (1987)].

Crouzeix, Ferland and Schaible (1985) related (3) to the following parametric programming problem (P_v):

(5) (P_v) $$F(v) = \inf_{x \in S} \{ \max_{1 \le i \le p} [f_i(x) - v\, g_i(x)) \}$$

where $v \in R$, and established the following very important result.

Theorem 3.1. (a) F is nonincreasing and upper semicontinuous:
(b) $F(v) < 0$ if and only if $v > v$; hence $F (v) \geq 0$;
(c) If (GFP_1) has an optimal solution then $F (v) = 0$;
(d) If $F (v) = 0$, then programs (GFP_1) and (P_v) have the same set of optimal solutions (which may be empty).

Using Theorem 3.1 and proving few more results, Crouzeix, Ferland and Schaible (1985) developed the following Dinkelbach type (1967) algorithm.

Algorithm:

Step 1. Start with some $x^0 \in S$. Let

$$v_1 = \max_i f_i(x^0)/ g_i(x^0)$$

Let $k = 1$

Step 2. Solve (P_{v_k}). Let $x^k \in M(v_k)$ where $M(v)$ is the (possibly empty) set of optimal solutions of (P_v)

Step 3. If $F(v_k) = 0$, stop. Then x^k and x^{k-1} are optimal solutions of (GFP_1), and $v_k = v$.

Step 4. If $F(v_k) \neq 0$, take

$$v_{k+1} = \max_i f_i(x^k)/ g_i(x^k).$$

Let $k = k+1$ and go to Step 2.

Later, Crouzeix, Ferland and Schaible (1986) modified the above algorithm to improve the speed of convergence.

Ferland and Potvin (1985) summarized several algorithms to solve (GFP_1) and compared them numerically in the linear case. Bhatt and Rosenbloom (1985) considered a special case of (GFP_1) in which each $\phi_i(x) = (c_i x + \alpha_i)/(d_i x + \beta_i)$, $i = 1, 2, \ldots, p$ and the constraint set is given by linear functions. They suggest a dynamic programming approach to solve such problems. Recently, Benadada and Ferland (1988) proposed an algorithm based on partial linearization to solve (GFP_1) and showed that their algorithm is equivalent to the generalized Dinkelbach's algorithm when applied to (GFP_1).
Duality is an other direction in which researchers have worked for (GFP_1). In addition to the work done by Jagannathan and Schaible (1983) and Crouzeix, Ferland and Schaible

(1983) reported in Schaible (1983), recently few more papers have appeared in different journals. Chandra, Craven and Mond (1986) using ratio game approach of Schroder (1970) showed that (GFP_1) is equivalent to the following equivalent program (EP):

$$\text{(EP)} \qquad \inf\ q \quad \text{Subject to} \quad f_i(x) - v_i\, g(x) \le q,\ i = 1, 2, \ldots, p, \quad h(x) \le 0, \quad x \in S,\ q \in R$$

Finding equivalent linearized program to (EP) they indicated that different duals can be formulated for (GFP_1) and using the linearized program Chandra, Craven and Mond (1986) discussed the duality of (GFP_1). Bector, Husain, Chandra and Bector (1988), Bector and Bector (1989), Bector and Suneja (1988), and Bector, Chandra and Bector (1989), used (P_v) to get down to (EP). Using Fritz John optimality conditions on (EP), Bector, Husain, Chandra and Bector (1988) and Bector and Bector (1989) introduced different duality models for (GFP_1). Using Kuhn-Tucker optimality conditions on (EP) and using the parameter v and Theorem 1, Bector, Chandra and Bector (1989) generalized Schaible (1976) type and Bector (1973) type duals for (GFP_1). Bector and Suneja (1988), using the Lagrangian approach and assuming the functions to be nondifferentiable obtained the following generalized fractional dual (GFD) as the dual to (GFP_1).

$$\text{(GFD)} \qquad \max \left[\inf \left(\frac{u^T F(x) + w^T h(x)}{u^T G(x)} \;\middle|\; u \ge 0,\ w \ge 0,\ \sum_{i=1}^{p} u_i = 1,\ u \in R^p,\ w \in R^m,\ x \in S,\ q \in R \right) \right]$$

where, $u^T F(x) = \sum_{i=1}^{p} u_i f_i(x)$, $u^T G(x) = \sum_{i=1}^{p} u_i g_i(x)$, $v^T h(x) = \sum_{j=1}^{m} w_j h_j(x)$.

4. Multiobjective Programming Problem

A multiobjective fractional program is stated as

$$\text{(6)} \qquad \text{(MOFP)} \quad \underset{x \in S}{\text{minimize}} \quad \phi(x) = (f_1(x)/g_1(x),\ f_2(x)/g_2(x), \ldots, f_p(x)/g_p(x))$$

where all the functions are as in (GFP_1) and S is as in (2), i.e. $S = \{ x \in \Gamma : h(x) \geq 0 \}$

In the literature the above problem is also called <u>fractional vector minimization problem</u>. In the present paper we shall deal with (MOFP) in Pareto Optimality' sense. We define $x^0 \in S$ to be a vector minimum (Pareto minimum or efficient solution) on S id there is no $x \in S$ such that

$$\phi(x) \leq \phi(x^0)$$

We shall mainly restrict ourselves to results that have been proved (MOFP) after 1982. For results derived up to 1982 the reader is referred to Schaible (1983).
Singh and Hanson (1986) established the association of (MOFP) with the following vector optimization problem.

(VOP) $$\min_{x \in S} \left(F_1(x, y_1^0), F_2(x, y_2^0), \ldots\ldots, F_p(x, y_p^0)\right)$$

where $(F_i(x, y_1^0) = f_i(x) - y_i^0 g_i(x)$ and the y_i^0 are some nonnegative fixed real numbers.
Using the association of (VOP) with (MOFP) and introducing the Fritz John and Kuhn-Tucker saddle point problems, Singh and Hanson (1986) established, under nondifferentiability conditions, necessary and sufficient condition for an efficient point to be a saddle point. Singh (1986), under differentiability assumptions, obtained necessary and sufficient optimality conditions and duality for a very special type of multiobjective fractional program.
Gulati and Islam (1987) considered a special case of (MOFP) when the functions in the objectives are linear. Relating the efficient solution of (MOFP) with an optimal solution of a nonlinear programming problem, Gulati and Islam, under generalized convexity assumptions, established necessary and sufficient conditions for the existence of an efficient point. Bector, Chandra and Durgaprasad (1988) introduced a dual problem for a special case of (MOFP) when all the functions are linear.
The special case of the bicriteria linear fractional problem has been treated by several authors when the feasible region is a compact set [Choo and Atkins (1982); Choo and Atkins (1983); Warburton (1984); Shiable (1983 b)].
Cambini and Martein (1989 b) staded a sequential method for solving the bicriteria linear fractional problem for any feasible region; the method generates the set E of all efficient points and, at the same time, points out that E is arcwise-connected when at least one of the objective functions is linear.

5. <u>Algorithms</u>

In (MOFP) assuming all the functions to be linear, Nykowski and Zolkiewski (1985) considered a multiobjective linear fractional program (MOLFP) and related it to an associated multiobjective program (MOP). Nykowski and Zolkiewski (1985) established certain results relating (MOP) and (MOLFP). Based on these results they developed a compromise procedure for solving (MOLFP) and provided a numerical example to show how the procedure works.

Conclusions and Directions for Research.

In the present paper we have reviewed a number of papers on fractional programming that appeared after the year (1982) as the pre-1983 papers have already been reviewed by Schaible (1983). From the review it is clear that not much work has been done in (MOFP). Very few solution procedures to solve such problems are available and the theory part also has tremendous scope for development. Once this is done and appropriate computer softwares are made available, then the (MOFP's) will have a tremendous scope as an important decision support system.

References

/1/ **Bector C.R.**, Strong Pseudoconvex Functions, Mathematical Analysis of Some Nonlinear Programming Problems, Ph. D. Thesis, Indian Institute of Technology, Kampur, India, Chapter IV, Section II, pp. 114-119 (1968).

/2/ **Bector C.R.**, Duality in Linear Fractional Programming, Utilitas Mathematica 4, pp.411-428 (1973)

/3/ **Bector C.R.**, Duality in Nonlinear Fractional Programming, ZOR - Zeitschrift fur Operations Research 17, pp.183-193 (1973)

/4/ **Bector C.R., Bector M.K. and Klassen J.E.**, Duality for a Nonlinear Programming Problem, Utilitas Mathematica 11, pp.87-89 (1977)

/5/ **Bector C.R. and Bhatia B.L.**, Nature of Renyi's Entropy and Associated Divergence Function, Naval Research Logistics Quarterly, 33, pp.741-746 (1986)

/6/ **Bector C.R. and Bhatt S.K.**, A Linearization Technique for Solving Interval Linear Fractional Programs, Congrassus Numeratium XVI, pp.221-229 (1976)

/7/ **Bector C.R. and Jolly P.L.**, Programming Problems with Pseudo-Monotonic Objectives, Math. Operationsforsch. und Statist., Ser. Optimization, 15, pp.217-229 (1984)

/8/ **Bector C.R. and Chandra S.**, (Generalized)-Bonvex Functions and Second Order Duality in Mathematical Programming, Department of Actuarial and Management Sciences, Res. Rep. 85-2, The University of Manitoba, Canada (1986a)

/9/ **Bector C.R. and Chandra S.**, First and Second Order Duality for a Class of Nondifferentiable Fractional Programming Problem, Journal of Information and

Optimization Sciences, 7, pp.335-348 (1986b)

[10] **Bector C.R. and Chandra S.**, Duality for Pseudolinear Minmax Programs, Asia-Pacific Journal of Operational Research, 3, pp.86-94 (1986c)

[11] **Bector C.R. and Chandra S.**, Second Order Duality for a Generalized Fractional Programming, Methods of Operations Research, 56, pp.11-28 (1986d)

[12] **Bector C.R. and Chandra S.**, (Generalized)-Bonvexity and Higher Order Duality for Fractional Programming, Opsearch 24, pp.143-154 (1987)

[13] **Bector M.K., Husain I., Chandra S. and Bector C.R.**, A Duality Model for a Generalized Minmax Program, Naval Research Logistics, 35, pp.493-501 (1988)

[14] **Bector C.R. and Suneja S.K.**, Duality in Nondifferentiable Generalized Fractional Programming, Asia-Pacific Journal of Operational Research, 5, pp.134-139 (1988)

[15] **Bector C.R., Chandra S. and Durgaprasad M.V.**, Duality in Pseudolinear Multiobjective Programming, Asia-Pacific Journal of Operational Research (to appear) (1988)

[16] **Bector C.R. and Singh C.**, Beevex Functions, Research Report # 88-2, Dept. of Math. St. Lawrence Univ., Canton, New York (1988)

[17] **Bector C.R. and Bector M.K.**, Fritz John Sufficient Optimality Conditions and Duality for a Generalized Min-max Program, Journal of Information and Optimization Sciences, 10, (to appear) (1989)

[18] **Bector C.R., Chandra S. and Bector M.K.**, Generalized Fractional Programming Duality: A Parametric Approach, Journal of Optimization Theory and Applications, 60, pp.245-262 (to appear)(1989)

[19] **Benadada Y. and Ferland J.A.**, Partial Linearization for Generalized Fractional Programming, ZOR-Zeitschrift fur Operations Research 32, pp.101-106 (1988)

[20] **Bhatia D. and Dutta N.**, Necessary Conditions and Subgradient Duality for Nondifferentiable and Nonconvex Multiobjective Programming Problem, Cahiers du Centre d'Etude de Recherche Operationelle 27, pp.131-139 (1985)

[21] **Bhatt S.K.**, Linearization Technique for Linear Fractional and Pseudo-Monotonic Programs Revisited, Cahiers du Centre d'Etudes de Recherche Operationelle 23, pp.53-56 (1981)

[22] **Bhatt S.K. and Rosenbloom E.S.**, A Dynamic Programming Approach to Generalized Linear Fractional Programs, Cahiers du Centre d'Etudes de Recherche Operationelle 27, pp.207-212 (1985)

[23] **Bhatt S.K.**, Equivalence of Various Linearization Algorithms for Linear Fractional Programming, ZOR-Zeitschrift fur Operations Research 33 (to appear) (1989)

[24] **Bitran G.R. and Novaes A.G.**, Linear Programming with a Fractional Objective Function, Operations Research 24, pp.675-699 (1976)

[25] **Cambini A. , Martein L. and Pellegrini L.**, A Decomposition Algorithm for a Particular Class of Nonlinear Programs, First Meeting AFCET-SMF on Applied Mathematics, Ecole Polytecnique Palaiseau, (Paris) 2, pp.179-189 (1978)

[26] **Cambini A.**, An Algorithm for a Special Class of Generalized Convex Programs, Generalized Concavity in Optimization and Economics, (ed.) Schaible S. and Ziemba W.T., Academic Press, pp. 491-508 (1981)

[27] **Cambini A. and Martein L.**, On the Fenchel-Like and Lagrangian Duality in Fractional Programming, Methods of Operations Research, 53, pp.21-32 (1985a)

[28] **Cambini A. and Martein L.**, A Modified Version of Martos's Algorithm for Linear Fractional Problem, Methods of Operations Research, 53, pp.33-44 (1985b)

[29] **Cambini A., Martein L. and Schaible S.**, On Maximizing a Sum of Ratios, Journal of Information and Optimization Sciences, vol. 10, n.1 (1989a)

[30] **Cambini A. and Martein L.**, Linear Fractional and Bicriteria Linear Fractional Programs, This volume (1989b)

[31] **Cambini A. and Sodini C.**, On parametric Linear Fractional Programs, Research Report n.6, Dept. of Statistics and Applied Mathematics, University of Pisa, Italy (1987)

[32] **Chandra S., Craven B.D. and Mond B.**, Symmetric Dual Fractional Programming, Pure Mathematics Research Report n. 83-12, La Trobe University, Australia (1983)

[33] **Chandra S., Craven B.D. and Mond B.**, Generalized Fractional Programming Duality: A Ratio Game Approach, Journal of Australian Mathematical Society, Series B, 28, pp.170-180 (1986)

[34] **Choo E.U. and Atkins D.R.**, Bicriteria Linear Fractional Programming, J.O.T.A., vol.36, pp.203-220 (1982)

[35] **Choo E.U. and Atkins D.R.**, Connectedness in Multiple Linear Fractional Programming, Management Science, vol. 29, pp. 250-255 (1983)

[36] **Crouzeix J.P., Ferland J.A. and Schaible S.**, Duality in Generalized Linear Fractional Programming, Mathematical Programming 27, pp.342-354 (1983)

[37] **Crouzeix J.P., Ferland J.A. and Schaible S.**, An Algorithm for Generalized Fractional Programs, Journal of Optimization Theory and Applications, 47, pp.35-49 (1985)

[38] **Crouzeix J.P., Ferland J.A. and Schaible S.**, A Note on an Algorithm for Generalized Fractional Programs, Journal of Optimization Theory and Applications, 50, pp.183-187 (1986)

[39] **Dutta N.**, Efficiency in Multiobjective Fractional Programming, Journal of Information and Optimization Sciences, 3, pp.262-268 (1982)

[40] **Dinklebach W.**, On linear Fractional Programming, Management Science 13, pp.492-498 (1967)

[41] **Ferland J.A. and Potvin J.Y**, Generalized Fractional Programming: Algorithm and Numerical Experimentation, European Journal of Operational Research, 20, pp.92-101 (1985)

[42] **Gulati T.R. and Islam M.A.**, Proper Efficiency in Linear Fractional Vector Maximum Problem with Generalized Convex Constraints, Research Report Roorkie Univ., Rooekie, India, European Journal of Operational Research (to appear)(1987)

[43] **Isbell J.R. and Marlow W.H.**, Attrition Games, Naval Research Logistics Quarterly 9, pp.181-186 (1956)

[44] **Ibaraki T.**, Parametric Approaches to Fractional Programming, Mathematical Programming, 26, pp.345-362 (1983)

[45] **Jagannathan R. and Schaible S.**, Duality in Generalized Fractional Programming via Farkas' Lemma, J. Opt. Theo. and Appl. 41, pp.417-424 (1983)

[46] **Jeyakumar V.**, First and Second Order Fractional Programming Duality, Opsearch, 22, pp.24-41 (1985)

[47] **Kaul R.N. and Kaur S.**, Optimality Criteria in Nonlinear Programming Involving Nonconvex Functions, J. Math. Anal. Appl. 105, pp.104-112 (1985)

[48] **Mangasarian O.L.**, Second and Higher Order Duality in Nonlinear Programming, J. Math. Anal. Appl. 51, pp.607-620 (1975)

[49] **Mangasarian O.L.**, Nonlinear Programming, McGraw-Hill, New York (1969)

[50] **Mangasarian O.L.**, Nonlinear Fractional Programming, J. of the O.R. Soc. of Japan 12, pp.1-10 (1969)

[51] **Mond B.**, Second Order Duality for Nonlinear Programs, Opsearch, 11, pp.90-99 (1974)

[52] **Mond B. and Weir T.**, Generalized Convexity and Higher Order Duality, Pure Mathematics Research Paper n. 81-16, Department of Mathematics, La Trobe Univ., Australia (1981)

[53] **Mond B. and Weir T.**, Duality for Fractional Programming with Generalized Convexity Conditions, J. of Inf. and Opt. Sci. 3, pp.105-124 (1982)

[54] **Martos B.**, Hyperbolic Programming, Nav. Res. Quart. 11, pp.135-155 (1964)

[55] **Nykowski I. and Zolkiezwski Z.**, A Compromise Procedure for the Multiple Objective Linear Fractional Programming Problem, European Journal of Operational Research, 19, pp.91-97 (1985)

[56] **Schaible S.**, Fractional Programming I, Duality, Management Science 22, pp.858-867 (1976)

[57] **Schaible S.**, Fractional Programming II, On Dinkelbach's Algorithm, Management Science 22, pp.858-867 (1976)

[58] **Schaible S.**, Fractional Programming, ZOR-Zeitschrift fur Operations Research 33, pp.39-54 (1983a)

[59] **Schaible S.**, Bicriteria Quasiconcave Programs, Cahiers du C.E.R.O. vol.25, pp.93-101 (1983b)

[60] **Schoeder R.G.**, Linear Programming Solution to Ratio Games, Operations Research 18, pp.300-305 (1970)

[61] **Singh C.**, Convex Programming with Set Inclusive Constraints and its Applications to Generalized Linear and Fractional Programming,J. Optim. Th. and Appl. 38, pp.33-42 (1982)

[62] **Singh C.**, Optimality Conditions for Fractional Minmax Programming, J. Math. Anal. Appl. 100, pp.409-415 (1984)

[63] **Singh C. and Hanson M.A.**, Saddle Point Theory for Nondifferentiable Multiobjective Programming, Journal of Information and Optimization Sciences 7, pp.41-48 (1986)

[64] **Singh C.**, A Class of Multiple-Criteria Fractional Programming Problem, Journal of Mathematical Analysis and Applications, 115, pp.202-213 (1986)

[65] **Sniedovich M.**, Fractional Programming Revisited, European Journal of Operational Research 33, pp.334-341 (1988)

[66] **Vartak M.N. and Gupta I.**, Duality Theory for Fractional Programming Problem Under η-Convexity, Opsearch, 24, pp.163-174 (1987)

[67] **Wagner H.M. and Yuan S.C.**, Algorithmic Equivalence in Linear Fractional Programming, Management Science 14, pp.301-306 (1968)

[68] **Warburton A.R.**, Parametric Solution of Bicriterion Linear Fractional Programs, Operations Research vol.33, n.1, pp.74-84 (1985)

[69] **Weir T.**, A Note on Duality for Fractional Programming Problems, Opsearch 22, pp.241-247 (1985)

RECENT RESULTS IN DISJUNCTIVE LINEAR FRACTIONAL PROGRAMMING

V. Patkar - I.M. Stancu-Minasian

Abstract

A short review of the recent results pertaining to linear fractional Programming problems under disjunctive constraints is presented in this paper. A few areas for further research are outlined.

1. Introduction

In the last few years, attempts have been made to generalize some of the results of disjunctive linear programming [1, 2, 3, 7, 8, 10] to the problems in linear fractional programming under disjunctive constraints. The purpose of this paper is to describe the major results and to outline areas for further research in disjunctive fractional programming. The present paper is based on our three studies [11, 12, 13].

2. Disjunctive linear fractional programming problems

Basically, a disjunctive linear fractional programming problem is of the form,

(DLFP) Minimize $\{ (CX + C_0)/(DX + D_0) : AX \geq b, X \geq 0, X \in L \}$,

where A is an mxn matrix, C, D are 1x n vectors, C_0, D_0 are scalars, b is an m x 1 vector and L is a set of logical conditions.

Following Balas [3], DLFP can be classified as belonging to a disjunctive normal form or conjuctive normal form according to the constraint set formulation as given below.

Disjunctive normal form:

$$(1) \qquad \bigvee_{h \in Q} \{ A^h X \geq b^h ,\ X \geq 0 \}$$

Conjuctive normal form:

$$(2) \qquad AX \geq b,\ X \geq 0 \quad \text{and} \quad \bigvee_{i \in Q_j} \{ E_i X \geq E_{io} \},\ j \in S,$$

where E_i is an $1 \times n$ vector and E_{io} is a scalar. The sets Q and Q_j, $j \in S$, may or may not be finite.

The five variants of disjunctive linear fractional programming problem considered in [11], fall under the category of problems with constraints in the form of (1) or (2). The general strategy followed to deal with disjunctive constraints is to introduce an indicator variable $\delta_i \in \{0, 1\}$, $i \in Q$, so that DLFP is transformed to an ordinary mixed-integer linear fractional programming problem for which many solution methods are available. To illustrate the method, let us consider the following typical problem (P1):

(3) $$\text{Maximize} \quad F(X) = (CX + C_0)/(DX + D_0)$$

subject to

(4) $$A_1X \le b_1 \cup A_2X \le b_2 \cup \ldots \cup A_pX \le b_p, \; X \ge 0$$

The indicator variable δ_i ($i = 1, 2, \ldots, p$) takes value 1 if the i-th constraint set is picked up in the solution procedure, otherwise its value is 0. The modality of introducing δ_i is as follows:

For all non-negative x_j, the inequality $\sum_{j=i}^{n} a^i_j x_j \le b_i$ (a^i_j is a column vector of the matrix A_i) holds then $\delta_i = 1$ i.e.

(5) $$\delta_i = 1 \rightarrow \sum_{j=i}^{n} a^i_j x_j \le b_i$$

Now (5) can also be written as,

(6) $$\sum_{j=i}^{n} a^i_j x_j + M_i \delta_i \le M_i b_i$$

where M_i is an upper bound of the expression $(\sum_{j=i}^{n} a^i_j x_j - b_i)$.

Thus the problem (P1) can be transformed to the following integer fractional programming problem (P2) [14]:

(7) $$\text{Maximize } F(X) = (CX + C_0)/(DX + D_0)$$

subject to

(8) $$A_i X + V_i \delta_i b^+_i, \quad i = 1, 2, \ldots, p$$

(9) $$\sum_{i=1}^{p} \delta_i \geq 1$$

(10) $$X \geq 0,\ \delta_i \in \{0, 1\},\ i = 1, 2, \ldots\ldots, p$$

where b^+_i denotes the modified value of b_i in the light of coefficient V_i of δ_i and (9) ensures that one of the constraint sets is chosen.
It can be proved that if (X_o, δ_o) is the optimal solution of (P2) then X_o is the optimal solution of (P1) [11].

Remark 2.1. The greater than or equal to type of constraints in (4) can be similarly treated.
In [12], a parametric algorithm was presented to solve a class of DLFP in conjunctive normal form, namely,

(11) $$\text{Maximize } F(x_1, x_2, \ldots\ldots, x_n) = \Big(\sum_{j=1}^{n} c_j x_j + c_o\Big)\Big/\Big(\sum_{j=1}^{n} d_j x_j + d_o\Big)$$

subject to

(12) $$\sum_{j=1}^{n} a_{ij} x_j \leq b_i,\quad i = 1, 2, \ldots\ldots, m$$

(13) $$x_j \geq 0,\quad j = 1, 2, \ldots\ldots, n$$

and logical constraints,

(14) $$\prod_{j\in J_1} x_j = 0, \text{---------}, \prod_{j\in J_k} x_j = 0$$

where c_j, d_j, a_{ij} and b_i are constants for all i and j, c_o and d_o are scalars and $J_1, J_2, \ldots\ldots, J_k$ are subsets of index set $N=\{1, 2, \ldots\ldots, n\}$. It is assumed that $\sum_{j=1}^{n} d_j x_j + do > 0$ for all values of x_j satisfying the constraints (12) to (14).
The approach is to solve an equivalent parametric disjunctive linear programming problem in the form,

(15) $$\text{maximize } G(x_1, x_2, \ldots\ldots, x_n, \mu) = \sum_{j=1}^{n} (c_j - \mu d_j)\, x_j + (c_o - \mu d_o)$$

subject to (12), (13) and (14), where

(16) $$\mu = \left(\sum_{j=1}^{n} c_j x_j + c_o\right) / \left(\sum_{j=1}^{n} d_j x_j + d_o\right).$$

The procedure is initiated by assigning any arbitrary value to μ. Now, for a fixed value of μ, the problem (15) reduces to a form where the solution method proposed by Owen can be applied [10]. The resulting values of $x_1, x_2, \ldots\ldots, x_n$ are to be used to compute new value of μ by (16). These steps are to be repeated till the value of (15) becomes 0 or less than a very small preassigned number [5, 6].

Remark 2.2. The solution is independent of the initial value of μ.

Remark 2.3. The parametric algorithm is bound to terminate in a finite number of steps because the function in (15) is monotonically decreasing over constraints (12) to (14).

Remark 2.4. The algorithm can be extended to solve a large scale form of the problem (11) to (14) if constraints can be suitably partitioned [12].

3. Duality

The duality theory in disjunctive linear programming [2] has been generalized to a disjunctive linear fractional programming problem (PP) defined as [13]:

(17) $$\text{Minimize} \quad F(x) = (c\,x + c_o)/(d\,x + d_o)$$

subject to

(18) $$\bigvee_{h \in Q} \{ A^h x \geq b^h, \ x \geq 0 \}$$

where A^h is an mxn matrix for all $h \in Q$, c, d are axn vectors, c_o, d_o are scalars and Q is a finite index set. It is assumed that $dx + d_o > 0$ over the feasible region and the Q constraint sets do not consist of further disjunctions.
Associated with the problem (PP), we define a dual problem (DP):

(19) $$\text{Maximize} \quad G(z) = z$$

subject to

$$(20) \qquad \bigwedge_{h \in Q} \{ u^h A^h + dz \leq c, -u^h b^h + d_0 z \leq c_0, u^h \geq 0 \}$$

where (DP) is feasible if there exists z and for each $h \in Q$, there exists u^h such that (u^h, z) satisfies corresponding bracketed system.
Let

$$X_h = \{x : A^h x \geq b^h, x \geq 0\}, \bar{X}_h = \{x : A^h x \geq 0, x \geq 0\},$$

$$U_h = \{(u^h, z) : u^h A^h + dz \leq c, -u^h b^h + d_0 z \leq c_0, u^h \geq 0\},$$

$$\bar{U}_h = \{(u^h, z) : u^h A^h + dz \leq 0, -u^h b^h + d_0 z \leq 0, u^h \geq 0\},$$

and

$$Q^* = \{ h \in Q : X_h \neq \emptyset \}, \; Q^{**} = \{ h \in Q : U_h \neq \emptyset \}.$$

The following regularity condition is assumed:

$$\text{(RC)} \qquad (Q^* \neq \emptyset, Q \setminus Q^{**} \neq \emptyset) \Rightarrow Q^* \setminus Q^{**} \neq \emptyset;$$

i.e. if (PP) is feasible and (DP) is infeasible, then there exists $h \in Q$ such that $X_h \neq \emptyset$ and $U_h = \emptyset$.
Making use of the result of disjunctive linear programming and the regularity condition (RC), the following main duality theorem is established [13].

Theorem: If the problems (PP) and (DP) satisfy the regularity condition (RC) then either both problems are feasible, each has an optimal solution and the optimal values are equal, or one of the problems is infeasible implying other one is either infeasible or has no finite optimum.

It is to be noted that the problem (DP) is an ordinary linear programming problem which could be easily solved.

Remark 3.1. Generally, the problems in practice would satisfy the regularity condition (RC), but there can be situations when it breaks down and a duality gap is formed as illustrated by an numerical example in [13].

4. Discussion

The results in disjunctive linear fractional programming reported so far can be treated as an extension of those in disjunctive linear programming. It is therefore necessary to establish algorithms and various dual formulations for DLFP on the lines of those available for ordinary linear fractional programming. An approach based on cutting plane theory [7, 8] appears promising in this respect.

It is found that certain special forms of fairly large disjunctive linear programming problems offer appreciable computational advantage over the others in terms of computer time requirement [9]. Thus, there is an urgent need for computer based numerical experimentation to compare various DLFP models.

A study of topics like post-optimality analysis, parametric programming, explict integer programming, multi-objective programming and max-min programming under disjunctive linear fractional programming is expected to generate many interesting results. Some of these topics, incidently, have also remained unexplored in disjunctive linear programming. Further, the area of nonlinear version of DLFP development needs the attention.

A few applications of DLFP in machine flow-shop scheduling, technology selection, regional planning etc. were identified in [11] and it seems possible to treat some of the problems in economic planning through this development. The potential for obtaining new results in disjunctive fractional progtramming is tremendous and it is likely to attract many researchers in future.

References

[1] **Balas E.**, Disjunctive Programming: Cutting Planes from Logical Conditions, in: O.L. Mangasarian, R.R. Meyer and S.M. Robinson (eds.), Non-Linear Programming 2, Academy Press, New York, N.Y., 1975, pp.279-312.

[2] **Balas E.**, A note on Duality in Disjunctive Programming, Journal of Optimization Theory and Applications, 21 (1979), pp.523-528.

[3] **Balas E.**, Disjunctive Programming, Annals of Discrete Mathematics, 5 (1979), 3-51.

[4] **Balas E., and Jeroslow, R.G.**, Strengthening Cuts for Mixed Integer Programs, European Journal of Operational Research, 4 (1980), pp.224-234.

[5] **Dinkelbach W.**, On linear Fractional Programming, Management Science, 12 (1967),pp. 492-498.

[6] **Jagannathan R.**, On Some Properties of Programming Problems in Parametric Form Pertaining to Fractional Programming, Management Sciences, 12 (1966), pp.609-615.

[7] **Jeroslow R.G.**, Cutting Plane Theory: Disjunctive Methods, Annals of Discrete Mathematics, 1 (1977), pp.293-330.

[8] **Jeroslow R.G.**, Cutting Plane for Complementary Constraints, SIAM Journal of Control and Optimization, 16 (1978), pp.56-62.

[9] **Jeroslow R.G.**, Private Communication.

[10] **Owen G.**, Cutting Planes for Programs with Disjunctive Constraint, Journal of Optimization Theory and Applications, 11 (1973), pp.49-55.

[11] **Patkar V. and Stancu-Minasian I.M.**, On Disjunctive Linear Fractional Programming, Economic Computation and Economic Cybernetics Studies and Research, 17 (1982), n.2, pp.87-96.

[12] **Patkar V. and Stancu-Minasian I.M.**, Parametric Algorithm for a Class of Disjunctive Linear Fractional Programs, Bulletin Mathematique de la Societe des Sciences Mathematiques de la Republique Socialiste de Roumania (N.S.), 29 (1985), pp.279-284.

[13] **Patkar V. and Stancu-Minasian I.M.**, Duality in Disjunctive Linear Fractional Programming, European Journal of Operational Research, 21 (1985), pp.101-105.

[14] **Williams H.P.**, Model Building in Mathematical Programming, John Wiley and Sons, N.Y., 1978.

AN INTERVAL-TYPE ALGORITHM FOR GENERALIZED FRACTIONAL PROGRAMMING

Y. Benadada - J.P. Crouzeix - J.A. Ferland

Abstract

The purpose of this paper is to introduce and analyze the convergence of a new interval-type algorithm for generalized fractional programming; This new algorithm has the advantage of being easier to implement than earlier algorithms of this type, especialy for nonlinear problem. The numerical results indicate that it is as efficient as other interval-type and Dinkelbach-type algorithms for these problems.

1. Introduction

Consider the generalized fractional program to determine

$$(P) \qquad \lambda^* = \operatorname*{Inf}_{x \in X} \left\{ \operatorname*{Max}_{1 \le i \le p} \left\{ \frac{f_i(x)}{g_i(x)} \right\} \right\}$$

Assume that $X \subset R^n$ is nonempty and compact, f_i and g_i are continuous real valued functions on an open set $S \subset R^n$ including the closure of X, and $g_i(x) > 0$ for all $x \in S, 1 \le i \le p$. Associate with (P) the parametric program

$$(P_\lambda) \qquad F(\lambda) = \operatorname*{Inf}_{x \in X} \{ \operatorname*{Max}_{1 \le i \le p} \{ f_i(x) - \lambda g_i(x) \} \}$$

In [6, 7, 11, 13], function F is analyzed, and solving (P) is shown to be equivalent to determine the root of the equation $F(\lambda) = 0$.

If $p = 1$, then (P) reduces to the classical fractional programming problem which has been extensively investigated in the last two decades. This type of problem occurs frequently in models where an efficiency measure expressed as a ratio is to be optimized. In numerical analysis, the eigenvalue problem is formulated as a fractional program. Fractional programs are also exhibited in stochastic programming. These applications and others are discussed in Schaible [17] where appropriate references are given.

An early application of generalized fractional programming ($m > 1$) is found in the Von Neumann's model of an expanding economy [19]. The best rational approximation problem [1] is another manifestation of generalized fractional programming. Furthermore, goal programming and multicriteria optimization where several ratios are considered and Chebychev's norm is used give rise to generalized fractional programming. More specific applications are given in [8, 12, 17].
The basic procedure to identify the root of $F(\lambda)=0$ is summarized as follows:

Step 0. Let $x^0 \in X$. Determine λ_1, and let $k = 1$.
Step 1. Determine $x^k \in X$ an optimal solution of (P_{λ_k}).
Step 2. If $F(\lambda_k) = 0$, then λ_k is an optimal value and x^k an optimal solution of (P); STOP.
Step 3. Determine λ_{k+1}, let $k = k + 1$, and return to Step 1.

□

Several variants of this algorithm have been proposed corresponding to different choices of λ_{k+1} in Step 3.
In [7], Crouzeix et al. generalize the approach proposed by Dinkelbach [9] for the case $p = 1$ by selecting

$$\lambda_{k+1} = \underset{1 \le i \le p}{\mathrm{Max}} \left\{ \frac{f_i(x^k)}{g_i(x^k)} \right\}$$

The rate of convergence of this variant is shown to be linear. To improve the rate of convergence, Crouzeix et al. [6] propose to replace (P_{λ_k}) in Step 1 of the algorithm by $(Q_k(\lambda_k))$ defined as follows:

$$(Q_k(\lambda_k)) \qquad F_k(\lambda_k) = \underset{x \in X}{\mathrm{Inf}} \left\{ \underset{1 \le i \le p}{\mathrm{Max}} \left\{ \frac{1}{g_i(x^{k-1})} (f_i(x) - \lambda_k g_i(x)) \right\} \right\}$$

It can be shown that the rate of convergence becomes superlinear when (P_{λ_k}) is replaced by $(Q_k(\lambda_k))$. This Dinkelbach-type algorithm is referred to as MAXMODM. Furthermore, the rate of convergence can be made equal to 1.618 [5, 11] or even quadratic [5] under more restrictive hypothesis on problem (P).
Interval-type variants of the algorithm are given in [2, 3, 4, 10]. In these cases, λ_{k+1} (in Step 3) is selected within an interval $[LB_k, UB_k]$ including λ^*. Furthermore, at each iteration, the bounds LB_k and UB_k are updated to generate a sequence of intervals with non-increasing length. In [4], Bernard and Ferland analyze the convergenze of BF II, an interval-type variant. The rate of convergence bahaves similarly to MAXMODM rate of convergence.
The purpose of this paper is to introduce and analyze the convergence of another interval-type variant BCF which is easier to implement than BF II. The numerical results indicate that BCF is as efficient as BF II and a little more efficient than MAXMODM.

In Section 2, the BCF procedure is presented. We also indicate how it differs from BF II and MAXMODM. The convergence of the procedure is analyzed in Section 3. First we show that the sequence $\{\lambda_k\}$ generated by procedure converges linearly to λ^*. Then, the rate of convergence can be made superlinear under additional conditions. Finally, in Section 4, we compare the numerical efficiency of MAXMODM, BF II, BCF and MINOS which is also used to solve the subproblems $(Q_k(\lambda_k))$.

2. BCF Procedure

It is easy to verify that F(x) is a non-increasing function of λ. Furthermore, since X is assumed to be compact, $F(\lambda)$ is continuous and $F(\lambda^*) = 0$ if and only if λ^* is the optimal value of (P) (see [7]).
Recall that in Dinkelbach-type variant [6,7]

$$\lambda_{k+1} = \operatorname*{Max}_{1\le i\le p} \left\{ \frac{f_i(x^k)}{g_i(x^k)} \right\}$$

and then,

$$\text{(2.1)} \qquad \lambda_{k+1} \ge \operatorname*{Min}_{x\in X} \left\{ \operatorname*{Max}_{1\le i\le p} \left\{ \frac{f_i(x)}{g_i(x)} \right\} \right\} = \lambda^*$$

Hence the sequence $\{\lambda_k\}$ has al its elements to the right of λ^*, and it follows that $F(x^k) \le 0$ for all k.
In BCF procedure, λ_{k+1} is determined as a linear combination of LB_k and UB_k using a scalar $\xi_{k+1} \in (0,1)$

$$\text{(2.2)} \qquad \lambda_{k+1} = \xi_{k+1}\, LB_k + (1 - \xi_{k+1})\, UB_k$$

Hence λ_{k+1} can either be to the right or to the left of λ^*.
To specific UB_k recall that given $x^k \in X$ determined in Step 1, it follows that

$$\operatorname*{Max}_{1\le i\le p} \left\{ \frac{f_i(x^k)}{g_i(x^k)} \right\} \ge \operatorname*{Inf}_{x\in X} \left\{ \operatorname*{Max}_{1\le i\le p} \left\{ \frac{f_i(x)}{g_i(x)} \right\} \right\} = \lambda^*$$

Hence

$$(2.3) \qquad UB_k \leq \max_{1\leq i\leq p} \frac{f_i(x^k)}{g_i(x^k)}$$

It is more difficult to specify the lower bound LB_k unless $F_k(\lambda_k) > 0$ (i.e. $\lambda_k < \lambda^*$) in which case

$$(2.4) \qquad LB_k \geq \lambda_k.$$

To initialize the procedure, a point $x^0 \in X$ is selected, and $UB_k = \max_{1\leq i\leq p} \left\{ \frac{f_i(x^0)}{g_i(x^0)} \right\}$

Since there is no information available to specify the lower bound, LB_0 is arbitrarely taken to the left of UB_0 ; i.e.

$$LB_0 < UB_0 .$$

The procedure accounts for the case where LB_0 is not a lower bound on λ^* (i.e. $LB_0 > \lambda^*$).
The BCF procedure can be summarized as follows:

Step 0. Let $u > 0$, and $\{\xi_k\}$ such that for all $k \geq 1$ $\xi_k \in (0,\xi]$ where $\xi < 1$.
Select $x^0 \in X$, and determine

$$UB_0 = \max_{1\leq i\leq p} \left\{ \frac{f_i(x^0)}{g_i(x^0)} \right\}$$

$$LB_0 < UB_0 .$$

Let $\lambda_1 = \xi_1 LB_0 + (1 - \xi_1)\, UB_0$, and $k = 1$.

Step 1. Determine $x^k \in X$ and optimal solution of $(Q_k(\lambda_k))$.
Step 2. If $F_k(\lambda_k) = 0$, then λ_k is an optimal value and x^k is an optimal solution of (P); STOP.
Step 3. To determine λ_{k+1}, two different cases have to be analyzed:

a) If $F_k(\lambda_k) > 0$ (i.e. $\lambda_k < \lambda^*$), then

$$(2.5) \qquad UB_k = \min \left\{ UB_{k-1} , \max_{1\leq i\leq p} \left\{ \frac{f_i(x^k)}{g_i(x^k)} \right\} \right\}$$

(2.6) $$LB_k = \lambda_k .$$

Go to Step 3 c).

b) Otherwise, $F_k(\lambda_k) < 0$ (i.e. $\lambda_k > \lambda^*$).

(2.7) $$UB_k = \underset{1 \le i \le p}{\text{Max}} \left\{ \frac{f_i(x^k)}{g_i(x^k)} \right\}$$

If $LB_{k-1} \ge \underset{1 \le i \le p}{\text{Max}} \left\{ \frac{f_i(x^k)}{g_i(x^k)} \right\} > \lambda^*$, then

(2.8) $$LB_k = \underset{1 \le i \le p}{\text{Max}} \left\{ \frac{f_i(x^k)}{g_i(x^k)} \right\} - u.$$

Otherwise

(2.9) $$LB_k = LB_{k-1}.$$

c) $\lambda_{k+1} = \xi_{k+1} LB_k + (1 - \xi_{k+1}) UB_k$, and $k = k + 1$.

Return to Step 1.

□

In Step 3 b), as long as $LB_{k-1} \ge UB_k > \lambda^*$ (i.e. LB_{k-1} is not a lower bound of λ^*) we specify LB_k as in (2.8) hoping to obtain $LB_k \le \lambda^*$.

Note that whenever $\lambda_{k+1} \ge \lambda^*$, then λ_{k+1} is closer to λ^* than $\underset{1 \le i \le p}{\text{Max}} \left\{ \frac{f_i(x^k)}{g_i(x^k)} \right\}$ which is the value of λ_{k+1} in MAXMODM. Indeed

$$\xi_{k+1} LB_k + (1 - \xi_{k+1}) UB_k \le UB_k \le \underset{1 \le i \le p}{\text{Max}} \left\{ \frac{f_i(x^k)}{g_i(x^k)} \right\}$$

BCF is easier to implement than BF II [3, 4] mostly because this last procedure requires a "real" lower bound at each iteration. In particular, it may require a huge amount of computation to generate $LB_0 < \lambda^*$.

3. Convergence of the procedure

To complete the analysis, we first introduce some preliminary results useful to show that the procedure converges with a linear rate of convergence. Then, additional assumptions are specified to obtain a superlinear rate of convergence.

3.1 Preliminary results

Some properties of the sequences $\{\lambda_k\}$, $\{LB_k\}$ and $\{UB_k\}$ are now derived before analyzing the convergence of the sequences $\{\lambda_k\}$ and $\{x^k\}$. The first two are straighforward to verify.

Property 3.1 For all $k \geq 1$, $UB_k \geq \lambda^*$.

Property 3.2 For all $k \geq 1$, $LB_k < UB_k$, and hence $LB_k < \lambda_{k+1} < UB_k$.

Property 3.3 The sequence $\{UB_k\}$ is non-increasing.

Proof. If $F_k(\lambda_k) > 0$, then it follows from (2.5) that $UB_k \leq UB_{k-1}$. If $F_k(\lambda_k) < 0$, then

$$F_k(\lambda_k) = \underset{1 \leq i \leq p}{\mathrm{Max}} \left\{ \frac{1}{g_i(x^{k-1})} (f_i(x^k) - \lambda_k\, g_i(x^k)) \right\} < 0$$

implies that for all $1 \leq i \leq p$,

$$\frac{f_i(x^k)}{g_i(x^k)} < \lambda_k.$$

Hence

$$UB_k = \underset{1 \leq i \leq p}{\mathrm{Max}} \left\{ \frac{f_i(x^k)}{g_i(x^k)} \right\} < \lambda_k < UB_{k-1} \tag{3.1}$$

□

Property 3.4 There exists a finite iteration index q such that for all $k \geq q$, $LB_k \leq LB_{k+1} \leq \lambda^*$.

Proof. For contradiction, suppose that such an index does not exist. Hence for all $k \geq 1$

$$\lambda^* < LB_{k-1} < \lambda_k < UB_{k-1}, \tag{3.2}$$

and via (3.1),

$$\lambda^* < LB_k < \lambda_{k+1} < UB_k < \lambda_k .$$

Referring to [6, Theorem 2.1], for all $k \geq 1$

$$F_k(\lambda_k) \leq (\lambda^* - \lambda_k) \operatorname*{Min}_{1\leq i\leq p} \left\{ \frac{g_i(x^*)}{g_i(x^{k-1})} \right\} \tag{3.3}$$

where x^* is an optimal solution of (P). Since x^k is an optimal solution of $(Q_k(\lambda_k))$, it follows that

$$F_k(\lambda_k) = \operatorname*{Max}_{1\leq i\leq p} \left\{ \frac{g_i(x^k)}{g_i(x^{k-1})} \left[\frac{f_i(x^k)}{g_i(x^k)} - \lambda_k \right] \right\}$$

Hence, it follows from (2.7) that

$$F_k(\lambda_k) \geq \frac{g_j(x^k)}{g_j(x^{k-1})} (UB_k - \lambda_k) \tag{3.4}$$

$$\text{for all } j \in J(x^k) = \left\{ j : \frac{f_j(x^k)}{g_j(x^k)} = \operatorname*{Max}_{1\leq i\leq p} \left\{ \frac{f_i(x^k)}{g_i(x^k)} \right\} \right\}$$

From (3.3) and (3.4) it follows that for all $j \in J(x^k)$

$$UB_k - \lambda^* \leq (\lambda_k - \lambda^*) \left[1 - \left\{ \frac{g_j(x^{k-1})}{g_j(x^k)} \right\} \operatorname*{Min}_{1\leq i\leq p} \left\{ \frac{g_i(x^*)}{g_i(x^{k-1})} \right\} \right] \tag{3.5}$$

Denote

(3.6) $$\alpha = \operatorname*{Inf}_{x,y \in X} \left\{ \operatorname*{Min}_{1 \le i \le p} \left\{ \frac{g_i(x)}{g_i(y)} \right\} . \operatorname*{Min}_{1 \le i \le p} \left\{ \frac{g_i(x^*)}{g_i(x)} \right\} \right\}$$

Since X is compact and g_i are continuous and positive on X, then $\alpha > 0$. Substituting in (3.5), it follows from (3.1) that

(3.7) $$\lambda_{k+1} - \lambda^* < UB_k - \lambda^* \le (\lambda_k - \lambda^*)(1 - \alpha).$$

Then, the sequence (λ_k) converges (linearly) to λ^*, and from (3.2) the sequence (BI_k) converges also to λ^*.
But referring to Step 3 b), it is easy to verify that the lower bound LB_k would become smaller than λ^* after a finite number of iterations where (2.8) is applied. Hence, there must exist an iteration index q^1 such that . LB_k = cte for all $k \ge q^1$ (i.e., (2.9) applies for all iterations $k \ge q^1$). Thus referring to (3.2), it follows that $LB_k = \lambda^*$ for all $k \ge q^1$. This contradiction shows the existence of a finite iteration index q such that $LB_q \le \lambda^*$.
Furthermore, for all $k \ge q + 1$,
whenever $\lambda_k > \lambda^*$, (2.9) implies that $LB_k = LB_{k-1}$;
whenever $\lambda_k < \lambda^*$, (2.6) implies that $LB_k = \lambda_k > LB_{k-1}$.
Hence, there exists an index q such that for all $k \ge q$ $LB_k \le LB_{k+1} \le \lambda^*$.

□

3.2 Linear rate of convergence

Now, we analyze the convergence of the sequences (λ_k) and (x^k) generated with BCF procedure. In order to do this, we partition (λ_k) into two subsequences (λ^l_j) and (λ^r_j) where:

$\lambda_k \in (\lambda^l_j)$ if and only if $\lambda_k < \lambda^*$ (i.e., $F_k(\lambda_k) > 0$).
$\lambda_k \in (\lambda^r_j)$ if and only if $\lambda_k > \lambda^*$ (i.e., $F_k(\lambda_k) < 0$).

Theorem 3.5. If it is not finite, the sequence (λ_k), generated with BCF procedure, converges at least linearly to λ^*.

Proof. Three different cases have to be studied:

i) (λ^l_j) is finite;
ii) (λ^r_j) is finite;
iii) (λ^l_j) and (λ^r_j) are not finite.

i) (λ^l_j) is finite.

Since $\{\lambda_k\}$ is not finite, it follows that $\{\lambda^r_j\}$ is not finite. Hence there exists an iteration index k_1, such that $\lambda_k \geq \lambda^*$ for all $k \geq k_1$.
Using the same argument as in the proof of Property 3.4, it follows that (3.7) is verified; i.e., for all $k \geq k_1$.

(3.8) $$0 \leq \lambda_{k+1} - \lambda^* < (\lambda_k - \lambda^*)(1 - \alpha_k) \leq (\lambda_k - \lambda^*)(1 - \alpha)$$

where

(3.9) $$\alpha_k = \underset{1\leq i\leq p}{\text{Min}} \left\{ \frac{g_i(x^{k-1})}{g_i(x^k)} \right\} \cdot \underset{1\leq i\leq p}{\text{Min}} \left\{ \frac{g_i(x^*)}{g_i(x^{k-1})} \right\}.$$

and $\alpha > 0$ is specified by relation (3.6). Hence $\{\lambda_k\}$ converges at least linearly to λ^*.

ii) $\{\lambda^r_j\}$ is finite.

Since $\{\lambda_k\}$ is not finite, it follows that $\{\lambda^1_j\}$ is not finite. Hence, there exists an iteration index k_2 such that $\lambda_k \leq \lambda^*$ for all $k \geq k_2$.
It follows from (2.6) that for all $k \geq k_2$

$$LB_k = \lambda_k.$$

Substituting in (2.2), it follows that for all $k \geq k_2$

$$\lambda_{k+1} = \xi_{k+1}\lambda_k + (1 - \xi_{k+1})\, UB_k$$

and

$$0 \leq \lambda^* - \lambda_{k+1} = \xi_{k+1}(\lambda^* - \lambda_k) - (1 - \xi_{k+1})(UB_k - \lambda^*).$$

But since $UB_k - \lambda^* \geq 0$ and $\xi_{k+1} \leq \xi < 1$, then

(3.10) $$0 \leq \lambda^* - \lambda_{k+1} \leq \xi_{k+1}(\lambda^* - \lambda_k) \leq \xi(\lambda^* - \lambda_k).$$

Hence, $\{\lambda_k\}$ converges at least linearly to λ^*.

iii) $\{\lambda^1_j\}$ and $\{\lambda^r_j\}$ are not finite.

a) Convergence of $\{\lambda^1_j\}$.

Let $\lambda^1_j = \lambda_k$ and $\lambda^1_{j+1} = \lambda_{k+\rho}$ where $\rho \geq 1$. From (2.6) and (2.9), it follows that

$$LB_k = \dots\dots = LB_{k+\rho-1} = \lambda_k.$$

Substituting in (2.2) $\lambda_{k+\rho} = \xi_{k+\rho}\lambda_k + (1 - \xi_{k+\rho})\, UB_{k+\rho-1}$.

Since $UB_{k+p-1} \geq \lambda^*$ and $\xi_{k+p} \leq \xi < 1$, then it follows that

$$\lambda^* - \lambda_{k+p} \leq \xi_{k+p}(\lambda^* - \lambda_k) \leq \xi(\lambda^* - \lambda_k) \tag{3.11}$$

or

$$\lambda^* - \lambda^l_{j+1} \leq \xi(\lambda^* - \lambda^l_j).$$

Hence, $\{\lambda^l_j\}$ converges at least linearly to λ^*.

b) Convergence of $\{\lambda^r_j\}$

Let $\lambda^r_j = \lambda_k$ and $\lambda^r_{j+1} = \lambda_{k+p}$ where $p \geq 1$. From (2.7) it follows that

$$UB_k = \operatorname*{Max}_{1 \leq i \leq p} \left\{ \frac{f_i(x^k)}{g_i(x^k)} \right\}.$$

Referring to the proof of Property 3.4, it follows from (3.7) that

$$0 \leq UB_k - \lambda^* \leq (\lambda_k - \lambda^*)(1 - \alpha_k) \leq (\lambda_k - \lambda^*)(1 - \alpha) \tag{3.12}$$

where $\alpha > 0$ is specified in (3.6). But from Property 3.3 the sequence $\{UB_k\}$ is non-increasing, and hence

$$\lambda_{k+p} - \lambda^* < UB_{k+p-1} - \lambda^* \leq UB_k - \lambda^* \leq (\lambda_k - \lambda^*)(1 - \alpha).$$

Then

$$\lambda^r_{j+1} - \lambda^* < (\lambda^r_j - \lambda^*)(1 - \alpha)$$

and $\{\lambda^r_j\}$ converges at least linearly to λ^*.

c) Since both $\{\lambda^l_j\}$ and $\{\lambda^r_j\}$ converge at least linearly to λ^*, it follows that $\{\lambda_k\}$ converges at least linearly to λ^*.

□

Now, we analyze the convergence of $\{x^k\}$.

Theorem 3.6. Each convergent subsequence of $\{x^k\}$, generated with BCF procedure, converges to an optimal solution of (P).

Proof. The argument is identical to the proof of [6, Theorem 2.1].

□

3.3 Superlinear rate of convergence

To improve the rate of convergence of $\{\lambda_k\}$, an additional condition on the sequence $\{\xi_k\}$ is required.

Theorem 3.7. Assume that the sequence $\{\xi_k\}$ converges to zero and that the sequence $\{x^k\}$ converges to x^*. Then, the sequence $\{\lambda_k\}$ generated with BCF procedure converges superlinearly to λ^*.

Proof. We refer to the three cases studied in Theorem 3.5 to complete the proof.

i) $\{\lambda^l_j\}$ is finite

To analyze the convergence of $\{\lambda^r_j\}$, recall that in this case (3.8) is verified; i.e., for all $k \geq k_1$

$$0 \leq \lambda_{k+1} - \lambda^* < (\lambda_k - \lambda^*)(1 - \alpha_k).$$

Since $\{x^k\}$ converges to x^*, it follows from (3.9) that $\lim_{k\to\infty} \alpha_k = 1$. Hence $\{\lambda^r_j\}$ converges superlinearly to λ^*.

ii) $\{\lambda^r_j\}$ is finite

To analyze the convergence of $\{\lambda^l_j\}$, recall that in this case (3.10) is verified; i.e., for all $k \geq k_1$

$$0 \leq \lambda^* - \lambda_{k+1} \leq \xi_{k+1}(\lambda^* - \lambda_k).$$

Since $\{\xi_k\}$ converges to 0, it follows that $\{\lambda^l_j\}$ converges superlinearly to λ^*.

iii) $\{\lambda^l_j\}$ and $\{\lambda^r_j\}$ are not finite

Referring to (3.11) and (3.12), similar arguments are used to show that $\{\lambda^l_j\}$ and $\{\lambda^r_j\}$ converges superlinearly to λ^*. Hence, $\{\lambda_k\}$ converges superlinearly to λ^*.

□

The following example shows that the condition requiring $\{\xi_k\}$ to converge to 0 is not only sufficient to guarantee superlinear convergence (Theorem 3.7), but also necessary. In this example $n = p = 1$ (i.e., one variable and one ratio):

$$f(x) = 2x, \; g(x) = 1, \text{ and } X = [0, 1].$$

It follows that

$$\lambda^* = \operatorname*{Inf}_{x\in X} \left\{ \frac{f(x)}{g(x)} \right\} = \operatorname*{Inf}_{x\in[0,1]} \; (2x) = 0,$$

$$F_k(\lambda_k) = \operatorname*{Inf}_{x\in[0,1]} \; (2x - \lambda x) = - \lambda_k,$$

and

$$x^k = 0.$$

Consider the sequence $\{\xi_k\}$ specified as follows:

$$\xi_k = \frac{1}{4} + \frac{1}{k+2}$$

Then $\lim_{k\to\infty} \xi_k = \frac{1}{4} \neq 0.$

Suppose that $x^0 = 0$. Then $UB_0 = 0$. Let $LB_0 < 0$.
It follows that

$$\lambda_1 = \xi_1 \, LB_0 < 0.$$

Furthermore, since $x^k = 0$ for all k, it follows that

$$UB_k = \frac{f(x^k)}{g(x^k)} = 0.$$

and inductively $F_k(\lambda_k) > 0$. Hence, from (2.6)

$$LB_k = \lambda_k,$$

and from (2.2)

$$\lambda_{k+1} = \xi_{k+1} \, LB_k + (1 - \xi_{k+1}) \, UB_k = \xi_{k+1} \lambda_k.$$

Thus

$$\frac{|\lambda_{k+1}|}{|\lambda_k|} = \xi_{k+1} < 1,$$

and (λ_k) converges linearly to 0, but not superlinearly since $\lim_{k\to\infty} \xi_k \frac{1}{4} \neq 0.$

4. Final Remarks

Like MAXMODM and BF II, BCF procedure was implemented on the CYBER 173 of the University of Montreal. Numerical results are reported in [3, 4, 10] to compare the efficiency of MAXMODM and BF II in solving generalized linear fractional programming problems. In these references, the authors use Land and Powell [14] computer code to deal with the linear programs (P_{λ_k}) or $(Q_k(\lambda_k))$.

Here, we are comparing the efficiency of the three procedures in solving generalized fractional programming problems. It follows that (P_{λ_k}) and $(Q_k(\lambda_k))$ are not linear in general, and we use MINOS [15] to deal with these. Furthermore, since MINOS is a general procedure capable of dealing with (P), we compare also its efficiency with that of the other three procedures.

A first set of generalized linear fractional programming problems is used. They are generated in the same way as in [10] (see [2] for details). Recall that we are using MINOS instead of Land and Powell code to deal with $(Q_k(\lambda_k))$. Nevertheless, the results are consistent with those in [3, 4, 10] to indicate that BF II is a little more efficient than MAXMODM. Furthermore, they also indicate that BCF and BF II have comparable efficiency. Finally, for small problems ($n \leq 100$) MINOS is more efficient than BCF and BF II, but for larger problems ($n \geq 150$) BCF and BF II become more efficient than MINOS.

The same relative efficiency of the four procedures is observed on a second set of problems where the function f_i are quadratic and convex and g_i are linear. MINOS is more efficient for problem where $n \leq 150$, but for larger problems ($n \geq 200$) BCF and BF II become more efficient.

These numerical results indicate that the interval-type algorithms BCF and BF II and the Dinkelbach-type algorithm MAXMODM have comparable numerical efficiency. Nevertheless, for any interval-type or Dinkelbach-type algoritms, the numerical results in [2, 3, 4, 10] indicate that the variant using $(Q_k(\lambda_k))$ is always numerically more efficient than the one using (P_{λ_k}).

References

[1] **Barrodale I, Powell M.J.D., Roberts F.D.K.**, The Differential Correction Algorithm for Rational l_∞ approximation, SIAM J.Numerical Analysis 9, (1972) pp.493-504.

[2] **Benadada Y.**, Approches de Résolution du Problème de Programmation Fractionnaire Généralisée, Thèse de doctorat Université de Montréal (Montréal, Canada, 1989).

[3] **Bernard J.C.**, Quelques Aspects Théoriques en Programmation Fractionnaire Généralisée et Algorithmes, Thése de doctorat, Université de Montréal (Montréal, Canada, 1986).

[4] **Bernard J.C., Ferland J.A.**, Convergence of Interval-Type Algorithms for Generalized Fractional Programming, Mathematical Programming (to appear).

[5] **Borde J., Crouzeix J.P.**, Convergence of a Dinkelbach-Type Algorithm in Generalized Fractional Programming, Zeitschrift fur Operations Research 31 (1987) A31-A54.

[6] **Crouziex J.P., Ferland J.A., Schaible S.**, A Note on an Algorithm for Generalized Fractional Programs, Journal of Optimization Theory and Applications 50 (1986) pp. 183-187.

[7] **Crouzeix J.P., Ferland J.A., Schaible S.**, An Algorithm for Generalized Fractional Programs, Journal of Optimization Theory and Applications 47 (1985) pp. 35-49.

[8] **Crouzeix J.P., Ferland J.A., Schaible S.**, Duality in Generalized Linear Fractional Programming, Mathematical Programming 27 (1983) pp. 342-354.

[9] **Dinkelbach W.**, On Non-Linear Fractional Programming, Management Science 13 (1967) pp. 492-498.

[10] **Ferland J.A., Potvin J.Y.**, Generalized Fractional Programming: Algorithms, and Numerical Experimentation, European Journal of Operational Research 20 (1985) pp. 92-101.

[11] **Flachs J.**, Generalized Cheney-Loeb-Dinkelbach-Type Algorithms, Mathematics of Operations Research 10 (1985) pp. 674-687.

[12] **Jagannathan R.**, An Algorithm for a Class of Nonconvex Programming Problems with nonlinear Fractional Objectives, Management Science 31 (1985) pp. 847-851.

[13] **Jagannathan R., Schaible S.**, Duality in Generalized Fractional Programming via Farkas'Lemma, Journal of Optimization Theory and Applications 41 (1983) pp. 417-424.

[14] **Land A.H. and Powell S.**, Fortran Code for Mathematical Programming, Wiley, New York, 1973.

[15] **Murtagh B.A. and Saunders M.A.**, MINOS AUGMENTED, User's Manual, Technical Report SOL 80-14, Systems Optimization Laboratory, Operations Research Department, Stanford University (Stanford, CA, 1980).

[16] **Schaible S.**, Bibliography in Fractional Programming, Zeitschrift fur Operations Research 26 (1982) pp. 211-241.

[17] **Schaible S.**, A Survey of Fractional Programming, in S.Schaible, W.T.Ziemba, eds., Generalized Concavity in Optimization and Economics, Academic Press, New York, (1981) pp.417-440.

[18] **Schaible S.**, Fractional Programming, II: on Dinkelbach's Algorithm, Management Science 22 (1976) pp. 868-873.

[19] **Von Neumann J.**, A Model of General Economic Equilibrium, Review of Economic Studies 13 (1945) pp. 1-9.

A MODIFIED KELLEY'S CUTTING PLANE ALGORITHM FOR SOME SPECIAL NONCONVEX PROBLEMS

E.A. Sideri

Abstract

We consider the three problems

P1) $\min F(x) - G(x)$

both $F(x)$ and $G(x)$ convex functions from R^p to R

P2) $\min \{ \max F_i(x), i \in I \}$

each function F_i is a pseudoconvex function from R^p to R

P3) $\min \max \{ F_i(x) / G_i(x), i \in I \}$

F_i is a convex function, G_i is a concave function, $G_i(x) > 0$ and $F_i(x) \geq 0$ if G_i is a non linear function.

We show that the classical Cutting Plane Algorithm for convex programming can be modified to obtain a rather general algorithm suitable both for pseudoconvex problems P2 and P3 and for the problem P1 having a particular duality theory (Toland's duality theory). The limit points of the algorithm are minimizer for P2 and P3 and critical points (eventually local or global minima) for P1.

The algorithm is p-step quadratically convergent in a neighborhood of the solution when the Haar condition is meet (i.e. if H is the objective function and x^* the solution there are $v_1, \ldots, v_{p+1}$ in $\partial H(x^*)$ and the set of vectors $\{ v_i \}, i = 1, \ldots, p+1$ has rank p).

Introduction

The Cutting Plane is a classical Algorithm for convex programming (see Kelley [9]). This Algorithm uses the point-convex separation property. In this form it cannot be applied to pseudoconvex problems and the Algorithm does not work in the framework of nonconvex dualities.

The aim of this paper is the discussion of a modified Cutting Plane Algorithm, suitable for the solution of the following problems

P1) $\min F(x) - G(x)$

$x \in R^p$, both $F(x)$ and $G(x)$ convex functions from R^p to R

P2) $\min \{ \max F_i(x), i \in I \}$

$x \in R^p$, the set I is a finite set of indices and each function F_i is a pseudoconvex function from R^p to R

P3) $\min \max \{ F_i(x)/ G_i(x) , i \in I \}$
$x \in R^p$, the set I is a finite set of indices and for all indices i
F_i is a convex function, G_i is a concave function, $G_i(x) > 0$ and $F_i(x) \geq 0$ if G_i is a non linear function.

The outline of the paper is as follow.Section 1 recalls the classical Cutting Plane Algorithm and presents the new one. Section 2 discuss the Algorithm for problem P1 using the Toland's duality theory. Section 3 present the results for the problems P2 and P3.
Some convergence theorems are given in these sections. The algorithm solves at every step a linear problem. The limit points are solutions for P2 and P3 and critical points for P1. Section 4 show a second order convergence result assuming the Haar condition. This result holds only for P1 and P2.
Let $H(x)$ be any function of the following kind:
$H(x) = \max \{ H_i(x) , i \in I \}$ and let I be a finite set. Let $J(x)$ be the set of indices such that $j \in J$ implies $H_j(x) = H(x)$.
We denote by $\partial H(x)$ the set of the vectors u, $u = \nabla H_j(x)$ for some j in $J(x)$.

1. The Algorithm

(For further information on Cutting Plane see [1], [9], [10], [11], [16], [17], [18]).
Let H be a convex function to minimize.
In the classical Cutting Plane Algorithm we assume to have $p+1$ points $x_1, \ldots, x_{p+1}$ and $p+1$ vectors $p_1, \ldots, p_{p+1}$, verifying for every i $p_i \in \partial H(x_i)$.
Moreover there are weights $\mu_1, \ldots, \mu_{p+1}$ satisfying

(1) $$\sum \mu_i p_i = 0$$

(2) $$\sum \mu_i = 1$$

(3) $$\mu_i \geq 0 \ \forall \ i$$

and the set of vectors $\{ p_i \}$ has maximum rank.
We compute the point (z , ξ), $z \in R^p$, $\xi \in R$ such that for every i

(4) $$H(x_i) + p_i^t(z - x_i) = \xi$$

By (4) and by convexity for all u in R^p it holds

$$H(u) \geq H(x_i) + p_i^t(u - x_i)$$

i.e.

(5) $$H(u) \geq H(x_i) + p_i{}^t(z - x_i) + p_i{}^t(u - z) = \xi + p_i{}^t(u - z)$$

We multiply by μ_i, we add the $p+1$ relationships and we obtain by using (1)-(3)

(6) $$H(u) \geq \xi$$

Thus z is optimal when $H(z) = \xi$
The new point (z^+, ξ^+) is computed in the following way.
Let $p_o \in \partial H(z)$; first we compute the weights $\sigma_1, \ldots, \sigma_{p+1}$

(7) $$\sum \sigma_i p_i = p_o$$

(8) $$\sum \sigma_i = 1$$

We define

(9) $$\rho = \min \{ \mu_i / \sigma_i, \sigma_i > 0 \}$$

We assume without lost of generality $\rho = \mu_j / \sigma_j$. We define for $i \neq j$

(10) $$\mu^+_i = (\mu_i - \rho\sigma_i), \; i \neq j$$

We add ρp_o and $(-1) \sum \sigma_i p_i$ to (1) and we add ρ and $(-1) \sum \sigma_i$ to (2).
We obtain

(11) $$\sum_{i \neq j} \mu^+_i p_i + \rho p_o = 0$$

(12) $$\sum_{i \neq j} \mu^+_i + \rho = 1$$

By (10) and (9) $\mu^+_i \geq 0$.
Thus (10), (11) and (12) tell us that (1) - (3) hold for the $p+1$ points and vectors $\{ x_1, \ldots, x_{p+1} \} \cup \{ z \} \setminus \{ x_j \}$, $\{ p_1, \ldots, p_{p+1} \} \cup \{ p_o \} \setminus \{ p_j \}$.
The computations are the same of a single simplex steps (see [8]).
The new point (z^+, ξ^+) satisfies the p equations

(13) $$H(x_i) + p_i{}^t(z^+ - x_i) = \xi^+ \qquad i = 1, \ldots, p+1, i \neq j$$

and

(14) $$H(z) + p_o{}^t(z^+ - z) = \xi^+$$

By (4) and (13) we have

(15) $$p_i^t(z^+ - z) = \xi^+ - \xi \qquad i = 1, \ldots, p+1, i \neq j$$

After multiplication by μ^+_i or by ρ we sum the equalities (14) and (15) yielding $\rho H(z) = \xi^+ - (1-\rho)\xi$ i.e.

(16) $$\rho\ (H(z) - \xi) = \xi^+ - \xi$$

Thus ξ^+ is an improved approximation to min $H(x)$.When H is not a convex function (6) does not hold.

We modify the Cutting Plane Algorithm by using only the subgradient vectors and by dropping any explicit use of the values $H(x_i)$. We assume to have $p+1$ points $x_1, \ldots, x_{p+1}$ and $p+1$ vectors $p_1, \ldots, p_{p+1}$, $p_i \in \partial H(x_i)\ \forall\ i$. We suppose that there are non negative weights μ_i satisfying (1) - (3). The point (z, η) is computed not by (4) but by using the $p+1$ equations

(17) $$p_i^t(z - x_i) = \eta \qquad \forall\ i.$$

Let p_0 be, $p_0 \in \partial H(z)$ then the equations and substitutions (7) - (12) hold and we can compute the new point (z^+, η^+) by solving

(18) $$p_i^t(z^+ - x_i) = \eta^+ \qquad \forall\ i, i \neq j$$

(19) $$p_i^t(z^+ - z) = \eta^+$$

Thus

(20) $$p_i^t(z^+ - z) = \eta^+ - \eta \quad i = 1, \ldots, p+1, i \neq j$$

and we obtain, similary (16)

$$\sum_{i \neq j} p_i^t(z^+ - z) + \rho p_0^t(z^+ - z) = (1-\rho)(\eta^+ - \eta) + \eta^+$$

and

(21) $$\eta^+ = (1-\rho)\,\eta$$

We have the Modified Cutting Plane Algorithm.

Algorithm

1) $j = 1$
2) Find a set $X^j = (x_1, \ldots, x_{p+1})$ and p+1 vectors $(p_1, \ldots, p_{p+1})$ such that for every i $p_i \in \partial H(x_i)$ and (1) - (3) hold for some weights μ_i.
3) Compute (z^j, η^j) by solving
   $p_i^t(z^j - x_i)\, \eta^j \quad \forall\, i$
4) If $\eta^j = 0$ STOP
   OTHERWISE
5) Compute a vector $p_0 \in \partial H(z)$ and by (7)-(12) the minimum ratio ρ^j, the point x_r to be dropped and the new values μ^+_i.
6) Set $X^{j+1} = X^j \cup \{z\} \setminus \{x_r\}$
7) $j = j + 1$
8) Compute (by (18)-(19) the new $((z^j, \eta^j)$
9) Go to 3

After some steps (p+1 at least) all the initial points x_i are deleted and only points z^i, $i < j$ are in X^j.
We assume that the vector $p_0 \in \partial H(z)$ computed in the step 5 be always a gradient vector. If $H(z) = \max\{ H_i(z), i \in I \}$ then $p_0 = \nabla H_j(z)$ for some index j such that $H_j(z) = H(z)$.
By (21) and (2)-(3) all the values η^j have constant sign hence the sequence $|\eta^j|$ is not increasing.
When for some positive ρ_0 holds $\rho^j \geq \rho_0$ the limit value of η^j is 0. The meaning of $\eta^j = 0$ and the relationships between $|\eta^j|$ and the distance of z^j of the solution are treated in the next sections for all the problems.
The following lemma explains the difference between classical and modified Cutting Plane.

Lemma 1.1 Let H be a convex function, let $x_1, \ldots, x_{p+1}$ and $p_1, \ldots, p_{p+1}$ be sets of points and vectors such that for every i $p_i \in \partial H(x_i)$ and (1) - (3) hold for some weights μ_i.
Let H1 and H2 be, $H1 = \min\{ H(x_i), i = 1, \ldots, p+1 \}$, $H2 = \max\{ H(x_i), i = 1, \ldots, p+1 \}$.
Let (z, ξ) be the point given by the Cutting Plane (equations (4)) and let (w, η) be the point given by the modified algorithm (equations (17). Then

(22) $$\xi - H1 \leq \eta \leq \xi - H2$$

Proof. By (4) and (17) we have for all i
$H(x_i) + p_i^t(z - x_i) = \xi, \; p_i^t(w - x_i) = \eta$ i.e.

(23) $$p_i^t(z - w) = \xi - \eta - H(x_i)$$

By using (1)-(3) it follows

$$0 = \sum \mu_i p_i{}^t(z - w) = \xi - \eta - \sum \mu_i H(x_i)$$

(24) $$\eta = \xi - \sum \mu_i H(x_i)$$

It follows (22).

□

From previous lemma when $\eta = 0$, $\xi - \sum \mu_i H(x_i) = 0$ and by (6) $\sum \mu_i H(x_i) = \min H(x)$. i.e. $\mu_i > 0$ implies $H(x_i) = \min H(x)$.
In the convex case given (w, η) z is not know in general but ξ can be computed by the equation (24).
The points z and w coincide iff $H(x_i) = H(x_j)$ holds for all i, j.

2. The Algorithm and Toland's Duality

A non convex duality theory has been developed for the problem P1. This is the Toland's duality theory (see [14], [15], [6]).
Given the problem P1
P1) $\min F(x) - G(x)$
$x \in R^p$, both $F(x)$ and $G(x)$ convex functions from R^p to R

we consider the problem

TD) $\min G^*(y) - F^*(y)$
$y \in R^p$
with F^* and G^* the convex conjugate functions of F and G

Then we have (see [6])
1) The problems P1 and TD have the same value i.e.
$\min (F(x) - G(x)) = \min (G^*(y) - F^*(y))$
2) A point x^* is said a critical point if $\partial F(x^*) \cap \partial G(x^*) \neq \emptyset$
3) If x^* is a critical point then there is y^* such that
$F(x^*) - G(x^*) + G^*(y^*) - F^*(y^*)$ and $y^* \in \partial F(x^*) \cap \partial G(x^*)$

Since $\max (F(x) - G(x)) = - \min (G(x) - F(x))$ and $G(x) - F(x)$ is always the difference of two convex functions all the max and min of $(F(x) - G(x))$ are critical points.
We consider the function $H(x) = F(x) - G(x)$. Then $p \in \partial H(x)$ iff $p = q - r$, $q \in \partial F(x)$, $r \in \partial G(x)$.
The algorithm can be applied without the explicit knowledge of $F(x)$, $G(x)$, and their subdifferentials.

The set of p+1 points and p+1 vectors in this case is

- a set of (p+1) points $x_1, \ldots, x_{p+1}$
- a set of (p+1) vectors $p_i = q_i - r_i$. The 2(p+1) vectors q_i, r_i, $q_i \in \partial F(x_i)$, $r_i \in \partial G(x_i)$ are not know.

For the problem P1 η has the meaning of a generalized duality gap

Theorem 2.1

$$(25) \qquad \sum \mu_i(G^*(q_i) - F^*(r_i)) - \eta = \sum \mu_i(F(x_i) - G(x_i))$$

Proof. Since $p_i^t(z - x_i) = \eta$ and

$$p_i^t x_i = (q_i - r_i)^t x_i = q_i^t x_i - r_i^t x_i = F(x_i) + F^*(q_i) - G(x_i) - G^*(r_i) \ \forall\ i$$

by (1) -(3)

$$\eta = (\sum \mu_i)\, \eta = \sum \mu_i p_i^t(z - x_i)$$

It follows

$$(26) \qquad -\eta = \sum \mu_i(F(x_i) - G(x_i)) + \sum \mu_i(F^*(q_i) - G^*(r_i))$$

and (25) □

$\sum \mu_i(F(x_i) - G(x_i)) = \sum \mu_i H(x_i)$ is an estimate of the primal value. This estimate uses p+1 points $x_1, \ldots, x_{p+1}$. $\sum \mu_i(G^*(q_i) - F^*(r_i))$ is an estimate of the dual (Toland's dual) problem.

The expression depends on 2(p+1) vectors q_i and r_i. This value may be computed as $\eta + \sum \mu_i H(x_i)$; $\eta \geq 0$ implies that the estimated dual value is greater that the primal one.

Theorem 2.2 Assume that $(z^j, \eta^j) \to (w, 0)$, the functions F^*, G^* are finite for every $y \in R^p$, and for each point z^j there is an integer s(j) such that z^j leaves the set $X^{j+s(j)}$. Then w is a critical point.

Proof. Let i be a fixed index. We consider the sequence of points x_i^j i.e. of the points at the place i when the index j is varying, $x_i^j \in X^j$.

The points x_i^j are not all the same because for some integer k(j) $x_i^j = z^{j-k(j)}$ and this point leaves the set X after some steps.

Thus $\lim_{j \to \infty} x_i^j = w$

Let q^* be a cluster point of the sequence q_i^j and let r^* be a cluster point of the sequence r_i^j. Thus $q_i^j \in \partial F(x_i^j)$, $r_i^j \in \partial G(x_i^j)$ and $q^* \in \partial F(w)$, $r^* \in \partial G(w)$.

If the cluster points are the same for all indices i, by (25)(26)

$$0 = \sum\mu_i(F(w) - G(w)) + \sum\mu_i(F^*(q^*) - G^*(r^*))$$

On the other hand $p_i = q_i - r_i$ and by (1) - (3) $q^* = r^*$; thus $(G^*(q^*) - F^*(q^*)) = F(w) - G(w)$ and $q^* \in \partial F(w) \cap \partial G(w)$ and w is a critical point.
If the cluster points of the sequence q_i^j, $i=1, p+1$, r_i^j, $i=1, p+1$ are not the same we have $0 \leq \mu_i \leq 1$ and the μ_i are compact.

If q^* and $q^\#$ are two cluster point of the sequence q_i^j then $q^* \in \partial F(w)$, $q^\# \in \partial F(w)$, $w \in \partial F^*(q^*) \cap \partial F^*(q^\#)$ and F^* is linear on the set of cluster points q^* and $q^\#$. A similar result holds for G^* (see [12], [13]). Let y be a cluster point of $\sum_{i=1}^{p+1} \mu_i q_i^j$; clearly y is also a cluster point of $\sum_{i=1}^{p+1} \mu_i r_i^j = y$. We assume $\lim_{j\to\infty} \sum \mu_i q_i^j = y$.

Thus by taking the limit in (25) we have by linearity of conjugate functions $F(w) - G(w) = G^*(y) - F^*(y)$, $y \in \partial F(w) \cap \partial G(w)$ and w is a critical point.

□

We assume the limit point w be max or min. In this case we know the kind of the limit point from the sign of η.

Theorem 2.3 If $(z^j, \eta^j) \to (w, 0)$, and w is a critical point but not a saddle point for $(F(x) - G(x))$ then if $\eta^j \leq 0$ w is a point of minimum while, if $\eta^j \geq 0$, w is a point of maximum.

Proof. Let η^j be, $\eta^j < 0$. For all x_i in X^j it holds $p_i^t(z^j - x_i) = \eta^j < 0$.
Each $H(x_i) + \eta^j$ is a first order approximation to $H(z^j)$. If H is a non smooth function (for instance $H(x) = \max\{H_j(x), j \in J\}$) we assume that at least an approximation be a first order approximation). Thus for at least an index r $H(z^j) < H(x_r)$.
The point z^j enter the set X^{j+1} thus

(27) $$\max\{H(x_i), x_i \in X^{j+1}\} \leq \max\{H(x_i), x_i \in X^j\}$$

Let $\{y^j\}$ be the sequence of points satisfying
i) $y^j \in X^j$
ii) $H(y^j) = \max\{H(x_i), x_i \in X^j\}$
Clearly y^j converges to w and $H(y^j)$ is not increasing. Thus w cannot be a local maximum and is either a saddle point or a local minimum. By assumption w is not a saddle point hence w is a (local) minimum. The proof that $\eta^j > 0$ implies (local) maximum is similar. All the previous inequalities are reversed and min substitutes max.

3. Pseudoconvex Problems

We consider the problem

P2) $\min(\max F_i(x), i \in I)$

$x \in R^p$, the set I is a finite set of indices and each function F_i is a pseudoconvex function from R^p to R

We recall that a function F is said pseudoconvex (see [1], [2]) if

$$\nabla F(x)^t h \geq 0 \Rightarrow F(x+h) \geq 0$$

and strictly pseudoconvex if

$$\nabla F(x)^t h \geq 0 \Rightarrow F(x+h) > 0 \quad \text{when } h \neq 0$$

This condition holds when the Hessian matrix of F is positive definite on the space orthogonal to the gradient (see [4]).We define $H(x) = (\max F_i(x), i \in I)$.
When we use the algorithm for this problem we have:

Theorem 3.1 Let x^* a solution of P2 then assuming all functions F_i pseudoconvex
i) $\eta^j \leq 0$ for all j
ii) if $\eta^j = 0$ then $H(x_i) = H(x^*)$ if $x_i \in X^j$ and the corresponding μ_i is strictly positive.

Proof. Let the index j given. If for some $x_i \in X^j$ we have $p_i^t(x^* - x_i) > 0$ then recalling $p_i = \nabla F_k(x_i)$ for some $k \in I$ it follows

(28) $$F_k(x^*) = F_k(x_i + (x^* - x_i)) > F_k(x_i)$$

and

(29) $$H(x^*) \geq F_k(x^*) > F_k(x_i) \geq H(x_i)$$

The strict inequality follows from Taylor's formula, since we know $H(x^*) \leq H(x_i)$ $p_i^t(x^* - x_i) \leq 0$ for all $x_i \in X^j$. On the other hand

$$p_i^t(z^j - x^*) = p_i^t(z^j - x_i) - p_i^t(x^* - x_i)$$

$$0 = \sum \mu_i p_i^t(z^j - x^*) = \sum \mu_i p_i^t(z^j - x_i) - \sum \mu_i p_i^t(x^* - x_i) = \eta^j - \sum \mu_i p_i^t(x^* - x_i)$$

thus

(30) $$\eta^j = \sum \mu_i p_i^t(x^* - x_i) \leq 0$$

ii) If in (30) $\eta^j = 0$ then

(31) $$\sum \mu_i \, p_i^t(x^* - x_i) = 0$$

hence $\mu_i > 0$ implies $p_i^t(x^* - x_i) = 0$. In this case the observations made in i) lead to $H(x_i) = H(x^*)$.

□

A similar result holds when η^j converges to 0.

Theorem 3.2 If $(z^j, \eta^j) \to (w, 0)$ then w is optimal

Proof. Let (x_i^j) be a sequence of points, depending on j, $(x_i^j) \in X^j$; the index i being fixed. Since $z^j \to w$ $\lim x_i^j = w$ holds for at least a subsequence $i=1, p+1$.
By (30) $\lim \mu_i (p_i^j)^t(x^* - x_i^j) = 0$ i.e. $\lim \ (\mu_i (p_i^j)^t(x^* - w) = 0$ and $(p_i^j) \to q \in \partial H(w)$
If $q \neq 0$ and $\lim \mu_i \neq 0$ since $\lim H(x_i^j) = H(w)$ by (28) - (31) $H(w) = H(x^*)$.
Let ρ^j be the ratio (9) computed when we move from X^j to X^{j+1}. By (21) $\eta^{j+1} = (1 - \rho^j)\eta^j$.
Thus $\lim_{j\to\infty} \eta^j = \lim \eta^j \prod_{k<j}(1-\rho^k)$
We assume $\rho \geq \rho_0 > 0$. Since $\rho^j = \mu_i^+$ for some index i at the next step for some i $\mu_i \geq \rho_0 > 0$.
Thus $\lim \mu_i \neq 0$.
If $(p_i^j) \to q \in \partial H(w)$ and q is the null vector $0 \in \partial H(w)$, in both cases w is optimal

□

For the problem P3
P3) $\min \max \{ F_i(x)/G_i(x) , i \in I \}$

$x \in R^p$, the set I is a finite set of indices and for all indices i, F_i is a convex function, G_i is a concave function,
$G_i(x) > 0$ and $F_i(x) \geq 0$ if G_i is a non linear function

again each function $F_i(x)/G_i(x)$ is pseudoconvex and the corresponding max function

$$H(x) = \max \{ F_i(x)/G_i(x) , i \in I \}$$

is also pseudoconvex

Lemma 3.3 Let $F(x)/G(x)$ be a function such that either A or B holds then $F(x)/G(x)$ is a pesudoconvex function
A) G positive, F convex, G linear
B) G positive, F positive, F convex, G concave
then $F(x)/G(x)$ is a pseudoconvex function

Proof. Let y an arbitrary but fixed point in R^p. We define

(31) $$\theta = F(y)/G(y)$$

and the function from R^p to R

(32) $$S(x) = F(x) - \theta G(x)$$

Clearly $S(y) = 0$, S is convex because is either the sum of a convex function and of a linear function (if A holds) or the sum of two convex functions (if B holds) and

$S(x) > 0$ iff $F(x)/G(x) > \theta = F(y)/G(y)$. Thus

(33) $$\nabla(F(y)/G(y)) = (G(y))^{-1} (\nabla F(y) - ((F(y)/G(y))\nabla G(y)$$

and

(34) $$\nabla S(y) = \nabla F(y) - \theta \nabla G(y) = G(y)\ \nabla(F(y)/G(y))$$

If $\nabla(F(y)/G(y))^t h \geq 0$ then $\nabla S(y)^t h \geq 0$, $S(y+h) \geq S(y) + \nabla S(y)^t h = \nabla S(y)^t h$ and

(35) $$F(y+h)/G(y+h) \geq F(y)/G(y) = \theta$$

□

For the problem P3 we can consider the vectors $t = G_i(x)\ \nabla(F_i/G_i)$; these vectors are the gradients of the function S corresponding to the point x. We start from a set of $(x_1, \ldots, x_{p+1})$ points and $(t_i, \ldots, t_{i+1})$ vectors

(36) $$G(x_i)\ t_i \in \partial(F(x_i)/G(x_i))$$

The algorithm can be modified and we can solve

(37) $$t_i^t(z - x_i) = \eta \quad i=1, p+1$$

instead of (17) $p_i^t(z - x_i) = \eta$

Since $t_i = G_i p_i$ and $G_i > 0$ the theorem 3.1 holds with similar computation and the following modification: the equality (30) became

(38) $$\eta^j = \sum \mu_i\, t_i^t(x^* - x_i) \leq 0$$

Thus $\eta^j = 0$ in (38) implies if $\mu_i > 0$ then $t_i^t(x^* - x_i) = 0$ and by (33) - (35) $H(x_i) = H(x^*)$.
There is a great difference between the p_i and t_i possible algorithms. Using the p_i

vectors $H(x_i)+\eta^j$, $x_i \in X^j$ is a first order approximation to $H(z^j)$,while using the t_i vectors for every $x_i \in X^j$ we build implicitely the function (depending on the point x_i)

$$S_k^{\,i}(x) = F_k(x) - \theta\, G_k(x),\quad \theta = F_k(x_i)/G_k(x_i) = H(x_i)$$

and η^j is a first order approximation to the value $S_k^{\,i}(z^j)$.

This algorithm is different but similar to others algorithm suggested for fractional programming ([3], [5], [7]).

4. Convergence

The Cutting Plane Algorithm has in general a slow rate of convergence (see [1], [16], [17], [18]). In some particular cases the convergence is p-steps quadratic. This result can be extended to the modified algorithm for non smooth problems assuming the Haar condition.
We consider the non smooth problem

NSP) $\min H(x)$, $\quad H(x) = \{ \max\ H_i(x),\ i \in I \}$.

We assume that either $H(x) = F(x) - G(x)$ (i.e. problem P1) or that each function $H_i(x)$ be pseudoconvex (i.e. problem P2)

Definition (property H - Haar condition)
Let x^* be a solution of the problem NSP. The Haar condition holds at x^* if
i) the set of vectors $\{ p_i \in \partial H(x^*) \}$ has rank p
ii) $\sum \mu_j\, p_j(x^*) = 0$, $p_j \in \partial H(x^*)$, $\sum \mu_j = 1$, $\mu_j \geq 0$ implies that at least $p+1$ multipliers μ_j are nonzero.
The meanings are: maximum rank of the subgradients at the solution and strict positivity of the dual variables (Lagrange multipliers). Clearly the Haar condition implies that there are at least $p+1$ functions $H_i(x)$.

Lemma 4.1. If t_i, $i=1,\ldots,p+1$ are vectors in R^p and
i) $\sum \mu_i\, t_i = 0$, $\sum \mu_i = 1$, $\mu_i > 0$
ii) $\operatorname{rank}\{ t_i \} = p$
then for every nonnull $s \in R^p$ there are $\alpha(s)$, $\beta(s)$ $\beta(s) < 0 < \alpha(s)$ such that $\beta(s) = \min t_i^{\,t}s$, $\alpha(s) = \max\ t_i^{\,t}s$; moreover $\alpha(s)$, $\beta(s)$ are continuous functions from R^p in R.

Proof. Clearly $\beta(s) = \min t_i^{\,t}s$, $\alpha(s) = \max\ t_i^{\,t}s$ are continuous functions because the min and the max are taken with respects to linear (hence both convex and concave) functions. The condition $\alpha(s) > 0$ holds because $\alpha(s) \leq 0$ implies $\max t_i^{\,t}s \leq 0$, i.e. $\sum \mu_i\, t_i^{\,t}s \leq 0$ and by the assumptions it follows $t_i^{\,t}s = 0$ for all i and s is the null vector.

In the same way we obtain $\beta(s) < 0$. □

Assuming the property H we can apply the lemma to the set of vectors $\{p_i(x^*), p_i \in \partial H(x^*)\}$

Let α^* be,

$$\alpha^* = \min \{ \alpha(s), \| s \| = 1 \}$$

i.e.

(39) $$\alpha^* = \min \{ \max \{ p_i^t s, p_i \in \partial H(x^*) \}, \| s \| = 1 \}$$

clearly $\alpha^* > 0$

Lemma 4.2. If property H holds then in a neighbourhood of x^*

(40) $$H(x) - H(x^*) \geq \alpha^* \| x - x^* \|$$

hence the solution x^* posseses local uniqueness

Proof. We have

$$\liminf (H(x^* + v) - H(x^*)) \| v \|^{-1} = \liminf (\max \{(H_i(x^* + v) - H_i(x^*)) \| v \|^{-1}\})$$

and

$$\max \{(H_i(x^* + v) - H_i(x^*)) \| v \|^{-1}\} = \max \| v \|^{-1}(\nabla H_i(x^*)^t v + 0 (\| v \|))$$

hence

$$\liminf (H(x^* + v) - H(x^*)) \| v \|^{-1} \geq \alpha^* > 0.$$

Thus $H(x^* + v) - H(x^*) \geq \alpha^* \| v \|$, with the previously defined α^*.

□

We recall that assuming bounded partial derivatives holds

(41) $$H(x^* + v) - H(x^*) \leq \beta^* \| v \|$$

for some β^*.

Lemma 4.3. Let x^* be the solution of the NSP. If $z^j \rightarrow x^*$ and the Haar condition holds at the solution then definitively holds

i) $x_i \in X^j$, $x_k \in X^j$, $k \neq j$, then $H(x_i) = H_r(x_i)$, $H(x_k) = H_s(x_k)$, $s \neq r$;

ii) the weights μ_i are strictly positive.

Proof. We consider the sequence of equalities

(42) $$\sum \mu_i p_i = 0, \; p_i \in \partial H(x_i), \; x_i \in X^j$$

By our assumption each function H_i is smooth and $p_i \in \partial H(x_i)$ is equivalent to $p_i =$

$\nabla H_{t(i)}(x_i)$ for some index $t(i) \in I$.Thus by taking the limit with respect to j it follows

$$\lim \sum \mu_i p_i = \lim \sum \mu_i \nabla H_{t(i)}(x_i) = \sum \mu_i^* \nabla H_{t(i)}(x^*) p_i$$

Since the Haar condition holds, in the sum all the μ_i^* are positive and there are p+1 different subgradients. Hence there are p+1 different vectors $\nabla H_{t(i)}(x^*)$ and $i \neq j$ implies $t(i) \neq t(j)$.
Thus the same considerations holds for $\sum \mu_i \nabla H_{t(i)}(x_i) - \sum \mu_i p_i$ and definitively holds if $x_i \in X^j$, $x_k \in X^j$, $k \neq j$, then $H(x_i) = H_r(x_i)$, $H(x_k) = H_s(x_k)$, $s \neq r$, and all the weights μ_i are strictly positive.

□

We shall proof the p-steps quadratic convergence by making two other assumptions A2 and A3.
A2) There are only p+1 functions H_i involved (This assumption is technical and can be relaxed, see lemma 4.10).
A3) There is an integer t such that every non optimal point in X^j is replaced after at most t steps.

Clearly if both properties H and A2 hold from the equation $X^{j+1} = X^j \cup \{z^j\} \setminus \{x_k\}$ it follws $H(z^j) = H_s(z^j)$, $H(x_k) = H_s(x_k)$ for the same index s because the vector $p_k = \nabla H_s(x_k)$ is replaced by $p_0 = \nabla_{t(i)} H(z)$, there are only p+1 functions and Lemma 4.3 holds. Clearly if z^j enter X^{j+1} and x_k leaves X^j then z^j is the k-th point of the set X^{j+1}.
To avoid too many indices we assume that $x_i \in X^j$ implies $H(x_i) = H_i(x_i)$ i.e. that the index of the point in the set X^j is the same of the function. We assume that z^j converges to x^*. By lemma 4.3 $|\eta^j|$ is strictly deecreasing and all the points z^j lie in a neighbourhood of x^* by maximum rank assumption.

Lemma 4.4. There are two positive constant C_1, C_2 such that in the last steps of the algorithm if $X^{j+1} = X^j \cup \{z\} \setminus \{x_k\}$

(43) i) $|\eta^j| \geq C_1 \| z^j - x^* \|$

(44) ii) $C_2 \| x_k - x^* \| \geq |\eta^j|$

Proof. i) We have $p_i^t(z^j - x_i) = \eta^j < 0$ and from Taylor's formula we can assume that

(45) $p_i^t(x^* - x_i) = \nabla H_i(x_i)^t(x^* - x_i) \leq 0$

since $H_i(x_i) > H_i(x^*)$ holds for the optimal point x^*.

Thus

$$p_i^t(x^* - x_i) - p_i^t(x^* - z^j) + p_i^t(z^j - x_i) \le 0, \quad p_i^t(x^* - z^j) \le -\eta^j$$

Let s_j be, $s_j = (x^* - z^j) / \| x - z^j\|$, $\| s_j\| = 1$.
Since $\nabla H_i(x_i) \to p^\# \in \partial H(x^*)$ by the maximum rank assumption

$$\alpha^* \le \lim p_i^t(x^* - z^j)/\| x^* - z^j\| \le | -\eta^j| /\| x^* - z^j\|$$

i.e.

$$\| x^* - z^j\| \, C_1 \le | \eta^j| \text{ with } C_1 = \alpha^*/2$$

ii) x_k leaves X^j hence

$$H_k(x_k) = H(x_k) \quad \text{and} \quad H_k(z) = H(z)$$

We have near x^*

$$H(z) = H_k(z) = \sup \{ H_i(z) \} = \sup \{ H_i(x^*) + \nabla H_i(x^*)^t(z - x^*) + o(\| z - x^*\|) \}$$

and it follows

$$\nabla H_k(x^*)^t(z^j - x^*)/\| z^j - x^*\| \ge \alpha^*/2$$

By continuity, by observing that $\nabla H_k(x_k) \to \nabla H_k(x^*)$

$$\nabla H_k(x_k)^t(z^j - x^*) \ge \sigma \| z^j - x^*\| \text{ , for some } \sigma > 0$$

Hence

$$p_k^t(z^j - x^*) > 0$$

and

$$\eta^j = p_k^t(z^j - x_k) > p_k^t(x^* - x_k).$$

The first order Taylor's formula, starting from x_k yelds

$$H_k(x_k) + \eta^j + o(\| x^* - x_k\|) \ge H_k(x^*) , \quad H(x_k) - H(x^*) \ge -\eta^j - o(\| x^* - x_k\|)$$

thus by (41) for some $k > 0$

$$k \| x^* - x_k\| \ge -\eta^j - o(\| x^* - x_k\|)$$

and

$$k \ge \lim (-\eta^j)/\| x^* - x_k\| \quad \text{i.e.}$$

$$2k \ge (-\eta^j)/\| x^* - x_k\| \quad \text{and } (2k)\| x^* - x_k\| \ge (-\eta^j)$$

i.e. $C_2 = 2k$

□

Corollary 4.5. If (A3) holds then $\| z^j - x^* \| \neq 0$ implies $\| z^j - x^* \| \geq k(-\eta^j)$.

Proof. The point z^j enter the X^{j+1} but leaves the set after at most t steps, hence z^j cannot be in the set X^{j+t+1}.We assume (w.l.o.g.) that z^j leaves the set X^{j+t} and by lemma4.4

$$C_2 \| z^j - x^* \| \geq (-\eta^{j+t+1})$$

It is enough to prove that $| \eta^j / \eta^{j+t+1} |$ is bounded and does not converge to 0.
By (21) $\eta^{j+1} = (1-\rho^j)\eta^j$ and $\eta^{j+t+1} = (1-\rho^{j+t})(1-\rho^{j+t-1})...(1-\rho^j)\eta^j$
The multipliers ρ^i are the weights of the new vectors p. By lemma 4.3 all the weights are strictly positive and converge to some optimal non zero multipliers μ^*.
Thus

$$(1-\mu^*_M)^t \leq | \eta^j / \eta^{j+t+1} | \leq (1-\mu^*_m)^t$$

with $\mu^*_M = \max \mu^*_i \;\; i \in I(x^*)$, $\mu^*_m = \min \mu^*_i \;\; i \in I(x^*)$

□

Remark. When corollary 4.5 holds we have that for some positive C_3, C_4,

(46) $$C_3 (-\eta^j) \leq \| z^j - x^* \| \leq C_4 (-\eta^j)$$

The left inequality follows from corollary 4.5.The other inequality follows from lemma 4.4 with $C_4 = (C_1)^{-1}$.
Let t be a fixed integer then $| \eta^{j+t} | / | \eta^j |$ is bounded and for some positive number C_5 holds

(47) $$\| z^{j+t} - x^* \| \leq C_5 \| z^j - x^* \|$$

Clearly C_5 depend on the fixed integer t.

For every step of the algorithm we define $\alpha^j = \min (H(x_i), x_i \in X^j)$, $\beta_{ij} = H(x_i) - \alpha^j$ (thus $\alpha^j = H(x_i) - \beta_{ij} = H_i(x_i) - \beta_{ij}$) and the new problem

$$\min H_\beta(x)$$

(NSP$_\beta$)

$$H_\beta(x) = \max (H_i(x) - \beta_{ij}, i \in I)$$

The gradient vectors $\nabla H_i(x)$ are unchanged for every $x \in R^k$ and for every $i \in I$.
The problem (NSP$_\beta$) is a perturbation of the original problem (NSP) hence by continuity (assuming the values β_{ij} bounded and near 0)
- (NSP$_\beta$) has an unique solution (say x_β)

- (NSP_β) possesses property H
- if $\max(\beta_{ij}) \to 0$ then $\| x_\beta - x^* \| \to 0$
- at the solution x_β holds $H_i(x_\beta) = H_\beta(x_\beta)$ for all $i \in I$ and $\sum \lambda_i^* \nabla H_i(x_\beta) = 0$ with p +1 positive weights λ_i^*.

We remark that the point (z^j, η^j) obtained from the algorithm for the problem NSP can be used as starting point for the algorithm applied to problem NSP_β and

$$p_i^t(x_\beta - x_i) \le 0 \text{ for all } x_i \in X^j \text{ (see (45))}, \quad \eta^j \le \max\ (p_i^t(x_\beta - x_i))$$

Lemma 4.6. If $x_\beta \neq z^j$ then for some $K_3, K_4 > 0$

$$K_3 \le \| z^j - x^* \| / \| z^j - x_\beta \| \le K_4 \tag{48}$$

Proof. The point $(z^j - \eta^j)$ can be identified as a point produced from the algorithm applied to problem (P_β). We can use the previous results and we have

$$\| z^j - x^* \| \le C_4(-\eta^j)$$

For small perturbation β the behaviour of the algorithm is the same for thetwo problems. The constants obtained in the previous lemmas depends only on the optimal point x^* and on the vectors $\nabla H_i(x^*)$. By the Haar condition all the sets of vectors have maximum rank. Thus the multipliers μ_i^* and α^* are very close for the two problems and

$$\| z^j - x_\beta \| \ge C_3(-\eta^j)/\ 2 \quad \text{thus} \quad \| z^j - x^* \| / \| z^j - x_\beta \| \le 2(C_4/C_3) = K_4$$

The constant K_3 is computed in a similar way

□

We don't know if really for all the steps of the algorithm $x_\beta \neq z^j$ holds.
But if for all the steps and for all the sets X^j we have $x_\beta = z^j$ then all the functions H_1 are linear. The assumption $x_\beta \neq z^j$ seems to be quite reasonable if all the functions H_1 are nonlinear.

Theorem 4.7. If

i) $x_\beta \neq z^j$

ii) all the functions H_1 are C^2 and non linear

then there is $K^*, K^* > 0$ such that for all j there is a point $x_k = z^{j-s}$, $x_k \in X^j$ satisfiing

$$\| z^j - x^* \| / \| x_k - x^* \|^2 \le K^*. \tag{49}$$

Proof. For all indices i the first order approximation to $H_\beta(x_\beta) = H_i(x_\beta) - \beta_{ij}$ is given by

$$H_i(x_i) + p_i^t(x_\beta - x_i) - \beta_{ij} = H_i(x_i) + p_i^t(x_\beta - z^j) + r_i^t(z^j - x_i) - \beta_{ij} = \alpha^j + \eta^j + p_i^t(x_\beta - z^j)$$

If $p_i^t(x_\beta - z^j) < 0$ for all indices then $\nabla H_i(x^*)^t(x_\beta - z^j) \le 0$ for all i; but the vector $(x_\beta - z^j)$ is fixed and this contradicts lemma 4.1.

For at least an index i, $p_i^t(x_\beta - z^j) \ge 0$.

For this index we have $p_i^t(x_\beta - z^j) \ge \eta^j$ and it holds

$$H_\beta(x_\beta) = \alpha^j + \eta^j + p_i^t(x_\beta - z^j) + \tau \| x_\beta - x_i\|^2 \ge \alpha^j + \eta^j - |\tau| \, \| x_\beta - x_i\|^2$$
$$H_\beta(x_\beta) \ge \alpha^j + \eta^j - |\tau| \, \| x_\beta - x_i\|^2$$

for some real number τ

On the other hand $H_\beta(z^j) \ge H_\beta(x_\beta)$ and $H_\beta(z^j) \ge H_\beta(x_\beta) + \alpha \| z^j - x_\beta\|$

By Taylor's formula and assumption A2 for some index k

$$H_\beta(z^j) = H_k(z^j) - \beta_{kj} = H_k(x_k) - \beta_{kj} + p_k^t(z^j - x_k) + \omega \| z^j - x_k\|^2$$
$$H_\beta(z^j) = \alpha^j + \eta^j + \omega \| z - x_k\|^2$$

It is follows

$$|\omega| \, \| z^j - x_k\|^2 + |\tau| \, \| x_\beta - x_i\|^2 \ge \alpha \| z^j - x_\beta\|$$

Clearly τ and ω depends on the second derivative of H_i, H_k and α is given by lemma 4.2

For some σ, $-1 \le \sigma \le 1$, holds the equality

$$\| x_\beta - x_i\|^2 = \| z^j - x_i\|^2 + 2\sigma \| x_\beta - z^j\| \, \| z^j - x_i\| + \| x_\beta - z^j\|^2$$

Since $x_i \to x^*$, $z^j \to x^*$, $x_\beta \to x^*$

$$\lim \| z^j - x_i\|^2 / \| z^j - x_\beta\| = \lim \| x_\beta - x_i\|^2 / \| z^j - x_\beta\|$$

Thus $$\liminf \| z^j - x_\beta\|^{-1} (|\omega| \, \| z^j - x_k\|^2 + |\tau| \, \| z^j - x_i\|^2) \ge \alpha$$

and by lemma 4.6

$$\liminf \| z^j - x^*\|^{-1} (|\omega| \, \| z^j - x_k\|^2 + |\tau| \, \| z^j - x_i\|^2) \ge \alpha K_3$$

The point x_k, x_i are previous computed z i.e. for some integer s, r $x_k = z^{j-s}$, $x_i = z^{j-r}$ thus

$$\limsup \| z^j - x^*\| (|\omega| \, \| z^j - z^{j-s}\|^2 + |\tau| \, \| z^j - z^{j-r}\|^2)^{-1} \le (K_3 \alpha)^{-1}$$

Hence either $\limsup \| z^j - x^*\| / \| z^j - z^{j-s}\|^2 < +\infty$

or $\qquad \limsup \| z^j - x^* \| / \| z^j - z^{j-r} \|^2 < +\infty$

or both. Since

$$\| z^j - y \|^2 = \| z^j - x^* \|^2 + 2\sigma \| z^j - x^* \| \, \| y - x^* \| + \| y - x^* \|^2 \qquad -1 \le \sigma \le 1$$

either

$$\limsup \| z^j - x^* \| / \| z^{j-s} - x^* \|^2 < +\infty$$

or

$$\limsup \| z^j - x^* \| / \| z^{j-r} - x^* \|^2 < +\infty$$

or both.
We assume (w.l.o.g.) that the point x_k be the first introduced one ($s > r$).
If corollary 4.5 holds $\| z^{j-r} - x^* \| \le C_4 | \eta^{j-r} | \le C_4 | \eta^{j-s} | \le C_4 C_3 \| z^{j-s} - x^* \|$
thus $\| z^{j-r} - x^* \|^2 \le C_4 C_3 \| z^{j-s} - x^* \|^2$
and $\| z^{j-r} - x^* \|^2 \ge (C_4 C_3)^{-1} \| z^{j-s} - x^* \|^{-2}$ and $\limsup \| z^j - x^* \| / \| z^{j-s} - x^* \|^2 < +\infty$
i.e. for some $K^* > 0$ $\limsup \| z^j - x^* \| / \| x_k - x^* \|^2 < +\infty$

□

The proof of theorems depend on the assumption A2) used to obtain the Taylor's expansion for $H(z^j)$ starting from x_k.
If for the same index k $\max \{ H_i(z^j) - \beta_{ij}, i \in I \} = H_k(z^j) - \beta_{kj}$, $\max \{ H_i(z^j), i \in I \} = H_k(z^j)$ the point $x_k = z^{j-s}$ is the point leaving the set X^j.
Assuming that each point x_i is in no more that t set X^j the integer r and s are uniformely bounded and using the lemma 4.5 a), we can always obtain an inequality

$$\| z^j - x^* \| / \| z^{j-s} - x^* \|^2 \le K^*$$

with a sligthy different number K^*, $z^{j-s} = x_k$ (the point leaving the set X^j). We recall that point x_k is the point leaving the set X^j for the problems NSP_β but not necessarily for the original problem NSP).
If we assume further
- $z^j \ne x_\beta$ holds for every step of the algorithm
- for the same index (say k)
  $\max \{ H_i(z^j) - \beta_{ij}, i \in I \} = H_k(z^j) - \beta_{kj}$, $\max \{ H_i(z^j), i \in I \} = H_k(z^j)$
- the point x_k in the equation $\quad X^{j+1} = X^j \cup \{ z^j \} \setminus \{ x_k \}$

is the first entered one, the preceeding theorem is a proof of k step quadratic convergence.
We cannot assume in general that the point leaving the set X^j be the first entered one.
We show some further results: theorem 4.8 and corollary 4.9 assume that x_k be the point really leaving the set X^j, lemmas 4.10 and 4.11 hold in general.

Theorem 4.8. Given a set X^j, we define
dist $(X^j, x^*) = \min \{ \| x^* - x_i \|, x_i \in X^j \}$ then for some $k > 0$
dist $(X^{j+t}, x^*) / (\text{dist}(X^j, x^*))^2 \le K$
if every points leave X^j after at most t iteration

Proof. Let $x_i \in X^j$, satisfiing dist $(X^j, x^*) = \| x^* - x_i \|$. Clearly if $x_i \in X^j$ cannot be in X^{j+t} thus for some steps, $j \le k < j+t$ the point x_i leaves the set X^k and is substituded by z^k. If $z^k \in X^{j+t}$ then
dist $(X^{j+p}) \le \| x^* - z^k \| \le k \| x^* - x_i \|^2 = k\, (\text{dist}(X^j, x^*))^2$
If z^k is not in X^{j+t} then the strongest result holds
dist $(X^{j+t}) \le \| x^* - z^{k+r} \| \le k \| x^* - z^k \|^2 \le k^2 \| x^* - x_i \|^4 = k^2 (\text{dist}(X^j, x^*))^4$
where z^{k+r}, is the point z entering the set X when z^k leaves this set

□

Observing only the $p+1$ subsequences $z^{i(p)}$, identified by the relationships:
$z^{i(p)}$ leaves the set X^j when $z^{i(p+1)}$ enter the set X^{j+1} i.e. by the equations

$$X^{i(p+1)+1} = X^{i(p+1)} \cup \{ z^{i(p+1)} \} \setminus \{ z^{i(p)} \}$$

we obtain, from the previous results

Corollary 4.9. The algorithm builds $p+1$ sequences, every sequences quadratically convergent to the solution x^*.
The assumption A2) i.e. there are only $k+1$ functions involved can be easily relaxed exploiting the

Lemma 4.10. When the Haar condition holds and for infinitely many j we have

$$X^{j+1} = X^j \cup \{ z^j \} \setminus \{ x_i \}, \quad H(x_i) = H_i(x_i), \quad H(z^j) = H_k(z^j)$$

then $| H_i(z^j) - H(x^*) / H(z^j) - H(x^*) | > \alpha$ for some nonzero α.

Proof. The equation $\sum \mu_i p_i = 0$, $p_i = \nabla H(x_i)$, $x_i \in X^j$ gives for X^j and X^{j+1}
$\sum \mu_r \nabla H_r(x_r) + \mu_i \nabla H_i(x_i) = 0$, $\sum (\mu_r)^+ \nabla H_r(x_r) + \mu_k \nabla H_k(z^j) = 0$
with $x_r \in X^j \cap X^{j+1}$ and all $\mu_r, \mu_i, \mu_k \ge 0$
All the points x_r, x_i, z^j converge to x^* hence in x^* holds
$\sum \mu_r^* \nabla H_r(x^*) + \mu_i^* \nabla H_i(x^*) = 0$, $\sum (\mu_r)^{+*} \nabla H_r(x^*) + \mu_k^* \nabla H_k(z^*) = 0$
Clearly $\lim H_i(x_i) = \lim H(x_i) = H(x^*)$ and $\lim H_k(z^j) = H(x^*)$
and from Haar condition it follows $\mu_i^* > 0, \mu_k^* > 0$.

Let J be, $J = \{ r \in I \,/\, H_r(x_r) = H(x_r) \text{ for some } x_r \in X^j \cap X^{j+1} \}$ and let J^* be, $J^* = J \cup \{i\}$.
From the previous analysis the reduced problem

NSPR) $\min H_r(x)$

$$H_r(x) = \max \{ H_j(x), j \in J^* \}$$

fulfills the Haar condition, has only the solution x^* and $H_r(x^*) - H(x^*)$.
Starting from X^j we can compute a different X^{j+1} for the problem NSPR). In this case we must introduce, at least definitively, $\nabla H_i(z^j)$: taking the limit we observe that for the others vectors p_i the Haar condition does not hold in x^*.
Hence $H_r(z^j) = H_i(z^j)$ and for some real numbers $\tau > 0$, $\omega > 0$

$$H_i(z^j) - H_r(x^*) = H_r(z^j) - H_r(x^*) \geq \tau \| z^j - x^* \|$$

and

$$H(z^j) - H(x^*) = H_k(z^j) - H(x^*) \geq \omega \| z^j - x^* \|$$

while from the boundness of the gradient it follows

$$H_r(z^j) - H_r(x^*) \leq A \| z^j - x^* \|, \; H(z^j) - H(x^*) \leq B \| z^j - x^* \|$$

Thus

$$| H_i(z^j) - H(x^*) / H(z^j) - H(x^*) | \geq \alpha$$

with $\alpha \geq (\tau / B)$

□

Hence all convergence theorems hold assuming only the Haar condition and A3).
When the Haar condition holds but we cannot prove that the point z^{j-s} given from Theorem 4.7 is the point leaving X^j we have a different form of theorem 4.8

Theorem 4.11. Given a set X^j, we define
$\text{dist}(X^j, x^*) - \max \{ \| x^* - x_i \|, x_r \in X^j \}$ then for some $k^* > 0$ $\text{dist}(X^{j+t}, x^*) / (\text{dist}(X^j, x^*))^2 \leq k$ where X^{j+t} is the first set with the property $X^j \cap X^{j+1} = \emptyset$

Proof. Let z^i the point $\text{dist}(X^{j+t}, x^*) = \| x^* - z^i \|$
Clearly if $x_i \in X^{j+t}$ then x_i is the point z^i for some index $i > j$ thus $\| x^* - z^i \| \leq k \| x^* - y_i \|^2$ for some point $y_i \in X^{j+t-r}$.
If y_i does not belongs to X^j we can obtain a further inequality $\| x^* - y_i \| < k \| x^* - w_i \|^2$
After a finite number of steps we have for some point $x_i \in X^j$ and for some integer m.
$\| x^* - z^i \| \leq (k)^m \| x^* - x_i \|^{2m}$
Thus

$$\text{dist}(X^{j+t}, x^*) - \| x^* - z^i \| \leq (k)^m \| x^* - x_i \|^{2m} \leq (k)^t (\text{dist}(X^j, x^*))$$

□

References

/1/ **Bazaara M.S., Shetty C.M.**, Nonlinear Programming Theory and Algorithm, Wiley, New York, 1979.

/2/ **Ben-Israel A., Ben-Tal A., Zlobec S.**, Optimality in nonlinear programming: A feasible directions approach, Wiley, New York, 1981.

/3/ **Chandras S., Craven B.D., Mond B.**, Generalized Fractional Programming Duality: A ratio game approach, Journ. Austr. Math. Soc. ser B, vol. 28 (1986)

/4/ **Crouzeix J.P., Ferland J.A.**, Criteria for Quasiconvex and Pseudoconvexity: Relationship and comparison, Math. Progr. vol. 23 (1982), pp. 193-205.

/5/ **Crouzeix J.P., Ferland J.A., Schaible S.**, An Algorithm for Generalized Fractional Programming, J.O.T.A., vol. 47 (1985) pp. 35-49.

/6/ **Ekeland I., Turnbull T.**, Infinite-dimensional Optimization and Convexity, University of Chicago Press, Chicago 1983.

/7/ **Ferland J.A., Potvin J.V.**, Generalized Fractional Programming: Algorithms and numerical Experimentation, E.J.O.R. vol. 20 (1985), pp.92-101.

/8/ **Hadley G.**, Linear Programming, Addison-Wesley, Reading Ma, 1962.

/9/ **Kelley J.E. jr.**, The Cutting Plane Method for Solving Convex Programming, S.I.A.M., journal vol. 8 (1960), pp. 703-712.

/10/ **Luenberger D.G.**, Introduction to Linear and non Linear Programming, Addison Wesley, Reading, 1973.

/11/ **Martos B.**, Nonlinear Programming: Theory and Methods, North-Holland, Amsterdam, 1975.

/12/ **Rockafellar R.T.**, Convex Analysis, Princeton University Press, Princeton, 1970.

/13/ **Stoer J., Witzgall C.**, Convexity and Optimization in Finite Dimension, Springer, Berlin, 1970.

/14/ **Toland J.F.**, Duality in Nonconvex Optimization, Jour. Math An Appl. 66 (1978), pp. 399-415.

/15/ **Toland J.F.**, A Duality Principle for Nonconvex Optimization and the Calculus of Variations, Arch. Rat. Mech. Anal. 71 (1979), pp. 41-61.

/16/ **Topkis**, A Cutting Plane Method With Linear and Geometric Rates of Convergence, J.O.T.A., vol. 36 (1982) n. 1, pp.1-22.

/17/ **Wolfe P.**, Convergence Theory in Nonlinear Programming-Integer and Nonlinear Programming, J. Abadie editor, pp. 1-36, North Holland, Amsterdam, 1970.

/18/ **Zoutendijk**, Mathematical Programming Methods, North-Holland, Amsterdam, 1976.

Equivalence and parametric analysis in linear fractional programming

C. Sodini

Abstract

In this paper a survey of some recent results in equivalence and parametric analysis for linear fractional programs with feasible region not necessarily bounded, is given. Furthermore an algorithm to solve a parametric linear fractional program with the parameter in the numerator of the objective function is suggested.

1. Introduction

This paper contains a survey of some recent results concerning equivalence and parametric analysis for linear fractional programs [5,6,7,9,10,13]. In section 2 the linear fractional program is defined and discussed. In section 3, for the case where the feasible region is bounded, the algorithms of Isbell-Marlow, Martos, Charnes-Cooper, Cambini are briefly described and compared in order to evidentiate differences, weaknesses and equivalences. In section 4 it is shown that only the algorithms of Cambini and Charnes-Cooper (this last one in a modified version suggested by Schaible) are able to solve a linear fractional program when the feasible region is unbounded. It is also shown that it is possible to modify the algorithm of Martos in order to work efficiently even for the unbounded case. Finally, in sections 5 and 6 some difficulties in studying parametric analysis are pointed out, and an algorithm to solve a parametric linear fractional program with the parameter in the numerator of the objective function is proposed.

2. The linear fractional problem

We consider the following linear fractional program:

$$\text{(LFP)} \quad \begin{aligned} &\sup\ f(x) = (c^T x + c_0)/(d^T x + d_0) \\ &x \in X = \{x \in R^n : Ax = b\ ,\ x \geq 0\} \end{aligned}$$

where A is a mxn matrix, c, $d \in R^n$, $b \in R^m$, c_0, $d_0 \in R$, $d^T x + d_0 > 0$, $\forall x \in X$.

For problem LFP the following three cases can be verified:

1) there exists an $x' \in X$ such that $\max \{f(x), x \in X\} = f(x')$; in this case at least a vertex of X is optimal;
2) it results $\sup \{f(x), x \in X\} = +\infty$; in this case there exists a vertex x' and an extreme ray r such that $\lim \{f(x'+tr), t \to +\infty\} = +\infty$;
3) 1) does not hold and it results $\sup \{f(x), x \in X\} = L < +\infty$; in this case there exists a vertex x' and an extreme ray r such that $\lim \{f(x'+tr), t \to +\infty\} = L$.

Clearly if X is a compact set only case 1) holds.

3. Analysis of the algorithms for the bounded case

In this section some algorithms for problem LFP, with bounded feasible region, are examined and compared in order to state possible equivalences.

The algorithm of Isbell - Marlow [14]

Step 0 Compute a feasible basic solution x'; set i=0, $x^i = x'$ and *go to* **Step 1**.

Step 1 Solve the linear program $\max \{f^i(x) = c^T x + c_0 - f(x^i)(d^T x + d_0), x \in X\}$. Let $f^i(x^*) = \max \{f^i(x), x \in X\}$; if $f^i(x^*) = 0$ *stop* (x^* is optimal); otherwise *go to* **Step 2**.

Step 2 Set i=i+1, $x^i = x^*$ and *go to* **Step 1**.

Let us note that the algorithm of Isbell - Marlow can examine the same vertex more than once, as it is shown in the following numerical example (see also [16]):

$\max f(x) = (2x_1 - x_2 + 1)/(x_1 + x_2 + 1)$

$-3x_1 + x_2 \le 1, x_1 - x_2 \le 2, x_2 \le 4, x_1 \ge 0, x_2 \ge 0$

which is depicted in fig. 1.

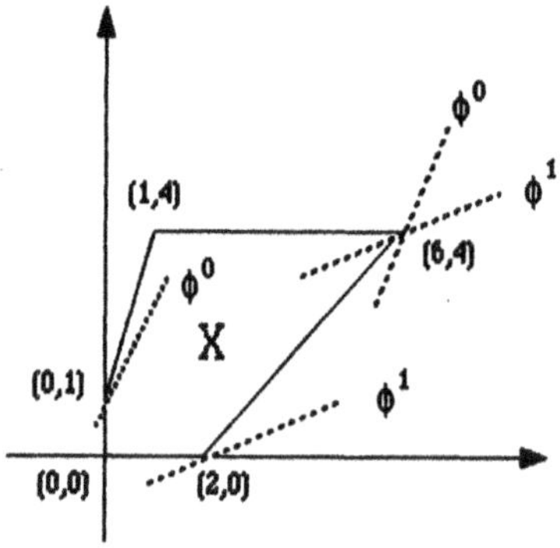

fig. 1

Starting from the vertex $x^0=(0,1)$ we have $f(x^0)=0$ and $f^0(x)=2x_1-x_2+1-f(x^0)(x_1+x_2+1) = 2x_1-x_2+1$. If we solve the linear subprogram the vertex $x^1=(6,4)$ is obtained by examining the vertices (0,0), (2,0) and (6,4). As x^1 is not optimal the linear subproblem with objective function $f^1(x)=2x_1-x_2+1-f(x^1)(x_1+x_2+1) = 13/11x_1-20/11x_2+2/11$ must be solved. Starting from the vertex $x^1=(6,4)$ we obtain the vertex (2,0) which is the optimal vertex. Clearly the vertex (2,0) has been examined twice.

Given a vertex x' of X and the corresponding base A_B let:

- $c'_N{}^T = c_N{}^T - c_B{}^T A_B{}^{-1} A_N$, $d'_N{}^T = d_N{}^T - d_B{}^T A_B{}^{-1} A_N$;
- $g' = c'_N - f(x')\, d'_N$.

The algorithm of Martos [15]

Step 0 Compute a feasible basic solution x'; *go to* **Step 1.**
Step 1 Compute c'_N, d'_N, g'. If $g' \leq 0$ *stop* (x' is optimal); otherwise select k such that $g'_k = \max\{g'_k\}$; *go to* **Step 2.**
Step 2 Do a simplex iteration with x_{N_k} as entering variable. Let x' be the new basic solution; *go to* **Step 1.**

Remark - In the algorithm of Martos the vector $g=d'_0\, c'_N - c'_0\, d'_N$, where $d'_0=d_B{}^T A_B{}^{-1} b+d_0$ and $c'_0= c_B{}^T A_B{}^{-1} b+c_0$, is considered instead of $g'= g/d'_0$. Let us note that g' is the vector of the reduced costs for the function $f(x) = c^Tx + c_0 - f(x')\,(d^Tx + d_0)$ and it results $f(x') = 0$.

The algorithm of Charnes - Cooper [11]

By using the transformation of variables $t=1/(d^Tx + d_0)$, $y=tx$ Charnes-Cooper have shown that problem LFP is equivalent to the following one:

$$(LP)\qquad \begin{aligned} &\max\ c^Ty + c_0t\\ &Ay - bt = 0\\ &d^Ty + d_0t = 1\\ &y \geq 0\\ &t > 0 \end{aligned}$$

Since the feasible region X is a compact set, we have $d^Tx + d_0 \geq p > 0$, $\forall x \in X$, so that $t = 0$ cannot occur in problem LP. Consequently the constraint $t > 0$ is equivalent to the constraint $t \geq 0$ and thus problem LP turns out to be a linear program which can be solved by any one of the algorithms for the linear program.

The "Parametric Algorithm" of Cambini [5]

Starting from a basic solution x^0 optimal for the linear subproblem:

$$P(\xi_0 = d^T x^0 + d_0): \quad \begin{array}{l} \max\ (c^T x + c_0)/\xi_0 \\ Ax = b \\ d^T x = \xi_0 - d_0,\ x \geq 0 \end{array}$$

the optimal solution x^* is obtained by means of a sequence of basic solutions $x^0, x^1, \ldots, x^k$, x^* optimal for problems $P(\xi_0)$, $P(\xi_1 = d^T x^1 + d_0), \ldots, P(\xi_k = d^T x^k + d_0)$, $P(\xi^* = d^T x^* + d_0)$ respectively and such that $f(x^0) < f(x^1) < \ldots < f(x^k) < f(x^*)$. A point x' optimal for the linear subproblem $P(d^T x' + d_0)$ is called an "optimal level point". Starting from the point x^i the new point x^{i+1} is obtained through a sensitivity analysis of the parametric problem (see fig. 2):

$$P(\xi_i + \theta): \quad \begin{array}{l} \max\ (c^T x + c_0)/(\xi_i + \theta) \\ Ax = b \\ d^T x = (\xi_i - d_0) + \theta \\ x \geq 0 \end{array}$$

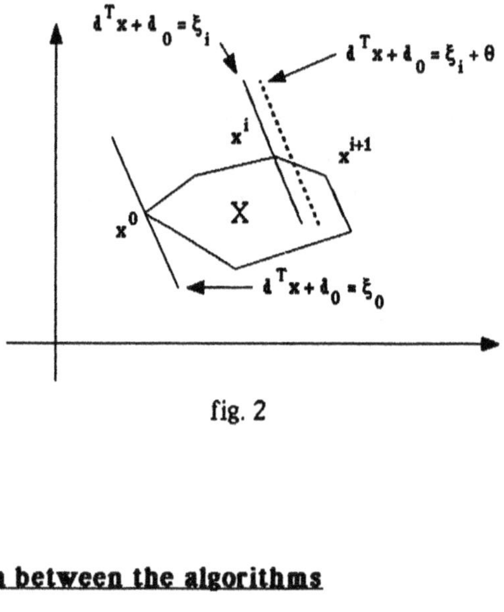

fig. 2

3.1 Comparison between the algorithms

As shown by Wagner and Yuan [19] Martos algorithm and Charnes - Cooper algorithm are equivalent. As we have just observed the algorithm of Isbell - Marlow can examine the same vertex more than once and so it is not equivalent to the other algorithms and it appears more "heavy" than the previous ones from a computational point of view.
The algorithm of Martos and the Parametric algorithm are not equivalent as it is shown by the following numerical example :

$\max\ (2x_1 + x_2 + 1)/(1000x_1 + x_2 + 1000)$

$x_1 + x_2 \leq 5,\ x_1, x_2 \geq 0$

Starting from the vertex (0,0) the algorithm of Martos obtains the optimal solution (0,5) by examining the vertex (5,0) and the vertex (0,5) while the Parametric algorithm consider only the vertex (0,5) (see fig. 3).

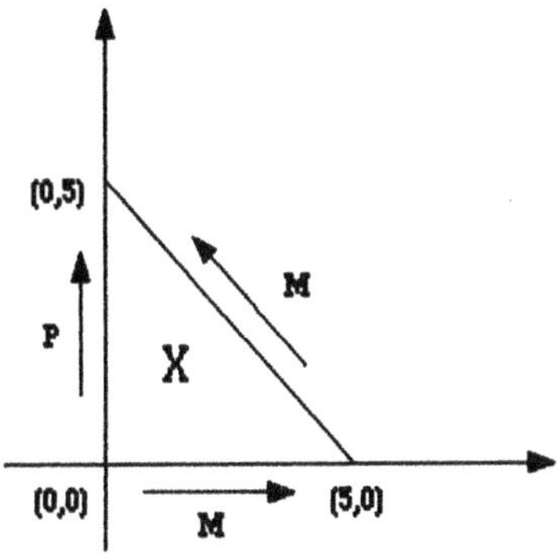

fig. 3

4. The unbounded case

Now we consider some weaknesses that the previously described algorithms can present when applied to problem LFP with an unbounded feasible region. Let us consider the following three problems:

Problem 1

$\max f(x) = (x_1 + 1)/(x_2 + 1)$

$-x_1 + x_2 \leq 2,\ x_1 - x_2 \leq 0,\ x_1, x_2 \geq 0;$

$R = \{(x_1, x_2): x_2 = x_1 + 2, x_1 \geq 0\}$, $\sup\{f(x), x \in R\} = 1$, (0,0) is a maximum and $f(0,0) = 1$ (see fig. 4).

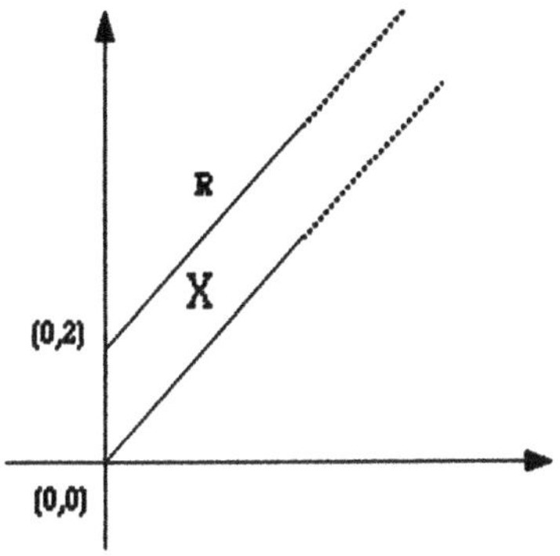

fig. 4

Problem 2

$\max f(x) = (x_1 + 1)/(x_2 + 1)$

$-x_1 + x_2 \leq 2,\ x_1 - 2x_2 \leq 0,\ x_1, x_2 \geq 0;$

$R = \{(x_1, x_2) : x_2 = x_1 + 2, x_1 \geq 0\}$, $\sup\{f(x), x \in R\} = 1$, $\sup\{f(x), x \in X\} = 2$. The maximum does not exist (see fig. 5).

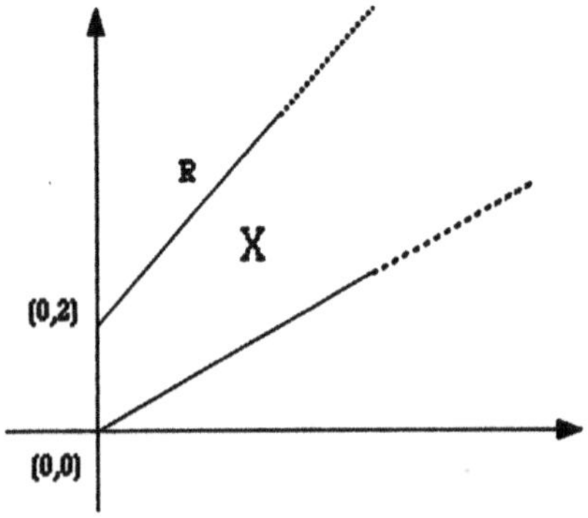

fig. 5

Problem 3

$\max f(x) = (x_1 + 1)/(x_2 + 1)$

$-x_1 + x_2 \leq 2,\ x_1, x_2 \geq 0;$

$R = \{(x_1, x_2) : x_2 = x_1 + 2, x_1 \geq 0\}$, $\sup\{f(x), x \in R\} = 1$, $\sup\{f(x), x \in X\} = +\infty$ (see fig. 6).

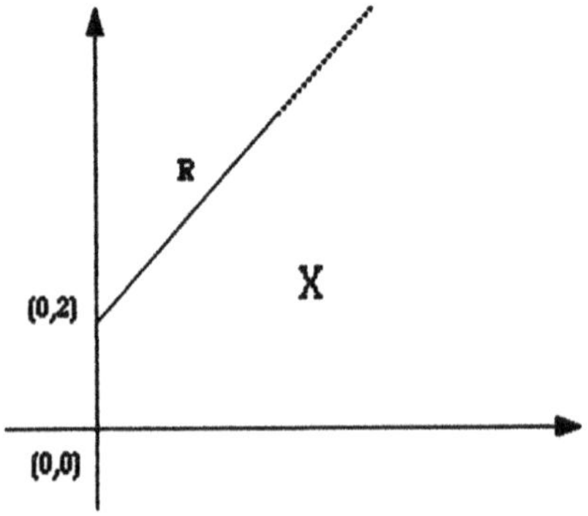

fig. 6

The algorithms of Isbell - Marlow, Martos and Charnes - Cooper do not guarantee the solution of an LFP if the feasible region is unbounded.

In fact for all the problems 1, 2 and 3, starting from the point (0,2):

- the algorithm of Isbell - Marlow generates a linear subproblem with unbounded objective function and this is true only for problem 3;
- the algorithm of Martos finds a supremum equal to 1, while problem 1 has maximum, problem 2 has a supremum equal to 2 and problem 3 has an unbounded objective function;
- the algorithm of Charnes - Cooper determines an optimal solution (y',t') where $t'=0$, which does not correspond to a feasible solution of the original LFP.

In the last case Schaible has suggested in [17] a way to find an extreme ray R such that $\sup\{f(x), x\in X\} = \sup\{f(x), x\in R\}$. The same result is obtained by Gasparotto, in his investigation on the equivalence between Martos and Charnes-Cooper algorithms [13], by interpreting the variable t as a parameter. The result given by Wagner-Yuan is extended to the unbounded case. A new proof of the above mentioned equivalence, in terms of optimal level points, is given by Cambini and Martein in [7].
The Parametric algorithm does not present the disadvantages of the other algorithms since it generates a sequence of optimal level vertices. This condition implies that when an increasing extreme ray R is detected:

- if $\sup\{f(x), x\in R\}=L<+\infty$, then $\sup\{f(x), x\in X\}=L$;
- if $\sup\{f(x), x\in R\}=+\infty$, then $\sup\{f(x), x\in X\}=+\infty$.

Note that in problems 1, 2 and 3 the point (0,2) is not an optimal level vertex.

4.1 A modified version of the algorithm of Martos (MVM)

In order to eliminate the disadvantages for the unbounded case Cambini and Martein [6] have proposed a new algorithm based on a modification of the algorithm of Martos. The algorithm, called MVM, differs from Martos algorithm in the fact that only optimal level vertices are examined.
Given a vertex x' of X and the corresponding base A_B let:

- $c'^T_N = c^T_N - c^T_B A_B^{-1} A_N$;
- $d'^T_N = d^T_N - d^T_B A_B^{-1} A_N$;
- $g' = c'_N - f(x')\, d'_N$.

The algorithm MVM

Step 0 Solve problem P_0 : $\min\ d^Tx + d_0$, $x\in X$. If the optimal vertex x' of P_0 is unique then *go to* **Step 1**; otherwise solve problem P_1 : $\max\ c^Tx + c_0$, $x\in X' = X\cap\{x: d^Tx = d^Tx'\}$; if $\sup\{c^Tx + c_0, x\in X'\}=+\infty$ then *stop* ($\sup\{f(x), x\in X\}=+\infty$); otherwise *go to* **Step 1** with x' equal to the optimal vertex of P_1.

Step 1 Compute c'_N, d'_N, g' and $J=\{j : g'_j > 0\}$. If $J=\Phi$ then *stop* (x' is an optimal solution); otherwise select k such that $c'_{Nk}/d'_{Nk} = \max\{c'_{Nj}/d'_{Nj} : j\in J\}$; *go to* **Step 2**.

Step 2 Compute $y^k = A_B^{-1}A_N^k$. If $y^k \leq 0$ then *stop* ($c'_{Nk}/d'_{Nk} = \sup\{f(x), x\in X\}$); otherwise do a simplex iteration with x_{Nk} as entering variable. Let x' be the new basic solution, *go to* **Step 1**.

The algorithm MVM does not present the disadvantages of the algorithm of Martos as it examines only optimal level vertices.

5. The parametric linear fractional problem [10]

Sensitivity and parametric analysis for a linear fractional problem whose feasible region is a compact set, is similar to the linear program one [1,2,4,18]. When the feasible region is unbounded, some difficulties arise in studying the supremum of the problem LFP as a function of a parameter, because of case 2) of section 2 which cannot happen in the linear case. Now we will point out these difficulties by means of the study of suitable cones, with respect to the variation of vector (c,c_0).
First we introduce the following notation:

- $V(X)$, set of vertices of X;
- $R(x^i)$, set of extreme rays starting from $x^i\in V(X)$;
- $C(x^i)=\{(c,c_0)\in R^{n+1}: \max\{f(x), x\in X\}=f(x^i), x^i\in V(X)\}$;
- $C(r_{ij})=\{(c,c_0)\in R^{n+1}: \sup\{f(x), x\in X\}=\lim \{f(x^i+tr_{ij}), t\to+\infty\}<+\infty, x^i\in V(X), r_{ij}\in R(x^i)\}$;
- $C^*(x^i)=C(x^i)\cup(\cup\{C(r_{ij}), r_{ij}\in R(x^i)\})$;
- $C=\cup\{C(x^i), x^i\in V(X)\}=\{(c,c_0)\in R^{n+1}: \exists x'\in V(X): \max\{f(x), x\in X\}=f(x')\}$;
- $C^*=\cup\{C^*(x^i), x^i\in V(X)\}=\{(c,c_0)\in R^{n+1}: \sup\{f(x), x\in X\}<+\infty\}$.

The following theorem holds:

Theorem 5.1

i) $C(x^i)$ is a convex polyhedral cone;
ii) $C(r_{ij})$ is a convex polyhedral cone;
iii) $C^*(x^i)$ is a cone (not necessarily convex);
iv) C is a cone (not necessarily convex);
v) C^* is a convex cone;
vi) $C^*=R^{n+1}$ iff X is linearly bounded with respect to $d^Tx + d_0$ [1].

---

[1] X is linearly bounded with respect to d^Tx+d_0 if the set $X\cap\{x\in R^n: d^Tx+d_0=k\}$ is a compact set for any $k\in R$.

Proof. i) $x^i \in V(X)$ is an optimal solution iff (*), (**) hold:

(*) $[(c^Tx^i + c_0)/(d^Tx^i + d_0)] \geq [(c^Tx^j + c_0)/(d^Tx^j + d_0)],\ x^j \in V(X);$

(**) $[(c^Tx^i + c_0)/(d^Tx^i + d_0)] \geq \lim \{[(c^T(x^j + tr_{jk})+c_0)/(d^T(x^j + tr_{jk})+ d_0)],\ t \to +\infty\} =$
$= (c^Tr_{jk}/d^Tr_{jk}),\ x^j \in V(X),\ r_{jk} \in R(x^j).$

Since (*),(**) are linear inequalities then $C(x^i)$ is a convex polyhedral cone. ii) the same proof of i). iii) and iv) follow directly from the definition of $C^*(x^i)$ and C. v) C^* is a cone as union of cones, it remains to prove that C^* is convex; let $(c',c'_0),(c'',c''_0) \in C^*$, it results:
$\sup \{[(c'^Tx + c'_0)/(d^Tx + d_0)],\ x \in X\} = L_1,$
$\sup \{[(c''^Tx + c''_0)/(d^Tx + d_0)],\ x \in X\} = L_2,$
$\sup \{[((\lambda c' + (1-\lambda)c'')^Tx + \lambda c'_0 + (1-\lambda)c''_0))/(d^Tx + d_0)],\ x \in X\} =$
$\sup \{[(\lambda(c'^Tx + c'_0) + (1-\lambda)(c''^Tx + c''_0))/(d^Tx + d_0)],\ x \in X\} \leq \lambda L_1 + (1-\lambda)L_2, 0 \leq \lambda \leq 1.$
vi) (if part) Ab absurdo suppose that $C^* \neq R^{n+1}$; then there exists $(c',c'_0) \in R^{n+1}$, $x^i \in V(X)$, $r_{ij} \in R(x^i)$ such that $\lim \{[(c'^T(x^i + tr_{ij})+c'_0)/(d^T(x^i + tr_{ij})+ d_0)],\ t \to +\infty\} = +\infty$; it follows that $d^Tr_{ij} = 0$ and $d^T(x^i + tr_{ij})+d_0 = d^Tx^i + d_0 =$ constant for each t, and this contradicts the hypothesis. (only if part) Ab absurdo suppose that there exist $x^i \in V(X)$, $r_{ij} \in R(x^i)$ such that $d^T(x^i + tr_{ij})+d_0 =$ constant for each t; then it follows that there exists $(c',c'_0) \in R^{n+1}$ such that $\lim \{[(c'^T(x^i + tr_{ij})+c'_0)/(d^T(x^i + tr_{ij})+ d_0)],\ t \to +\infty\} = +\infty$; the contradiction of the hypothesis is obtained.

Notice that the properties of Theorem 5.1 do not hold for the linear program; in fact in this problem it results $C(x^i) = C^*(x^i)$, $C = C^*$ and C is a convex cone.

Remark -Let us consider the following variation of the numerator of the objective function:

(+) $\qquad (c',c'_0) = (c,c_0) + \theta\ (u,u_0),\ \theta \in R.$

Relation (+) can be interpreted as a line in the space of variables (c,c_0), which intersects several of the above mentioned cones. It follows that the study of a linear fractional program as a function of a parameter in the numerator of the objective function, requires the capability of going from an extreme ray r_{ik} starting from the vertex x^i to a new extreme ray r_{jv} starting from the vertex x^j which may be not adjacent to the vertex x^i.

6. An algorithm for a parametric linear fractional problem [10]

In this section we consider the parameterization of the numerator of the objective function of a linear fractional program. Let us consider the following parametric problem:

$$z(\theta) = \sup [((c+\theta u)^T x + c_0 + \theta u_0)/(d^T x + d_0)]$$

PLF(θ):

$$x \in X = \{x \in R^n : Ax = b, x \geq 0\}, (u,u_0) \in R^{n+1}.$$

Now we will describe a procedure for solving problem PLF(θ).
Problem PLF(0) is solved by means of algorithm MVM. Let $x' = (x'_B = A_B^{-1}b, 0)$ be the solution obtained by the algorithm and

- $c'_N{}^T = c_N{}^T - c_B{}^T A_B{}^{-1} A_N$; $c'_0 = c_0 + c_B{}^T A_B{}^{-1} b$;
- $d'_N{}^T = d_N{}^T - d_B{}^T A_B{}^{-1} A_N$; $d'_0 = d_0 + d_B{}^T A_B{}^{-1} b$;
- $\gamma' = c'_N - f(x')\, d'_N = c'_N - (c'_0/d'_0)\, d'_N$.

Clearly x' is an optimal solution of the linear sub-problem PL(0); where:

$$\max\ (c+\theta u)^T x + c_0 + \theta u_0$$

PL(θ):

$$x \in X,\ d^T x = d^T x'.$$

Let $c^*_{N'}{}^T = c_{N'}{}^T - c_{B'}{}^T A^*_{B'}{}^{-1} A^*_{N'} \leq 0$ be the vector of the reduced costs of x' for PL(0).
Two cases can be verified:

1) $\gamma' \leq 0$, x' is a maximum of PLF(0);
2) γ' is not ≤ 0 and there exists an index $k : \gamma'_k > 0$, $c'_{Nk}/d'_{Nk} = \max \{c'_{Nj}/d'_{Nj}, j : \gamma'_j > 0\}$ and $y^k = A_B{}^{-1} A_N{}^k \leq 0$; in this case PLF(0) has a supremum (equal to c'_{Nk}/d'_{Nk}) on the extreme ray from x' along y^k.

Let "*the stability set* " of vertex x', refer to the set of values of θ such that x' is a maximum or the supremum is obtained on an extreme ray starting from x'. In order to find the stability set of x'; c'_N, γ', c'_0, $c^*_{N'}$ are considered as a function of the parameter θ. Such functions are denoted by $c'_N(\theta)$, $g'(\theta)$, $c'_0(\theta)$, $c^*_{N'}(\theta)$.
Let

$$H(\theta) = \{\theta : \gamma'(\theta) \leq 0\};$$
$$H'(\theta) = \{\theta : c^*_{N'}(\theta) \leq 0\};$$

clearly $H'(\theta) \supseteq H(\theta)$.

- If case 1) holds $H(\theta) \neq \Phi$ and for $\theta \in H(\theta)$ x' is a maximum of PLF(θ) with $z(\theta) = c'_0(\theta)/d'_0$.
- If case 2) holds let us consider the set: $H''(\theta) = \{\theta : c'_{Nk}(\theta)/d'_{Nk} = \max \{c'_{Nj}(\theta)/d'_{Nj}, j : \gamma'_j > 0\}\}$; for $\theta \in H'(\theta) \cap H''(\theta)$ the supremum is obtained along y^k with $z(\theta) = c'_{Nk}(\theta)/d'_{Nk}$.

Suppose that $H'(\theta) = \{\theta : \theta' \leq \theta \leq \theta''\}$. We will solve problem PL(θ^*) where $\theta^* = \theta' - \varepsilon$ ($\theta^* = \theta'' + \varepsilon$), $\varepsilon > 0$. Starting from x', by means of one simplex iteration, the optimal solution x^* of PLF(θ^*) is obtained. x^* lies on an edge of X and is not, in general, a vertex of X. We will

find the best vertex (in term of the value of the objective function) which belongs to the edge containing x^*. Let x'' such a vertex. The stability set of vertex x'' is adjacent to the stability set of vertex x'. Clearly the described procedure allows us to find the function $z(\theta)$ through a vertex following examination.
With regard to the properties of the parametric function $z(\theta)$ it is easy to prove that $z(\theta)$ is a convex piecewise linear function.

References

/1/ **Aggarwal S.P.**, Parametric linear fractional functional programming, Metrika, 1968, vol. 12, pp. 106-114.
/2/ **Aggarwal S.P.**, Analysis of the solution to a linear fractional functional programming, Metrika, 1970, vol. 16, pp. 9-26.
/3/ **Bitran G.**, Experiments with linear fractional problems, Nav. Res. Log. Quart., 1979, vol. 26, pp. 689-693.
/4/ **Bitran G., Magnanti T.L.**, Duality and sensitivity analysis for fractional programs, Opns. Res., 1976, vol. 24, pp. 675-699.
/5/ **Cambini A.**, Un algoritmo per il massimo del quoziente di due forme affini con vincoli lineari, Dip. Ric. Oper., Univ. di Pisa, 1977.
/6/ **Cambini A., Martein L.**, A modified version of Martos's algorithm, Methods of Oper. Res.,1985, vol. 53.
/7/ **Cambini A., Martein L.**, Equivalences in linear fractional programming, Dip. di Stat. e Mat. Appl. all'Ec., Univ. di Pisa, 1989, Report n. 28.
/8/ **Cambini A., Martein L., Sodini C.**, An algorithm for two particular non linear fractional programs, Methods of Operations Research, 1983, vol. 45, pp. 61-70.
/9/ **Cambini A., Sodini C.**, Sulla equivalenza di alcuni algoritmi di programmazione lineare frazionaria, Dip. di Mat., Univ. di Pisa, 1983, Report A-105.
/10/ **Cambini A., Sodini C.**, On parametric linear fractional programs, Dip. di Stat. e Mat. Appl. all'Ec., Univ. di Pisa, 1987, Report n. 6.
/11/ **Charnes A., Cooper W.W.**, Programming with linear fractional functionals, Nav. Res. Log. Quart., 1962, vol. 9, pp. 181-186.
/12/ **Dinkelbach W.**, On non linear fractional programming, Management Science, 1967, vol. 13, n. 7.
/13/ **Gasparotto G.**, On the Charnes - Cooper transformation in linear fractional programming, Dip. di Stat. e Mat. Appl. all'Ec., Univ. di Pisa, 1987, Report n.3.
/14/ **Isbell J.R.,Marlow W.R.**, Attrition games, Nav. Res. Log. Quart., 1956, vol. 3, pp. 71-93.
/15/ **Martos B.**, Hyperbolic programming, Nav. Res. Log. Quart., 1964, vol.11, pp. 135-155.
/16/ **Mond B.**, On Algorithmic Equivalence in Linear Fractional Programs, Mathematics of Computations, 1981, vol. 37, pp. 185-187.
/17/ **Schaible S.**, Analyse und anwendungen von quotientenprogrammen ein beitrag zur planung mit hilfe der nichtlinearen programmierung, Mathematical Systems in Economics, 1978, vol. 42, Hain-Verlag, Meisenheim.

[18] **Swarup K.**, On varying all the parameters in a linear fractional functionals programming problems, Metrika, 1968, vol. 13, pp. 196-205.
[19] **Wagner H.M., Yuan J.S.C.**, Algorithm equivalence in linear fractional programming, Management Science, 1968, vol. 14, pp. 301-306.

LINEAR FRACTIONAL AND BICRITERIA LINEAR FRACTIONAL PROGRAMS

A. Cambini - L. Martein

Abstract

In this paper we will restate the sequential methods suggested by the Authors [8] for solving a linear fractional problem for any feasible region using the concept of optimal level solutions.
By following this approach, sequential methods for solving the bicriteria linear fractional problem are proposed; these methods generate the set E of all efficient points and, at the same time, point out that E is arcwise-connected when at least one of the objective functions is linear.

1. Introduction

Bicriteria maximization problems, that is simultaneous maximization of a pair of functions in the sense given by Pareto, has been studied by several authors (see [1-6]). The aim of these contributions is to investigate the connectedness of the set E of all efficient points and to suggest sequential methods for generating E.

Problems with two linear fractional objective functions are important and have potentially broad applications. In fact, linear fractional objective functions occur in optimization problems involving criteria that are rates, such as return on investement, dividend coverage, margin on sales and residential density. Futhermore linear fractional functions are used as performance measures in many management situations such as production planning and scheduling, educational administration and analysis of financial enterprises.

In [7], the set E of all efficient points has been characterized in terms of the union of suitable sets of optimal solutions of a real-valued parametric problem. Recent results have enphasized the role of optimal level solutions in solving several fractional programs.

In this paper we will restate the sequential method suggested by Cambini-Martein [8] for solving a linear fractional problem for any feasible region (bounded or unbounded), using the concept of optimal level solutions.

By following this approach we are able to state sequential methods for solving the bicriteria linear fractional problem; these methods generate the set E of all efficient points and, at the same time, point out that E is arcwise-connected when at least one of the objective functions is linear. This last result extends the one given by some authors stated in the particular case of a bounded feasible region (see [2, 3, 6]).

2. Optimal level solutions

As we will see, the set E of all efficient points of the bicriteria linear fractional problem and the set of optimal solutions of the linear fractional problem are strictly related to the optimal solutions of a linear parametric problem of the following form :

$$P(\theta): \quad \begin{cases} \max cx+c_0, \; x \in R \triangleq \{ x \in \mathbb{R}^n : Ax=b, x \geq 0 \} \\ dx+d_0 = \theta \end{cases}$$

where A is a m x n real matrix and $b \in \mathbb{R}^m$.

We say that x^* is an optimal level solution (shortly o.l.s.) iff x^* is an optimal solution of the problem $P(\theta)$ for some θ (The reason for introducing such a definition will become clear later in the paper).

In the next sections, we will suggest sequential methods which are able to solve the linear fractional problem and the bicriteria linear fractional problem by generating a suitable finite sequence of o.l.s. of a linear parametric problem $P(\theta)$.

In order to characterize an o.l.s., we will introduce some notations.

Let x^0 be a vertex of the feasible region R with corresponding basis B; we partition the vectors c and d as $c=(c_B,c_N)$, $d=(d_B,d_N)$ and the matrix A as $A=[B|N]$. Set : $\bar{c}_0 = c_B B^{-1}b + c_0$, $\bar{d}_0 = d_B B^{-1}b + d_0$, $\bar{c}_N = c_N - c_B B^{-1}N$, $\bar{d}_N = d_N - d_B B^{-1}N$ and consider the following sets of indices: $J_1 = \{j: \bar{d}_j > 0\}$; $J_2 = \{ j: \bar{d}_j < 0 \}$; $J_3 = \{ j: \bar{d}_j = 0 \}$.

The following theorem holds:

Theorem 2.1 The vertex $x^0 \in R$ is an optimal level solution iff conditions i) and ii) both hold.

i) For any index $j \in J_3$, we have $\bar{c}_j \leq 0$.

ii) If J_1 and J_2 are not empty sets, then:

$$(\bar{c}_k / \bar{d}_k) \triangleq \max_{j \in J_1} (\bar{c}_j / \bar{d}_j) \leq \min_{j \in J_2} (\bar{c}_j / \bar{d}_j) \triangleq (\bar{c}_h / \bar{d}_h)$$

Proof. Let us note that, if we refer to the simplex tableau associated to the vertex x^0, x^0

turns out to be an o.l.s iff there exists an index s such that, by performing a pivot operation on d_s, the corresponding reduced costs $c^*_j = \bar{c}_j - (\bar{c}_s/\bar{d}_s)\bar{d}_j$ are non positive $\forall j$, and this is true iff (2.1) holds:

(2.1.a) $\bar{d}_j = 0$ implies $\bar{c}_j \le 0$
(2.1.b) $\bar{d}_j > 0$ implies $\bar{c}_s/\bar{d}_s \ge \bar{c}_j/\bar{d}_j$
(2.1.c) $\bar{d}_j < 0$ implies $\bar{c}_s/\bar{d}_s \le \bar{c}_j/\bar{d}_j$.

Condition (2.1) implies i) and ii).
Let us note that if J_1 and J_2 are not empty sets, then $s=k$ or $s=h$. This completes the proof.

Remark 2.1 Let us note that any feasible solution of the kind $x_B = B^{-1}b - B^{-1}N^{(k)} x_{N_k}$, $x_{N_k} \ge 0$, or of the kind $x_B = B^{-1}b - B^{-1}N^{(h)} x_{N_h}$, $x_{N_h} \ge 0$, is an o.l.s. with respect to the level $\theta_k = \bar{d}_0 + \bar{d}_k x_{N_k}$ or with recpect to the level $\theta_h = \bar{d}_0 + \bar{d}_h x_{N_h}$, where $N^{(k)}$ and $N^{(h)}$ are the columns of N associated with the indices k, h respectively.
This remark points out that it is possible to generate a sequence of vertices which are o.l.s. by performing a pivot operation which corresponds to the selected indices k, h.

3. The linear fractional problem.

For the linear fractional problem, that is, the problem of maximizing the quotient of two affine functions subject to linear inequalities, many algorithms have been proposed. Experiments performed by Bitran [9] have shown the superiority of Martos' procedure [10] over the others.
Unfortunately Martos' algorithm does not process the problem when the feasible region R is unbounded, in the sense that, in this case, the algorithm is not able to find a solution or show that no solution exists; this is the reason why the Authors put forward the so-called "modified version of Martos' algorithm" [8] based on the concept of optimal level solution. Technically, this differs from Martos'method on account of a particular choice of pivot, but it is capable of solving the problem whatever the feasible region.
In order to making the paper self-contained, in this section we will restate the sequential method suggested by Cambini- Martein [8]. Consider the problem

$$P_F : [\sup (f(x) = (cx+c_0)/(dx+d_0))] \triangleq L, \quad x \in R = \{x \in \mathbb{R}^n : Ax = b, x \ge 0\}$$

where A is a mxn real matrix, $b \in \mathbb{R}^m$, and $dx + d_0 > 0 \;\; \forall x \in R$.
The following theorem states some fundamental properties of a linear fractional problem:

Theorem 3.1 Consider the linear fractional problem P_F then i), ii), iii) hold.

i) $L = \max_{x \in R} f(x)$ iff there exists a vertex $x^o \in R$ such that $L = f(x^o)$.

ii) If $L \neq \max_{x \in R} f(x)$ then there exists a ray $r \subset R$ of equation $x = x^o + tu$, $t \geq 0$, such that

$$L = \sup_{x \in r} f(x) = \lim_{t \to +\infty} f(x^o + tu)$$

iii) Let D be the set of the optimal solutions of the linear problem: $\min_{x \in R} (dx + d_o)$. Then

$L = +\infty$ iff D is unbounded and $\sup_{x \in D} (cx + c_0) = +\infty$.

The sequential method for solving P_F that we are going to describe generates a finite sequence x^i, $i = 1, \ldots, l$, of o.l.s., the first of which is found in the following way:
the linear problem $P_0 : \min_{x \in R} (dx + d_0)$ has optimal solutions since the objective function is bounded below; let x^o be an optimal solution of P_0. If x^o is unique, then it is also an o.l.s., otherwise solve the linear problem:

P_1 : $\max (cx + co)$, $dx = dx^o$, $x \in R$.

If P_1 has no solutions, then we have $\sup_{x \in R} f(x) = +\infty$ for iii) of theorem 3.1; otherwise an optimal solution of P_1 is also an o.l.s.

Let us note that since we start solving P_0, we are interested in generating o.l.s. corresponding to increasing levels of the denominator, so that, taking into account remark 2.1, in order to obtain x^{i+1} from x^i, we must perform a pivot operation corresponding to the index k given by the following rule:

$$(3.1) \qquad (\bar{c}_{N_k} / \bar{d}_{N_k}) = \max_{j \in J_1} (\bar{c}_{N_j} / \bar{d}_{N_j})$$

Let us note that an optimal solution of P_F is also an o.l.s., but the converse is not true in general, so that the problem of establishing the optimality of x^i, $i = 1, \ldots, l$, arises.
Without loss of generality we can suppose that B is the basis associated to the vertex x^i.
The following theorem holds:

Theorem 3.2. If $J_1 = \emptyset$ or $\gamma_k \doteq \bar{d}_o \bar{c}_{N_k} - \bar{c}_o \bar{d}_{N_k} \leq 0$ then x^i is an optimal solution for the linear fractional problem P_F.

Proof. It is known (see Martos[10]) that a vertex is optimal iff $\gamma \leq 0$ where $\gamma \doteq \bar{d}_o \bar{c}_N - \bar{c}_o \bar{d}_N$.

Let us note that we generate the sequence x^i, $i=1,......,l$, in the following way: if in a vertex x^i, we have $\gamma_k > 0$, we find x^{i+1} by means of a pivot operation corresponding to the index k. As a consequence any vertex x^i is an optimal solution of the problem:

$$\max\,(cx + co),\ x\in R,\ dx \le dx^i.$$

Consequently for any $j\in J_2$, we have $\gamma_j \le 0$.

Now we will show that $\gamma_j \le 0$, $\forall j \in J_1$. With this aim let us note that $\gamma_k \le 0$ implies $(\bar{c}_k/\bar{d}_k) \le (\bar{c}_o/\bar{d}_o)$, so that for (2.1b) we have $(\bar{c}_j/\bar{d}_j) \le (\bar{c}_k/\bar{d}_k) \le (\bar{c}_o/\bar{d}_o)$. This completes the proof.

Taking into account the aforesaid results, the sequential method, suggested by Cambini-Martein in [8], for solving a linear fractional problem for any feasible region, can be restated in the following form:

Step 0 : Find the optimal level solution x^1; if such a solution does not exist, stop ($\sup_{x\in R} f(x) = +\infty$); otherwise set $i=1$ and go to step 1.

Step 1: If $J_1 = \emptyset$, stop, x^i is an optimal solution for P; otherwise let k be such that $(\bar{c}_{N_k}/\bar{d}_{N_k}) = \max_{j\in J_1} (\bar{c}_{N_j}/\bar{d}_{N_j})$, go to step 2.

Step 2: The non-basic variable x_{N_k} enters the basis by means of a pivot operation, set $i=i+1$ and go to step 1. If such an operation is not possible, stop ; we have $\sup f(x) = \bar{c}_{N_k}/\bar{d}_{N_k}$. Since :

$$\sup_{x\in R} f(x) = \sup_{\theta} \sup_{\substack{dx+d_0=\theta \\ x\in R}} f(x) = \sup_{x_{N_k}\ge 0} (\bar{c}_{N_k} x_{N_k} + \bar{c}_o)/(\bar{d}_{N_k} x_{N_k} + \bar{d}_o) = (\bar{c}_{N_k}/\bar{d}_{N_k})$$

4. The bicriteria linear fractional problem

In this section we will give some theoretical properties for a bicriteria linear fractional problem which allow us to suggest a sequential method for generating the set E of all efficient points.

The properties of E have been studied by several authors [2,3,4,5,6,11] who have pointed out that E is connected when the feasible region R is a compact set. The compactness of R is a crucial assumption, since it is easy to show, by means of simple examples, that E may be disconnected when R is unbounded.

One of the main findings of this paper is to show that E is still connected when R is not bounded for the class of bicriteria linear fractional problems where at least one of the objective functions is linear.
With this aim, let us consider the following problem

$$P : \max (f_1(x), f_2(x)), x \in R = \{x \in IR^n : Ax = b, x \geq 0\}$$

where f_1, f_2 are linear fractional functions, A is a m x n real matrix and $b \in IR^m$.
A point $x^0 \in R$ is said to be efficient for P if there does not exist a point $x \in R$ such that $f_i(x) \leq f_i(x^0)$ $(i = 1,2)$ where at least one of these inequalities is strict.
As outlined in [2,6,7], the set E is related to a suitable union of optimal solution sets $S(\alpha)$ of the following parametric problem:

$$P_\alpha : \sup f_2(x) \triangleq z(\alpha), \quad x \in R(\alpha) \triangleq \{x \in R : f_1(x) \geq -\alpha, \alpha \in IR\}$$

Set : $H \triangleq \{\alpha : R(\alpha) \neq \emptyset\}$, $H_1 \triangleq \{\alpha \in H : z(\alpha) < +\infty\}$; $H_2 \triangleq \{\alpha \in H_1 : z(\alpha) > z(\bar{\alpha})\ \forall \bar{\alpha} < \alpha\}$
$H_3 \triangleq \{\alpha \in H_1 : z(\alpha)$ is reached as a maximum$\}$
Taking into account the results given in [7], it is easy to prove the following theorem:

Theorem 4.1 If $H_3 \neq \emptyset$ then $E = \bigcup_{\alpha \in H_2 \cap H_3} S(\alpha)$

Let us note that, in order to generate the set E of all efficient points, it is necessary to be able to solve P_α for any α (this is possible by means of the algorithm suggested in section 3) and, furthermore, to know if $\alpha \in H_2 \cap H_3$.
To this aim, the properties of the parametric function $z(\alpha)$ must be investigated. Obviously $z(\alpha)$ is a non decreasing function since $\alpha_1 < \alpha_2$ implies $R(\alpha_1) \subset R(\alpha_2)$. When R is a compact set, $z(\alpha)$ turns out to be a semistrictly quasiconcave function (see theorem 2.4 in [7]) and, furthermore, H_3 is an interval of the real line, so that E can be easily characterized (see [7]). Unfortunately, in the unbounded case, these properties do not hold, but they are still valid for the subclass of bicriteria linear fractional problems where at least one of the objective functions is linear. In order to prove these properties, we will introduce some notations. Set $\alpha_1 \triangleq \inf H_3, \alpha_2 \triangleq \sup H_4$, $H_4 \triangleq \{\alpha : z(\alpha)$ is increasing in $[\alpha_1, \alpha]\}$ (α_1 and/or α_2 may be $-\infty, +\infty$ respectively). The following theorem holds:

Theorem 4.2. Consider the bicriteria linear fractional problem P where f_1 is linear.
If $H_3 \neq \emptyset$, then conditions i) and ii) hold:
i) H_3 is a convex set;
ii) $E = \bigcup_{\alpha \in [\alpha_1, \alpha_2]} S(\alpha)$

Proof. i) Let x^1, x^2 be optimal solutions of problems $P(\alpha_1)$, $P(\alpha_2)$ respectively, where, α_1, $\alpha_2 \in H_3$, $\alpha_1 < \alpha_2$. We must show that $P(\alpha^*)$ has optimal solutions for any $\alpha^* \in]\alpha_1, \alpha_2[$; if not, there exists a halfline, whose equation is in the form $x = \bar{x} + t\,u$, $t \geq 0$, contained in $R(\alpha^*)$, such that $z(\alpha^*) = \lim_{t \to +\infty} f_2(\bar{x} + t\,u)$.

Since $z(\alpha)$ is a non-decreasing function, we necessarily have $z(\alpha^*) > z(\alpha_1)$. It is easy to verify that the halfline $x = x^1 + t\,u$, $t \geq 0$, is contained in $R(\alpha_1)$ so that $\lim_{t \to +\infty} f_2(x^1 + t\,u) = z(\alpha^*) > z(\alpha_1)$ and this contradicts the optimality of x^1.

ii) The proof follows easily by corollary 2.1 and theorem 2.4 in [7].

As we will see, condition ii) of theorem 4.2 will play a foundamental role in finding simple sequential methods for generating E when at least one of the objective functions is linear. Since, by means of Charnes-Cooper transformation, problem P can be reduced to one that has f_1 as a linear function, we will be able to suggest, in the general case as well, an algorithm which generates E for any feasible region.

5. On generating E

In this section we will suggest simple sequential methods for generating E in the following exhaustive cases:

[A] the objective functions f_1 and f_2 are linear;
[B] f_1 is a linear function and f_2 is a linear fractional function;
[C] f_1 and f_2 are linear fractional functions.

With this aim, we will transform P_α in to the following parametric problem:

$$P(\theta): \quad \sup f_2(x) \doteq z(\theta)\,; \quad x \in R(\theta) \doteq \{x \in R: f_1(x) \geq \beta - \theta,\ \theta \in IR\}$$

in such a way, solving P_α $\forall \alpha$ is equivalent to solving $P(\theta)$ $\forall \theta$ starting from a suitable value β corresponding to $\theta = 0$.

As we will see $P(\theta)$ will be solved by means of post-optimality analysis performed either by a dual simplex-like procedure or by the algorithm given in section 3.

Set $K_1 \doteq \{\theta : z(\theta)$ is reached as a maximum$\}$, $\theta_1 \doteq \inf K_1$;
$K_2 \doteq \{\theta\,; z(\theta)$ is increasing in $]\theta_1, \theta[\}$, $\theta_2 \doteq \sup K_2$ and let $S(\theta)$, $\theta \in K_1$, be the set of optimal solutions of $P(\theta)$.

From ii) of theorem 4.2, the set E of all efficient points can be characterized, in cases A] and B] and when $K_1 \neq \emptyset$ (obvionsly $K_1 = \emptyset$ implies $E = \emptyset$) as the following union:

$$(5.1) \qquad E = \bigcup_{\theta \in [\theta_1, \theta_2]} S(\theta)$$

[A] The bicriteria linear problem

Consider now problem P where $f_1(x) = dx + d_0$, $f_2(x) = cx + c_0$; the parametric problem $P(\theta)$ reduces to the following one:

$P(\theta)$: $\sup cx+c_0 \doteq z(\theta)$, $x \in R(\theta) = \{ x \in R: dx+d_0 \geq \beta - \theta , \theta \in \mathbb{R}\}$.

First of all we will describe a way of finding β such that $0 \in [\theta_1, \theta_2]$ or of establishing that E is empty.
With this aim, consider the linear problem:

$$P_L: \ \max f_1(x), x \in R.$$

Let us note that if P_L has a unique solution x^0 or set S_L of the optimal solutions of P_L is compact, then we obviously have $\beta = f_1(x^0)$ and consequently $E = \bigcup_{\theta \in [0, \theta_2]} S(\theta)$.

In the general case let x^0 be an optimal basic solution to P_L, if one exists, or a vertex which is the end-point of a ray $r \subset R$ such that

$$\sup_{x \in r} f_1(x) = +\infty$$

Consider the problem

(5.2) $\qquad \sup f_2(x) \doteq L, x \in R, \quad f_1(x) \geq f_1(x^0)$

The following cases arise:
a) L is reached as a maximum; in this case we have $\beta = L$.
b) $L = +\infty$; in such a case $K_1 = \emptyset$ so that $E = \emptyset$.
We get this result after taking into account that $L = +\infty$ implies the existence of a halfline $x = x^0 + tu, t \geq 0$, such that $L = \lim_{t \to +\infty} f_2(x^0 + tu)$.
Since any halfline of equation $x = \hat{x} + tu, t \geq 0, \hat{x} \in R(\theta)$, is contained in the feasible region $R(\theta)$ of problem $P(\theta)$, we have $\sup_{x \in R(\theta)} f_2(x) = +\infty, \forall \theta$.
In order to generate the set E of all efficient points, let us note that E is given by (5.1) and it collapses with the set of optimal level solutions of the problem $P(\theta)$; such solutions are binding to the parametric constraint $d x + d_0 = \beta - \theta$ so that, starting from $\theta = 0$, we generate a sequence of vertices, by applying the algorithm given in section 3, by performing a pivot operation corresponding to the index k given by (3.1) for increasing levels and by performing a pivot operation corresponding to the index h given by

(5.3) $\qquad \min_{j \in J_2} (\bar{c}_j / \bar{d}_j) \doteq (\bar{c}_h / \bar{d}_h)$

for decreasing levels.
Let us note that when L is reached as a maximum, starting from $\theta = 0$, we have only increasing levels.
The algorithm stops when the step-wise linear function $z(\theta)$ reduces to a constant function.

[B] **(f_1 is linear and f_2 is a linear fractional function).**

Consider now problem P where $f_1(x) = a\,x$, $f_2(x) = (cx+co)/(dx+do)$; the parametric problem $P(\theta)$ reduces to the following one:

$$P(\theta):\quad \begin{cases} \sup\,[f_2(x) = (c\,x+c_0)/(d\,x+d_0)] \\ A\,x = b \\ a\,x = \beta - \theta\ ,\ x \geq 0 \end{cases}$$

Let $\hat{x} = (\hat{x}_B = \hat{b},\ \hat{x}_N = 0)$ be a feasible basic solution for $P(\hat{\theta})$, with corresponding basis $\hat{B}$.
We partition the vectors c and d as $c = (c_B, c_N)$, $d = (d_B, d_N)$ and the matrix

$$\hat{A} = \begin{bmatrix} A \\ -- \\ a \end{bmatrix} \quad \text{as } \hat{A} = [\hat{B} : \hat{N}].$$

Set

$$\bar{c}_N \triangleq c_N - c_B\,\hat{B}^{-1}\hat{N}\,,\quad \bar{d}_N \triangleq d_N - d_B\,\hat{B}^{-1}\hat{N}\,,\quad \bar{c}_0 \triangleq c\,\hat{x} + c_0$$

$$\bar{d}_0 \triangleq d\,\hat{x} + d_0\ ,\quad \gamma \triangleq \bar{d}_0\,\bar{c}_N - \bar{c}_0\,\bar{d}_N$$

and introduce the following notations:

$$\gamma(\theta) = \gamma - \theta\,w\ ,\quad w = \lambda_0\,\bar{d}_N - \mu_0\,\bar{c}_N\ ,\ \hat{x}_B(\theta) = \hat{x}_B - \theta\,h$$

where λ_0 and μ_0 are the last components of the vectors $c_B\,\hat{B}^{-1}$ and $d_B\,\hat{B}^{-1}$ respectively and h is the last column of the matrix $\hat{B}^{-1}$.
The parametric analysis is performed by studying the optimality condition $\gamma(\theta) \leq 0$ and the feasibility condition $\hat{x}_B(\theta) \geq 0$. With regard to the optimality condition, let us set: $I_1 = \{ i : w_i < 0 \}$; if $I_1 = \emptyset$, then $\gamma(\theta) \leq 0$, for any $\theta \geq 0$, otherwise $\gamma(\theta) \leq 0\ ,\forall\,\theta \in [0, \theta']$ where:

(5.4) $$\theta' = \min_{i \in I_1}\,(\gamma_i/w_i) \triangleq (\gamma_k/w_k)$$

With regard to the feasibility condition, let us set $I_2 = \{ i : h_i > 0 \}$; if $I_2 = \emptyset$, then $\hat{x}_B(\theta) \geq 0$ for any $\theta \geq 0$, otherwise $\hat{x}_B(\theta) \geq 0\ ,\forall \in [0, \theta'']$, where

(5.5) $$\theta'' = \min_{i \in I_2} (\hat{x}_{B_i}/h_i) \doteq (\hat{x}_{B_j}/h_j)$$

As a consequence for any $\theta \in [0, \bar{\theta}]$, where $\bar{\theta} \doteq \min(\theta', \theta'')$, $\hat{x}_B(\theta)$ is the optimal solution of the problem $P(\theta)$; when $\theta > \bar{\theta}$ and $\bar{\theta} = \theta''$, we can restore the feasibility by means of dual-simplex like algorithm; when $\theta > \bar{\theta}$ and $\bar{\theta} = \theta'$, we can restore the optimality by means of the algorithm given in section 3.
Now we will describe a sequential method for solving problem $P(\theta)$, $\theta \in [0, \theta_2]$ starting from an optimal basic solution to $P(0)$.

Step 0 (not iterative) Solve $P(0)$ and let $\hat{x}_B^o$ be an optimal basic solution; set $i=0$ and go to step 1.

Step 1 Consider $P(\theta^i + \theta)$, $\theta \geq 0$; calculate $\gamma(\theta)$, $\hat{x}_B^i(\theta)$, $\bar{\theta}$ and set $\theta^{i+1} = \theta^i + \bar{\theta}$; $\hat{x}_B^{(i+1)} = \hat{x}_B^{(i)}(\theta)$ is an optimal basic solution for $P(\theta^{i+1})$. If $z(\theta^{i+1}) = z(\theta^i)$, then $\theta_2 = \theta'$ STOP; otherwise go to step 2.

Step 2 If $\bar{\theta} = \theta' < +\infty$, we apply (3.1) and $\hat{x}_{N_k}$ enters the basis by means of a simplex-like pivot operation; set $i=i+1$ and return to step 1. If $\bar{\theta} = \theta'' < +\infty$, then $\hat{x}_{B_j}$ must leave the basis and a pivot operation is performed on a_{tj} such that $\gamma_t(\bar{\theta})/a_{tj} \doteq \min_{\substack{a_{ij}<0 \\ i \in I_2}} \gamma_i(\bar{\theta})/a_{ij}$; set $i=i+1$ and return to step 1. Otherwise $\hat{x}_B^{i+1}(\theta)$ is optimal for $P(\theta^{i+1}+\theta)$, $\theta_2 = +\infty$; STOP.

Remark 5.1 Let us note that, in order to find the set $S(\theta)$ of optimal solutions, $\forall\, \theta \in [\theta_1, 0]$, we can apply the same algorthm where (3.1) is now replaced by (5.3) in step 2.

In order to find β or to establish that $E = \emptyset$ we must repeat the same considerations in case A]; let us note that now in problem (5.2), $L < +\infty$ cannot reached as a maximum; in such a case solving the linear fractional problem $P(0)$ with $\beta_1 = f_1(x^o)$, by means of the algorithm given in section 3, we find a basic solution and an index t such that $\gamma_t > 0$ and the corresponding column has nonpositive coefficients.
Consider the parametric problem $P(\theta)$, $\theta \geq 0$, and calculate θ'', setting $\theta'' = +\infty$ if $I_2 = \emptyset$.
Let $\theta' > 0$ be the positive root of the linear equation $\gamma_t(\theta)$ and set $\theta' = +\infty$, if such a positive root does not exist. If $\theta' = \theta'' = +\infty$, then $K_1 = \emptyset$ so that $E = \emptyset$. If $\theta'' < \theta'$, set $\theta = \theta''$, consider problem $P(\theta'')$, and repeat the above considerations. If $\theta' \leq \theta''$, set $\theta = \theta'$ and consider problem $P(\theta')$; if $P(\theta')$ has optimal solutions, then $E \neq \emptyset$ and $\beta = \beta_1 - \theta'$, otherwise repeat the above considerations. After a finite number of such iterations we are able to find β or to establish that $E = \emptyset$.

[C] (The bicriteria linear fractional problem)

Now we will consider problem P, where f_1, f_2 are both linear fractional functions; in such a case the sequential method suggested in B] can be used in order to generate E since, by means of the Charnes-Cooper transformation, problem P reduces to a problem where one of the objective functions is linear.
Suppose that

$$f_1(x)=(a\,x+a_0)/(b\,x+b_0)$$

and consider the Charnes-Cooper transformation

$$t=1/(b\,x+b_0),\quad y=t\,x\,;$$

problem P is transformed in the problem

$$P^*:\begin{cases}\max\,(f_1(y/t),f_2(y/t))\\ (y/t)\in R\\ b\,y+b_0t=1\,,\,t\geq 0\end{cases}$$

where $f_1(y/t)$ and $f_2(y/t)$ are a linear function and a linear fractional real-valued function respectively.
It is easy to prove the following theorem which establishes a relationship between the efficient points of the problems P and P^*.

Theorem 5.1 i) If x^0 is an efficient point for P then (y^0, t_0) is an efficient point for P^*, where

$$t_0=1/(b\,x^0+b_0)\;,\quad y^0=t_0\,x^0$$

ii) If (y^0, t_0) is an efficient point for P^*, then $x^0=y^0/t_0$ is an efficient point for P.

Remark 5.2 Since P^* is a bicriteria linear fractional problem having, at least one linear objective function, the sequential method described in B], can be used to solve it.

6. On the connectedness of E.

The algorithms, given in the previous section, generate a finite sequence of parameters and a finite number of connected line segments whose endpoints are optimal basic solutions and also efficient points for P. Taking into account that $S(\theta)$ is a convex set for any $\theta\in[\theta_1,\theta_2]$, the following theorem is established:

Theorem 6.1 The set E of all efficient points of the bicriteria linear fractional problem P is connected when at least one of the objective functions is linear.

Remark 6.1 Let us note that E turns out to be disconnected (when f_1, f_2 are both linear fractional functions) if and only if in the algorithm described in C], the variable t leaves the basis and subsequently enters the basis again at a positive value.

References

/1/ **Benson H.P.**, Vector Maximization with two Objective Functions, J.O.T.A.,vol.28,1979, pp. 253-257.

/2/ **Choo E.U.-Atkins D.R.**, Bicriteria Linear Fractional Programming, J.O.T.A., vol.36, 1982,
pp. 203-220.

/3/ **Choo E.U.-Atkins D.R.**, Connectedness in Multiple Linear Fractional Programming, Management Science, vol. 29, 1983, pp.250-255.

/4/ **Choo E.U.-Schaible S.- Chew K.P.**, Connectedness of the Efficient Set in Three Criteria Quasiconcave Programming , Cahiers du C.E.R.O. ,vol.27, 1985 pp.213-220

/5/ **Martein L.**, On generating the set of all efficient points of a bicriteria linear fractional problem, Technical report n.13, Dep. of Statistics and Applied Mathematics, University of Pisa, 1988.

/6/ **Schaible S.**, Bicriteria Quasiconcave Programs, Cahiers du C.E.R.O., vol.25, 1983, pp.93-101

/7/ **Martein L.**, On the bicriteria maximization problem, This volum.

/8/ **Cambini A.-Martein L.**, A Modified version of Martos's Algorithm for Fractional Programs, Methods of Operation Research, vol.53, pp.33-44 (1985)

/9/ **Bitran G.R.**, Experiments with linear fractional problems, Naval Res.Logistic Quarterly, vol.26, 1979, pp. 689-693.

/10/ **Martos B.**, Hyperbolic programming, Naval Res. Logistic Quaterly, vol.11, 1964, pp. 135-155.

/11/ **Naccache P.H.**, Connectedness of the Set of Non-dominated Outcomes in Multicriteria Optimization, J.O.T.A., vol.25, 1978, pp.459-467.

/12/ **Bitran G.R.-Magnanti T.L.**, The structure of Admissible Points with Respect to Cone Dominance, J.O.T.A., vol.29, 1979, pp.573-614.

/13/ **Geoffrion A.M.**, Solving Bicriteria Mathematical Programs, Operations Research,vol. 15, 1967, pp.39-54.

/14/ **Kornbluth J.S.H.-Steuer R.E.**,Multiple Objective Linear Fractional Programming, Management Science, vol. 27, 1981, pp. 1024-1039.

/15/ **Yu P.L.**, Cone convexity , Cone Extreme Points and Nondominated Solutions in Decision Problem with Multiobjectives, J.O.T.A., vol.14, 1974, pp. 319-377.

/16/ **Warburton A.R.**, Parametric Solution of Bicriterion Linear Fractional Programs, Operation Research, vol. 33, n.1, 1985, pp.74-84.

CHAPTER III

DUALITY AND CONJUGATION

GENERALIZED CONJUGATION AND RELATED TOPICS

J. E. Martínez-Legaz

Abstract

An axiomatic approach to generalized conjugation theory, following [27], based on dualities between complete lattices, is outlined. Moreover, Toland-Singer type duality results for arbitrary conjugation operators, as well as a related formula for the Φ-conjugate of the difference of functions, which generalizes that of B. N. Pschenichnyi for the convex case, are studied in detail, examining some consequences for Moreau-Yosida and Lipschitz approximates of functions. Similar developments are made in connection with a duality formula due to M. Volle involving level set conjugations.

Introduction

In convex analysis and mathematical programming, the theory of Fenchel conjugation plays an important role [30, 36, 46, 23]. In order to extend it to the nonconvex situation a concept of generalized conjugation was introduced by J.J.Moreau [31], which has found useful applications to duality theory in optimization (see [43]). An axiomatic approach to these conjugation operators was given by I.Singer [41] and extended later to a wider class of operators in [27]. We will recall this abstract viewpoint for conjugation theory in Section 1.

In section 2, we will concentrate on Fenchel-Moreau conjugation operators. Moreover, a particular conjugation theory for functions defined on $\{0, 1\}^n$ will be studied.

Ten years ago, J.F. Toland [47, 48, 49] and I.Singer [38, 39] obtained, independently, approximately at the same time, the following duality formula:

$$(0.1) \qquad \inf_{x \in R^n} \{ f(x) \dot{-} h(x) \} = \inf_{x^* \in R^n} \{ h^*(x^*) \dot{-} f^*(x^*) \},$$

where $f, h : R^n \to \bar{R}$ are convex functions, * denotes convex conjugate and $\dot{-}$ is the upper subtraction of J.J.Moreau [28, 31] (we shall also use the operations $\dot{+}$, $\underset{\cdot}{+}$ and $\underset{\cdot}{-}$). Theorems of surrogate duality, closely related to Toland-Singer formula (0.1), were previously given in [13], in the context of approximation theory. The duality formula (0.1) was exploited by J.F. Toland in the analysis of some problems of Calculus of Variations. Formula (0.1) is almost characteristic of the convex conjugation operator *, since this is the only operator which satisfies (0.1), for any linear function h and arbitrary f, and the equality $\langle \cdot, x^* \rangle^* = \chi_{\{x^*\}}$ $(x^* \in R^n)$, where $\langle \cdot, \cdot \rangle$ represents the Euclidean scalar product in R^n and χ denotes indicator function (replace h by $\langle \cdot, x^* \rangle$ in (0.1)). This, in a sense, shows the fundamental character of (0.1). Moreover, it may be useful to give simple proofs of some formulas in analysis, as in the case of the formula of Hörmander [22, 4] on the Hausdorff distance between two nonempty closed convex subsets C, D of R^n (or, more generally, of a reflexive Banach space X), which is valid whenever the left side is finite:

(0.2) $$\mathrm{haus}\,(C, D) = \sup_{\|x^*\| \leq 1} | \chi^*_C(x^*) \dot{-} \chi^*_D(x^*) | .$$

Here, $\| \ \|$ denotes the Euclidean norm in R^n (or the dual norm in X^*, in the general case of a reflexive Banach space X) and one has to take $|-\infty| = -\infty$. Let us recall that the Hausdorff distance is defined by

$$\mathrm{haus}\,(C, D) = \max \{ \sup_{x \in D} \mathrm{dist}\,(x, C), \ \sup_{x \in C} \mathrm{dist}\,(x, D) \} \quad (C, D \subset R^n),$$

where $\mathrm{dist}\,(x, C) = \inf_{y \in C} \| x-y \| \ (x \in R^n, C \subset R^n)$. To prove (0.2), consider the equivalent

formula for the

Hausdorff distance [4, p.50] :

$$\mathrm{haus}\,(C, D) = \sup_{x \in R^n} | \mathrm{dist}\,(x, C) - \mathrm{dist}\,(x, D) | \qquad (C, D \subset R^n).$$

Using it and (0.1) (with inf and $\dot{+}$ replaced by sup and $\dot{-}$, respectively), we obtain:

$$\mathrm{haus}\,(C, D) = \max \{ \sup_{x \in R^n} \{ \mathrm{dist}\,(x, C) - \mathrm{dist}\,(x, D) \}, \ \sup_{x \in R^n} \{ \mathrm{dist}\,(x, D) - \mathrm{dist}\,(x, C) \} \}$$

$$= \max \{ \sup_{x^* \in R^n} \{ \mathrm{dist}(\cdot, D)^*(x^*) \dot{-} \mathrm{dist}(\cdot, C)^*(x^*) \}, \ \sup_{x^* \in R^n} \{ \mathrm{dist}(\cdot, C)^*(x^*) \dot{-} \mathrm{dist}(\cdot, D)^*(x^*) \} \}.$$

Let us compute the conjugate of dist $(\cdot, C)$:

$$\mathrm{dist}\,(\cdot, C)^* = (\| \cdot \| \nabla \chi_C)^* = \| \cdot \|^* + \chi^*_C = \chi_B + \chi^*_C$$

Here, ∇ denotes the inf-convolution operator of convex analysis [29, 30, 22] and B is the unit ball in R^n. Thus, we have:

$$\mathrm{haus}\,(C, D) = \max \{ \sup_{x^* \in R^n} \{ (\chi^*_B(x^*) + \chi^*_D(x^*)) \dot{-} (\chi_B(x^*) + \chi^*_C(x^*)) \} ,$$

$$\sup_{x^*\in R^n} \{ (\chi_B(x^*) + \chi^*_C(x^*)) \dot{-} (\chi_B(x^*) + \chi^*_C(x^*)) \} \} ,$$

$$= \max \{ \sup_{x^*\in B} \{ (\chi^*_D(x^*) \dot{-} \chi^*_C(x^*) \} , \sup_{x^*\in R^n} \{ \chi^*_C(x^*) \dot{-} \chi^*_D(x^*) \} \} =$$

$$= \sup_{x^*\in B} | \chi^*_C(x^*) \dot{-} \chi^*_D(x^*) | .$$

In section 3, we will study Toland-Singer duality for a general conjugation operator and its consequences in connection with Moreau-Yosida and Lipschitz approximates of functions. The Moreau-Yosida approximates f_r of $f : R^n \to \overline{R}$ are defined [8] by

$$f_r(x) = \inf_{y\in R^n} \{ f(y) + 1/2r \ \| x - y \|^2 \} \qquad (r > 0, x \in R^n).$$

It is evident that f_r depends nonincreasingly on r; moreover, it is well known that for any lower semicontinuous (l.s.c.) proper convex function f, f_r is also convex and of class C^1 and one has $\lim_{r\to 0} f_r(x) = f(x)$ $(x \in R^n)$. For any Hölder (in particular, Lipschitz) function f, f_r converges uniformly to f, since, for any $x \in R^n$, one has:

$$f(x) - f_r(x) = f(x) - \inf_{y\in R^n} \{ f(y) + 1/2r \ \| x - y \|^2 \} =$$

$$= \sup_{y\in R^n} \{ f(x) - f(y) - 1/2r \ \| x - y \|^2 \} \leq$$

$$\leq \sup_{y\in R^n} \{ N \| x - y \|^\alpha \ 1/2r \ \| x - y \|^2 \} =$$

$$= \sup_{t\geq 0} \{ N t^\alpha - 1/2r \ t^2 \} = N^{2/2-\alpha} \ \alpha^{\alpha \ 2-\alpha} (1 \ 1/2 \ \alpha) \ r^{\alpha/2-\alpha} ,$$

where we have assumed that f is α-Hölder with constant N. Another way of regularizing a function f is by means of its Lipschitz approximates $f_{[r]}$ [29, 54, 16, 17, 12, 25], which are defined in the following way:

$$f_{[r]}(x) = \inf_{y\in R^n} \{ f(y) + 1/r \ \| x - y \| \} \qquad (r > 0, x \in R^n).$$

These regularizations are usually expressed in terms of the inf-convolution operation, but they can also be obtained in the framework of generalized conjugation theory. In Section 3, we will adopt this viewpoint.

Let us also mention that, among other applications, Moreau-Yosida and Lipschitz approximates have been useful to define distance on the cone S of l.s.c. proper convex functions on R^n (or, more generally, a separable Hilbert space). In [4], H. Attouch and R. Wets have defined:

$$d_{r,\rho}(f,g) = \sup_{\|x\|\leq\rho} |f_r(x) - g_r(x)|$$

$$(r, \rho > 0 ; f, g \in S).$$

$$d_{[r],\rho}(f,g) = \sup_{\|x\|\leq\rho} |f_{[r]}(x) - g_{[r]}(x)|$$

A reasoning similar to that we have employed to derive the Hormander formula gives

$$d_{r,\rho}(f,g) = \sup_{\|x\|\leq\rho} |f_r(x) - g_r(x)| =$$

$$= \sup_{x\in R^n} |(f_r + \chi_{\rho B})(x) \dot{-} (g_r + \chi_{\rho B})(x)| =$$

$$= \sup_{x^*\in R^n} |(f_r + \chi_{\rho B})^*(x^*) \dot{-} (g_r + \chi_{\rho B})^*(x^*)| =$$

$$= \sup_{x^*\in R^n} |[(f_r)^* \nabla \chi^*_{\rho B}](x^*) \dot{-} [(g_r)^* \nabla \chi^*_{\rho B}](x^*)| =$$

$$= \sup_{x^*\in R^n} |[(f_r)^* \nabla (\rho \|\cdot\|)](x^*) \dot{-} [(g_r)^* \nabla (\rho \|\cdot\|)](x^*)| =$$

$$= \sup_{x^*\in R^n} |((f_r)^*)_{[\rho^{-1}]}(x^*) - ((g_r)^*)_{[\rho^{-1}]}(x^*)| ;$$

the same equalities hold when one replaces r by $[r]$ everywhere. In this way, Toland's duality provides new expression for the distances $d_{r,\rho}$ and $d_{[r],\rho}$.

In the last years, much effort has been devoted to the study of d.c. (i.e., difference of convex) functions (see [18] and the references contained therein). Among the applications of d.c. functions, we want to mention that the convex multiobjective programming problem can be formulated as the minimization of a d.c. function. Indeed, given $f_1,\ldots\ldots,f_k$:

$X \to R$, with $\emptyset \neq X \subset R^n$, it is easy to prove that, for any $\bar{x} \in X$, $\bar{x}$ is a Pareto minimum of $(f_1, \ldots, f_k)$ on X if and only if $g(\bar{x}) = 0$, where $g : X \to \bar{R}$ is defined by

$$(0.3) \quad g(x) = \sum_{j=1}^{k} f_j(x) - \inf \{ \sum_{j=1}^{k} f_j(y) \mid y \in X,\ f_i(y) \le f_i(x)\ (i = 1, \ldots, k) \};\ (x \in X)$$

(observe that g is nonnegative on X). This is also equivalent to say that $\bar{x}$ is a fixed point of the multivalued mapping $x \in X \mapsto M(x)$, where $M(x)$ denotes the solution set of the minimization problem appearing in the right hand side of (0.3). When X is a convex set and the objective functions f_j are convex, each term g^1, g^2 in the definition (0.3) of $g = g^1 - g^2$ is convex (and $M(x)$ is a convex set). Let us further assume that X is closed and g^2 is finite (this holds when $\sum_{j=1}^{k} f_j$ is bounded from below on X) and l.s.c.. Then (see (3.2) below), the minimization of g on X is equivalent to that of $\hat{g}_r^{\,1} - \hat{g}_r^{\,2}$ on R^n, where $\hat{g}^1$ and $\hat{g}^2$ are the extensions of g^1 and g^2, respectively, to R^n such that $\hat{g}^1(x) = g^2(x) = + \infty$ $(x \in R^n \backslash X)$. The last is an unconstrained C^1 d.c. problem. In the case of a linear problem (i.e., when each f_j is linear and X is a convex polyhedron), it is easy to see that the minimization of g on X can be formulated as a min-max linear problem or, using linear programming duality to transform the computation of g^2 into a maximization problem, as the minimization of a bilinear function under linear constraints.

A useful formula in the duality approach to d.c. functions is that of B.N. Pschenichnyi [34]:

$$(0.4) \qquad (f - h)^* (x^*) = \sup_{y^* \in R^n} \{ f^* (x^* + y^*) - h^* (y^*) \} \qquad (x^* \in R^n),$$

which he proved to be valid for convex functions $f, h : R^n \to R$. R. Ellaia and J.B. Hiriart-Urruty [11] relaxed the assumption on f, proving that (0.4) holds for arbitrary $f : R^n \to R \cup \{ + \infty \}$ and convex $h : R^n \to R$. The last author [19] obtained again (0.4) for proper extended real valued functions defined on a locally convex space, under the only assumption that h is convex and l.s.c.. In the same paper, he proved that (0.4) is equivalent to Toland-Singer duality formula (0.1). Recently, M. Volle [53] has given a very simple and elegant proof of (0.4). The right hand side of (0.4) coincides with the "deconvolution" of f^* and h^*, an operation introduced by J.B. Hiriart-Urruty and M.L. Mazure [20, 28], which plays the role of an inverse to the inf-convolution and can also be regarded as a generalization of the parallel subtraction of operators in the sense of J.R. Anderson, W. N. Duffin and G.E. Trapp [1]. The deconvolution operation has been also studied (without this name) by M. Volle [53], who related it to the star difference of sets, in the sense of B. N. Pschenichnyi [34]. Recently, J. B. Hiriart-Urruty and Ph. Plazanet [21] have proved a sort of converse to Moreau's decomposition theorem, using deconvolution.

The fundamental character of (0.4) is shown by the fact that the convex conjugation operator * is the only operator $S \overset{*}{\to} S$ which satisfes (0.4) for any $h \in S$ and

$$(0.5) \qquad (\chi_{\{.\}} \dot{-} h)^* (x^*) = h + \langle ., x^* \rangle = h^{**} + \langle ., x^* \rangle \qquad (h \in S, x^* \in R^n).$$

Indeed, if these conditions hold, for any $h \in S$ one has:

$$(0.6) \qquad h^* = h^{***} = (\chi_{\{.\}} - h^*)^* (0) = \sup_{x \in R^n} \{ \chi_{\{.\}}^* (x) \dot{-} h^{**}(x) \} = \sup_{x \in R^n} \{ \langle x, . \rangle - h(x) \}$$

In Section 4, we show the validity of (0.4) in the abstract setting of Φ-conjugation [9, 10], under the only assumption that Φ is closed under addition. We also give several necessary and sufficient conditions for the generalized version of (0.4) to hold.
Finally, in Section 5 we make similar developments on a related duality formula due to M. Volle [52], which is specific to level set conjugation.
Inless otherwise specified, the notation we shall use in the remaining of the paper will be the same we have used in this introduction.

1. Dualities between complete lattices [27]

In this Section, we present some results obtained jointly with I. Singer in [27].
Let E, F be two complete lattices. A mapping $\Delta : E \to F$ is called a duality (or a polarity [7]), if for each $X \subset E$ (including $X = \emptyset$) we have

$$(1.1) \qquad \Delta (\inf X) = \sup \Delta (X).$$

As the following theorem indicates, for Δ to be a duality it suffices (1.1) to be satisfied for each subset X of a given family of infimal generators of E. We recall that $Y \subset E$ is said to be a family of infimal generators of E, if for each $x \in E$ there exists $Y_x \subset Y$ such that $x = \inf Y_x$. In [45], I. Singer has given some applications of this concept to the study of dualities between complete lattices.

Theorem 1.1. Let E, F be two complete lattices, $\Delta : E \to F$ and Y a family of infimal generators of E. The following statements are equivalent:

1° Δ is a duality
2° For each $Y' \subset Y$ (including $Y' = \emptyset$) we have $\Delta (\inf Y') = \sup \Delta (Y')$.
These statements imply
3° We have
$$\Delta (x) = \sup \{ \Delta (y) \mid y \in Y, x \le y \} \qquad (x \in E).$$

Let us consider the particular case when $E = A^X$, X being an arbitrary set and $A \subset R$ being a complete lattice (with the usual order). Then, E is indeed a complete lattice (with the usual pointwise order). A convenient family of infimal generators of E is $(\chi^A_{x,a})_{x\in X, a\in A}$, where $\chi^A_{x,a}$ is defined by

$$(1.2) \qquad \chi^A_{x,a}(x') = \begin{cases} a & \text{if } x' = x \\ +\infty^A & \text{if } x' \neq x, \end{cases}$$

$+\infty^A$ denoting the greatest element of A. This is an immediate consequence of the following obvious equality:

$$f = \inf^E_{x\in X} \chi^A_{x,f(x)} \qquad (f \in E)$$

(by $\inf^E$ (or $\sup^E$) we denote infimum (resp. supremum) in E). Let us observe that, when $A = R$, we have $\chi^A_{x,a} = \chi_{\{x\}} + a$. Note also that, by (1.2), for any $f \in E$ and $(x, a) \in X \times A$, we have

$$f \le \chi^A_{x,a} \Leftrightarrow (x, a) \in \operatorname{epi} f,$$

where $\operatorname{epi} f = \{(x, a) \in X \times A \mid f(x) \le a\}$ is the epigraph of f. Combining these results with Theorem 1.1, we obtain:

Theorem 1.2. Let X be a set and let $(A, \le) \subset (\overline{R}, \le)$ and F be complete lattices. For a mapping $\Delta : f \in A^X \to f^\Delta \in F$, the following statements are equivalent:

1° Δ is a duality.

2° There exists a mapping $\Gamma : X \times A \to F$, satisfying, for every $A' \subset A$ (including $A' = \emptyset$),

$$\Gamma(x, \inf^A A') = \sup \Gamma(x, A') \qquad (x \in X),$$

and such that

$$f^\Delta = \sup_{x\in X} \Gamma(x, f(x)) \qquad (f \in A^X).$$

Moreover, in this case, Γ is uniquely determined by Δ, namely, we have

$$\Gamma(x, a) = (\chi^A_{x,a})^\Delta \qquad (x \in X, a \in A).$$

When $A = \{0, +\infty\} \subset \overline{R}$, which is obviously a complete lattice, the set A^X can be identified with 2^X, through the bijection

$$\begin{aligned} A &\to 2 = \{0, 1\}, \\ 0 &\mapsto 1 \\ +\infty &\mapsto 0. \end{aligned}$$

In other words, any $f \in A^X$ can be regarded as a subset of X, since one has $f = \chi_{f^{-1}(0)}$; conversely, any $X' \subset X$ can be represented by $f = \chi_{X'}$. It is clear that the mappings $f \mapsto f^{-1}(0)$, $X' \mapsto \chi_{X'}$ are mutually inverse. Using Theorem 1.2 we get:

Theorem 1.3. Let X be a set and F a complete lattice. For a mapping $\Delta : f \in \{0, +\infty\}^X \to f^\Delta \in F$, the following statements are equivalent:

1° Δ is a duality.
2° There exists a mapping $\gamma : X \to F$, such that

$$f^\Delta = \sup \gamma(f^{-1}(0)).$$

Moreover, in this case, γ is uniquely determined by Δ, namely, we have

$$\gamma(x) = (\chi_{\{x\}})^\Delta \qquad (x \in X).$$

In the particular case when $F = B^W$, W being another set and $(B, \leq) \subset (\overline{R}, \leq)$ a complete lattice, from Theorem 1.2, using the identification $(B^W)^{X \times A} \simeq B^{X \times W \times A}$, we obtain (compare with the F-conjugates of [6]):

Theorem 1.4. Let X and W be two sets and let $(A, \leq) \subset (\overline{R}, \leq)$ and $(B, \leq) \subset (\overline{R}, \leq)$ be complete lattices. For a mapping $\Delta : f \in A^X \to f^\Delta \in B^W$, the following statements are equivalent:

1° Δ is a duality
2° There exists a mapping $G: X \times W \times A \to B$, satisfying, for every $A' \subset A$ (including $A' = \emptyset$),

$$G(x, w, \inf{}^A A') = \sup{}^B G(x, w, A') \qquad (x \in X, w \in W),$$

and such that

$$f^\Delta = \sup_{x \in X}{}^B G(x, \cdot, f(x)) \qquad (f \in A^X).$$

Moreover, in this case, G is uniquely determined by Δ, namely, we have

$$G(x, w, a) = (\chi^A{}_{x,a})^\Delta (w) \qquad (x \in X, w \in W, a \in A).$$

Similarly, from Theorem 1.3 one gets:

Theorem 1.5. Let X and W be two sets, and let $(B, \leq) \subset (\overline{R}, \leq)$ be a complete lattice. For a mapping $\Delta : f \in \{0, +\infty\}^X \to f^\Delta \in B^W$, the following statements are equivalent:

1° Δ is a duality.
2° There exists a mapping $\varphi : X \times W \to B$, such that

$$f^{\Delta} \sup^{B} \varphi \, (f^{-1}(0), \cdot) \qquad (f \in \{0, +\infty\}^{W}).$$

Moreover, in this case, φ is uniquely determined by Δ, namely, we have

$$\varphi \, (x, w) = (\chi_{\{x\}})^{\Delta} (w) \qquad (x \in X, w \in W)$$

In the particular case when one also has $B = \{0, +\infty\}$, it is easy to show, by the identifications $\{0, +\infty\}^X \simeq 2^X$ and $\{0, +\infty\}^W \simeq 2^W$, that Theorem 1.5 is equivalent to the following result of I.Singer [42, Thm. 1.1]: "for a mapping $\Delta : (2^X, \supset) \to (2^W, \supset)$, the following statements are equivalent:

1° Δ is a duality.
2° There exists a set $\Omega \subset X \times W$ such that

$$\Delta(M) = \{ w \in W \mid (x, w) \in \Omega \ (x \in X) \} \qquad (M \subset X).$$

Moreover, in this case, Ω is uniquely determined by Δ, namely, we have

$$\Omega = \{ (x, w) \in X \times W \mid w \in \Delta(\{x\}) \} ".$$

Let us recall that if E and F are two complete lattices, the dual $\Delta^* : F \to E$ of any mapping $\Delta : E \to F$ is defined by

$$(1.3) \qquad \Delta^*(z) = \inf \{ x \in E \mid \Delta(x) \le z \} \qquad (z \in F).$$

When Δ is a duality, Δ^* is also a duality and for any $x \in E$ and $z \in F$ we have:

$$\Delta(x) \le z \Leftrightarrow \Delta^*(z) \le x.$$

From this, it follows that $\Delta^{**} = \Delta$. Moreover, in this case, $\Delta^* \Delta : E \to E$ is a hull operator (see [12], [51]).

Theorem 1.6. Let X and W be two sets, $(A, \le) \subset (\overline{R}, \le)$ and $(B, \le) \subset (\overline{R}, \le)$ two complete lattices, $\Delta : A^X \to B^W$ a duality, with dual $\Delta^* : B^W \to A^X$, and $G : X \times W \times A \to B$, $G^* : W \times X \times B \to A$ the mappings corresponding to them by Theorem 1.4.
Then, we have

$$G^*(w, x, b) = \min \{ a \in A \mid G(x, w, a) \le b \} \qquad (w \in W, x \in X, b \in B).$$

The next theorem explains the meaning of the hull operator $\Delta^* \Delta : A^X \to A^X$.

Theorem 1.7. Under the assumptions of Theorem 1.6, we have

$$f^{\Delta\Delta^*} = \sup{}^{A^X} \{ G^*(w, ., b) \mid w \in W, b \in B, G^*(w, ., b) \leq f \} \qquad (f \in A^X).$$

This result says that $f^{\Delta\Delta^*}$ coincides with the supremum (in A^X) of the elementary minorants of f, an elementary function being one of the form $G^*(w, ., b)$ for some $w \in W$ and $b \in B$. The following theorem says that, given any set $W \subset R^X$ (of "elementary" functions on X), one can find a duality Δ for which one has $f^{\Delta\Delta^*} = \sup \{ w \in W \mid w \leq f \}$.

Theorem 1.8. If X is a set and $W \subset \overline{R}^X$, then for $\Delta : f \in \overline{R}^X \to f^\Delta \in \{0, +\infty\}^W$ defined by

$$f^\Delta = \chi_{\{x \in W \mid w \leq f\}} \qquad (f \in \overline{R}^X),$$

we have

$$f^{\Delta\Delta^*} = \sup \{ w \in W \mid w \leq f \} \qquad (f \in \overline{R}^X).$$

2. Conjugation operators [41, 27]

In this Section we present some results of [41] and [27].
We recall that a duality $c : f \in \overline{R}^X f^c \in \overline{R}^W$ (where X and W are two sets) is called a conjugation, if

$$(2.1) \qquad (f + d)^c = f^c - d \qquad (f \in \overline{R}^X, d \in R).$$

The following theorem, which was proved by I. Singer [41, Thm. 3.1], shows that an operator $c : \overline{R}^X \to \overline{R}^W$ is a conjugation if and only if it is the Fenchel-Moreau conjugation [31] corresponding to some uniquely determined coupling function.

Theorem 2.1. Let X and W be two sets. For a mapping $c : f \in \overline{R}^X \to f^c \in \overline{R}^W$ the following statements are equivalent:

1° c is a conjugation
2° There exists a mapping $\varphi : X \times W \to \overline{R}$, such that

(2.2) $$f^c(w) = \sup_{x \in X} \{ \varphi(x,w) \dot{-} f(x) \} \qquad (f \in \overline{R}^X, w \in W).$$

Moreover, in this case φ is uniquely determined by c, namely, we have

(2.3) $$\varphi(x,w) = (\chi_{\{x\}})^c (w) \qquad (x \in X, w \in W).$$

In fact, this is an immediate consequence of Theorem 1.4. (with $A = B = \overline{R}$), since for G as in 2° of Theorem 1.4. by (2.1) one has:

$$G(x,w,a) = (\chi^R_{x,a})^c (w) = (\chi_{\{x\}} \dot{+} a)^c (w) = (\chi_{\{x\}})^c (w) \dot{-} a.$$

The conjugation operators have an important property of simmetry. If $c : \overline{R}^X \to \overline{R}^W$ is a conjugation, then the dual operator $c^* : \overline{R}^W \to \overline{R}^X$ (in the sense of (1.3)) is again a conjugation [41. Thm. 4.1] and one has [41. Thm. 4.2 and Cor. 4.3]:

Theorem 2.2. Let X and W be two sets, $c : \overline{R}^X \to \overline{R}^W$ a conjugation, with dual conjugation $c^* : \overline{R}^W \to \overline{R}^X$, and let $\varphi : X \times W \to \overline{R}^X$ and $\varphi^* : W \times X \to \overline{R}^X$ be the mappings associated with c and c^* respectively, by Theorem 2.1. Then

$$\varphi^*(w,x) = \varphi(x,w) \qquad (x \in X, w \in W).$$

This follows from Theorem 1.6, since, for $G^* : W \times X \times B \to A$, the mapping corresponding to c^* by Theorem 1.4, using (2.2) one obtains:

$$\begin{aligned} G^*(w,x,b) &= \min \{ a \in R \mid G(x,w,a) \leq b \} = \min \{ a \in R \mid \varphi(x,w) \dot{-} a \leq b \} = \\ &= \varphi(x,w) - b \qquad (w \in W, x \in X, b \in \overline{R}). \end{aligned}$$

This simple reasoning constitutes also a proof of the above mentioned Theorem 4.1 of [41].

Let us now esamine the particular case when $X = \{0,1\}^n$, $W = C_1 \times \ldots \times C_n$, where $C_i \subset R$ $(i = 1, \ldots, n)$ are unbounded from above and from below and φ is the restriction to $X \times W \subset R^n \times R^n$ of the usual scalar product.

Theorem 2.3. Let X, W and φ be as above and let c be the conjugation operator corresponding to the coupling function φ. For a function $f \in \overline{R}^X$, the following statements are equivalent:

1° $f = f^{cc^*}$.

2° Either $f(x) > -\infty$ $(x \in X)$ or $f \equiv -\infty$.

An important notion related to conjugation theory is that of subdifferential. Given two arbitrary sets X, W and a coupling function $\varphi : X \times W \to \overline{R}$, the subdifferential $\partial_\varphi f(x_0)$ of $f : X \to \overline{R}$ at x_0 is the set defined by [5]:

$$\partial_\varphi f(x_0) = \{ w \in W \mid \varphi(x, w) - \varphi(x, w_0) \le f(x) - f(x_0) \} \qquad (x \in X).$$

Theorem 2.4. Let X, W and φ be as in Theorem 2.3 and let $f \in \overline{R}^X$ and $x_0 \in X$.The following statements are equivalent:

1° $\partial_\varphi f(x_0) \neq \emptyset$.
2° $f(x) > -\infty \ (x \in X)$ and $f(x_0) \in R$.

The preceding theorem was proved by S. Fujishige [13] under the assumption of submodularity of f, for the case of X being a distributive sublattice of $\{0, 1\}^n$ and $W = R^n$ Since any function defined on a subset of $\{0, 1\}^n$ can be extended to the whole of $\{0, 1\}^n$, by assigning the value $+\infty$ to the points in the complememnt of this subset, and this procedure does not change the sets $\partial_\varphi f(x_0)$ for any x_0 belonging to the domain of f, it follows from Theorem 2.4 that the assumption of submodularity of f was superfluous. From the Theorem 2.4, we get

Corollary 2.1. Under the assumptions of Theorem 2.4, for $x_0 \in X$, the following statements are equivalent:

1° $\partial_\varphi f(x_0) \neq \emptyset \quad (x_0 \in X)$.
2° $f(x_0) \in R \quad (x_0 \in X)$.

Finally, let us mention that, in the example we have just studied, one has

$$f^c = \hat{f}^* |_W, \ \partial_\varphi f(x_0) = \partial \hat{f}(x_0) \cap W \qquad (f \in \overline{R}^X, x_0 \in X),$$
$$g^{c^*} = \tilde{g}^* |_X, \ \partial_{\varphi^*} g(w_0) = \partial \tilde{g}(w_0) \cap X \qquad (g \in \overline{R}^W, w_0 \in W),$$

where $\hat{f}$ (resp. $\tilde{g}$) denotes the extension of f (resp. g) to R^n such that $\hat{f}|_{R^n \setminus X} \equiv +\infty$ (resp. $\tilde{g}|_{R^n \setminus W} \equiv +\infty$) and ∂ represents the usual subdifferential mapping.

3. Toland-Singer Duality and Generalized Conjugation

In this Section, we will study Toland-Singer type duality results [47, 38, 48, 49, 39] for a general conjugation operator $c : f \in \overline{R}^X \to f^c \in \overline{R}^W$, defined (see the preceding Section) by

$$f^c(w) = \sup_{x \in X} \{ \varphi(x, w) \mathbin{\dot{-}} f(x) \} \qquad (f \in \overline{R}^X, w \in W),$$

where X and W are two sets and $\varphi : X \times W \to \overline{R}$ (for other definitions and notations concernings conjugation operators which we shall use in the sequel, we refer to the preceding sections). Some results of this kind were first given by I. Singer [44] and, independently, by M. Volle [52].

Lemma 3.1. [44, Theorem 5.2 and 52, p.117]. Let $f, h : X \to \overline{R}$. Then

$$\inf_{x \in X} \{ f(x) \mathbin{\dot{-}} h^{cc^*}(x) \} = \inf_{w \in W} \{ h^c(w) \mathbin{\dot{-}} f^c(w) \}.$$

Corollary 3.1. [44, Theorem 5.2 and 52, p. 117]. Let $f, h : X \to \overline{R}$. Then

$$\inf_{x \in X} \{ f(x) \mathbin{\dot{-}} h(x) \} \le \inf_{w \in W} \{ h^c(w) \mathbin{\dot{-}} f^c(w) \}.$$

Corollary 3.2. [52, p.117]. Let $f, g : X \to \overline{R}$. Then

$$\inf_{x \in X} \{ f(x) \mathbin{\dot{-}} h^{cc^*}(x) \} = \inf_{x \in X} \{ f^{cc^*}(x) \mathbin{\dot{-}} h^{cc^*}(x) \}.$$

Since $h^{cc^*c} = h^c$, Lemma 3.1 above is equivalent to a result of I. Singer [43, Thm.5.2], stating that, if $h = h^{cc^*}$, then

$$\inf_{x \in X} \{ f(x) \mathbin{\dot{-}} h(x) \} = \inf_{w \in W} \{ h^c(w) \mathbin{\dot{-}} f^c(w) \}. \qquad (f \in R^X).$$

Taking $f = \chi_{\{x_0\}}$, for any $x_0 \in X$, one obtains that the converse to the last statement also holds. A different proof of this fact is given in the next theorem.

Theorem 3.1. Let $h : X \to \overline{R}$. The following statements are equivalent:

1° $h = h^{cc^*}$.

2° $\inf_{x \in X} \{ f(x) \mathbin{\dot{-}} h(x) \} = \inf_{x \in X} \{ f(x) \mathbin{\dot{-}} h^{cc^*}(x) \}. \qquad (f \in \overline{R}^X).$

3° $\inf_{x \in X} \{ f(x) \mathbin{\dot{-}} h(x) \} = \inf_{w \in W} \{ h^c(w) \mathbin{\dot{-}} f^c(w) \}. \qquad (f \in \overline{R}^X).$

4° $\inf_{x\in X} \{ f(x) \dot{-} h(x) \} = \inf_{x\in X} \{ f^{cc^*}(x) \dot{-} h^{cc^*}(x) \}$. $\qquad (f \in \overline{R}^X)$.

5° There exists a formula expressing $\inf_{x\in X} \{ f(x) \dot{-} h(x) \}$ in terms of f^c.

6° $\inf_{x\in X} \{ f(x) \dot{-} h(x) \} = \inf_{x\in X} \{ f^{cc^*}(x) - h(x) \}$. $\qquad (f \in \overline{R}^X)$.

Proof: The implications 1°⇒ 2° and 4° ⇒ 5° are obvious. The implication 2°⇒ 3° is a direct consequence of Lemma 3.1. The implication 3° ⇒ 4° follows from Lemma 3.1 and Corollary 3.2. The implication 5°⇒ 6° is a consequence of the equality $f^c = f^{cc^*c}$. Finally, to prove the implication 6°⇒ 1°, taking $f = h$ in 6°, we obtain

$$(0, +\infty) \ni \inf_{x\in X} \{ h(x) \dot{-} h(x) \} = \inf_{x\in X} \{ h^{cc^*}(x) \dot{-} h(x) \},$$

whence $h^{cc^*}(x) \dot{-} h(x) \le 0$ $(x \in X)$, i.e., $h^{cc^*} \ge h$, which is equivalent to $h^{cc^*} = h$.

Let us now examine some particular cases of the generalized Toland-Singer duality formula (i.e., of statement 3° of Theorem 3.1). First, we consider the case when $X = Y = R^n$ and c is the conjugation operator corresponding to the coupling function φ defined by

$$(3.1) \qquad \varphi(x, y) = -\rho \| x-y \|^2 \qquad (x, y \in R^n),$$

where $\rho > 0$. This conjugation was studied by S. Dolecki and S. Kurcyusz [10], who obtained that the equality $f = f^{cc^*}$ means, in this case, that $f + \rho \| \cdot \|^2$ is either a l.s.c. proper convex function or identically $\pm\infty$. In the sequel, for the sake of abbreviation, we shall regard the constant functions $\pm\infty$ as being proper. The conjugation formula becomes now:

$$f^c(y) = \sup_{x\in R^n} \{ -\rho \| x-y \|^2 - f(x) \} = -\inf_{x\in R^n} \{ f(x) + \rho \| y-x \|^2 \} \qquad (f : R^n \to \overline{R},\ y \in R^n).$$

Thus, we have $f^c = -(f \nabla \rho \| \cdot \|^2) = -f_r$, where $r = 1/(2\rho)$. Therefore, the generalized Toland-Singer duality formula is now equivalent to

$$(3.2) \qquad \inf_{x\in R^n} \{ f(x) \dot{-} h(x) \} = \inf_{x\in R^n} \{ f_r(x) \dot{-} h_r(x) \}.$$

Formula (3.2) was obtained by D. Gabay [15] (see [18]) for the case when f and h are convex. From our Theorem 3.1, implication 1°⇒ 3°, it follows that (3.2) holds for any f if

and only if $h + 1/(2r)\|\cdot\|^2$ is a l.s.c. proper convex function.
A similar argument proves that the equality

$$\inf_{x\in R^n} \{ f(x) \dot{-} h(x) \} = \inf_{x\in R^n} \{ f_{[r]}(x) \dot{-} h_{[r]}(x) \},$$

obtained by J.B. Hiriart-Urruty [18] (for convex f and h), holds for any f if and only if either h is finite and Lipschitz with constant $1/r$ or $h \equiv -\infty$. In this case, $h_{[r]} = h$.
In our opinion, the generalized conjugation approach to this kind of regularizations is particulary clarifying. To further illustrate this point, we will now answer the following question, posed by H. Attouch, D. Azé and R. Wets [3, p.16]: "what is the largest class of functions for which the correspondence $f \longmapsto f_r$ is one-to-one?". They have proved [3, Prop.3.3] that the class Γ_k of functions $f : R^n \to \overline{R}$ (in fact, they consider functions defined on a Hilbert space; this level of generality would not present any difficulty in our following developments) such that $f + k\|\cdot\|^2$ is proper, l.s.c. and convex satisfies this condition when $k < 1/(2r)$ and that f_r is of class C^1 for any $f \in \Gamma_k \setminus \{\pm\infty\}$. Now, using generalized conjugation theory, we will prove that, although no such a largest class exists, one could say, in a sense, that it is just $\Gamma_{1/(2r)}$. Note also that $\Gamma_k \subset \Gamma_{k'}$ whenever $k < k'$, since $f + k'\|\cdot\|^2 = f + k\|\cdot\|^2 + (k'-k)\|\cdot\|^2$.

Lemma 3.2. Let c be the conjugation operator corresponding to the coupling function φ defined by (3.1) and $f : R^n \to \overline{R}$. Then

$$f^{cc^*} = (f + \rho\|\cdot\|^2)^{**} - \rho\|\cdot\|^2. \tag{3.3}$$

Proof: By the above mentioned result of S. Dolecki and S. Kurcyusz [10], f^{cc^*} is the greatest minorant of f such that, added to $\rho\|\cdot\|^2$, gives a l.s.c. proper convex function. Hence, from the inequality $f^{cc^*} + \rho\|\cdot\|^2 \le f + \rho\|\cdot\|^2$, it follows that $f^{cc^*} + \rho\|\cdot\|^2 \le (f + \rho\|\cdot\|^2)^{**}$. To prove the opposite inequality, it is enough to observe that the left hand side of $(f + \rho\|\cdot\|^2)^{**} - \rho\|\cdot\|^2 \le f$ is a function such that, added to $\rho\|\cdot\|^2$, gives a l.s.c. proper convex function (in other words, it belongs to Γ_ρ), whence $(f + \rho\|\cdot\|^2)^{**} - \rho\|\cdot\|^2 \le f^{cc^*}$.

Formula (3.3) has been obtained, indenpendently, by J.P. Penot and M. Volle [32].

Proposition 3.1. Let Γ be a class of functions $f : R^n \to \overline{R}$. Then, the correspondence

$$f \in \Gamma \longmapsto f_r \tag{3.4}$$

is one-to-one if and only if the correspondence

(3.5) $$f \in \Gamma \longmapsto (f + (1/2)r \| \cdot \|^2)^{**} - (1/2)r \| \cdot \|^2$$

is one-to-one.

Proof: Let c be as in Lemma 3.2, with $\rho = (1/2)r$. We have

(3.6) $$f_r = - f^c = - (f^{cc^*})^c = (f^{cc^*})_r$$

By Lemma 3.2, the correspondence (3.5) can be written

(3.7) $$f \in \Gamma \longmapsto f^{cc^*}$$

Let us first assume that it is one-to-one and let $f, g \in \Gamma$. If $f^{cc^*} = g^{cc^*}$, then by (3.6) we have $f_r = g_r$, whence $f = g$. This proves that (3.5) is one-to-one. Conversely, assume that (3.5) is one-to-one and let $f, g \in \Gamma$ be such that $f_r = g_r$. Then, using the equalities $f^c = - f_r$, $g^c = - g_r$ and Theorem 2.2, we obtain

$$f^{cc^*} = (- f_r)^{c^*} = - (- f_r)_r = - (- g_r)_r = (- g_r)^{c^*} = g^{cc^*} ,$$

whence $f = g$, thus proving that (3.4) is one-to-one.

Remark 3.1. Formula (3.6) has its own interest. It shows that the Moreau-Yosida approximate f_r of any function f coincides with that of its $\Gamma_{1/2r}$-regularized f^{cc^*}. Hence, in particular, if $f^{cc^*} \not\equiv \pm \infty$ and it is convex, which is equivalent to say that there is no $g \in \Gamma_{1/2r} \setminus S$ with $f^{**} \leq g \leq f$ (note that one always has $f^{**} \leq f^{cc^*} \leq f$, since $S \subset \Gamma_{1/2r}$), then f_r is C^1.

Corollary 3.3. Let Γ be a class of functions $f : R^n \to \overline{R}$ such that the corespondence (3.4) is one-to-one. Then, Γ is maximal if and only if (3.5) is a bijection onto $\Gamma_{1/2r}$. In particular, $\Gamma_{1/2r}$ is maximal.

Proof: It is clear that the image set of (3.5) is contained in $\Gamma_{1/2r}$. Hence, the statement follows from Proposition 3.1, Lemma 3.2 and the fact that, for $\Gamma = \Gamma_{1/2r}$, (3.7) is the identity mapping (hence, a bijection onto $\Gamma_{1/2r}$).

However, there are functions f in $\Gamma_{1/2r} \setminus \{\pm \infty\}$ for which f_r is not C^1. For example, for $f = - 1/(2r) \| \ \|^2$ one has $f_r = \chi_{\{0\}}$, which is not a finite function.
The Moreau-Yosida approximates of a function $f : R^n \to \overline{R}$ can also be expressed in terms of the classical convex conjugation, since we have:

$$f_r(x) = \inf_{y\in R^n} \{ f(y) + (1/2r) \| x-y \|^2 \} = \inf_{y\in R^n} \{ f(y) + 1/2r (\| x \|^2 - 2\langle x, y \rangle + \| y \|^2) \}$$

$$= 1/2r \| x \|^2 - \sup_{y\in R^n} \{ 1/r \langle x, y \rangle - (f(y) + 1/2r \| y \|^2) \} =$$

$$= 1/2r \| x \|^2 - (f + 1/2r \| \cdot \|^2)^* (1/r\, x) \qquad (r > 0, x \in R^n)$$

Thus,

$$f_r = 1/2r \| \cdot \|^2 - (f + 1/2r \| \cdot \|^2)^* (1/r.) \qquad (r > 0), \tag{3.8}$$

whence f_r is a d.c. function, provided that it is everywhere. Moreover, $f_r - 1/2r \| \cdot \|^2$ is concave. From (3.8), we can also get an expression for the convex conjugate of f in terms of Moreau-Yosida approximates:

$$f^* = r/2 \| \cdot \|^2 - (f - 1/2r \| \cdot \|^2)_r (r.) \qquad (r > 0). \tag{3.9}$$

In particular, (3.9) shows that, for a strongly convex (may be extended real valued) function (in the sense of B.T.Poljak [33, 35]), f^* is of class C^1 (this also follows from the fact that there cannot exist a common subgradient for f at two diferent points). Furthermore, (3.8) and (3.9) also show that the image set of the correspondence (3.4) is contained in $-\Gamma_{1/2r}$ and coincides with it when $\Gamma \supset \Gamma_{1/2r}$. Ideed, it is clear that the right hand side of (3.8) belongs to $-\Gamma_{1/2r}$. Conversely, given any $f \in -\Gamma_{1/2r}$, we have, using (3.9):

$$-f + 1/2r \| \cdot \|^2 = (-f(r.) + r/2 \| \cdot \|^2)(1/r\,.) = (-f(r.) + r/2 \| \cdot \|^2)^{**} (1/r.) =$$

$$= 1/2r \| \cdot \|^2 - ((-f(r.) + r/2 \| \cdot \|^2)^* - 1/2r \| \cdot \|^2)_r =$$

$$= 1.2r \| \cdot \|^2 - ((-f + 1/2r \| \cdot \|^2)^* (1/r.) - 1/2r \| \cdot \|^2)_r ,$$

whence

$$f = ((-f + 1/2r \| \cdot \|^2)^* (1/r.) - 1/2r \| \cdot \|^2)_r .$$

J.P.Penot and M.Volle [32] have proved the equivalent result that the inf-convolution equation $f \nabla r/2 \| \cdot \|^2 = g$ is solvable if and only if $-g \in \Gamma_{1/2r}$.

Another interesting particular case of the generalized Toland's duality formula is obtained when one takes $X = R^n$, $W = \{ \rho \in R \mid \rho > 0 \} \times R^n$ and c is the conjugation operator corresponding to the coupling function φ defined by

$$\varphi(x, (\rho, y)) = -\rho \| x-y \|^2 \qquad (x, y \in R^n, \rho > 0). \tag{3.10}$$

This conjugation was studied by S.Dolecki and S.Kurcyusz [10], who obtained that the equality $f = f^{cc^*}$ means, in this case, that f is l.s.c. and either $f \equiv -\infty$ or it satisfies the quadratic growth condition of R.T.Rockafellar [36], i.e.,

$$f(x) \geq s_0 - \rho_0 \| x-y_0 \|^2 \qquad (x \in R^n) \tag{3.11}$$

for some $s_0 \in R$, $\rho_0 > 0$ and $y_0 \in R^n$. The conjugation formula gives

$$f^c(\rho, y) = \sup_{x \in R^n} \{ -\rho \| x-y \|^2 - f(x) \} = -\inf_{x \in R^n} \{ f(x) + \rho \| x-y \|^2 \} = -f_r(y),$$

where $r = 1/2\rho$. Therefore, the generalized Toland-Singer duality formula becomes now:

$$\inf_{x \in R^n} \{ f(x) \dot{-} h(x) \} = \inf_{r>0} \inf_{x \in R^n} \{ f_r(x) - h_r(x) \}. \tag{3.12}$$

From Theorem 3.1, implication $1° \Rightarrow 3°$, it follows that (3.12) holds for any f if and only if h either is l.s.c. and satisfies the quadratic growth condition or $h \equiv -\infty$.
Similarly, the equality

$$\inf_{x \in R^n} \{ f(x) \dot{-} h(x) \} = \inf_{r>0} \inf_{x \in R^n} \{ f_{[r]}(x) - h_r(x) \}$$

holds for any f if and only if h is a supremum of Lipschitz functions or, equivalently, if h is l.s.c. and either $h \equiv -\infty$ or it satisfies

$$h(x) \geq s_0 - \rho_0 \| x-y_0 \|^2 \qquad (x \in R^n)$$

for some $s_0 \in R$, $\rho_0 > 0$ and $y_0 \in R^n$ (see [25]).
It is well known that for a l.s.c. proper convex function $f : R^n \to \overline{R}$, one has $\lim_{r \to \infty} f_r(x) = f(x)$ $(x \in R^n)$. In fact, convexity is not essential for this result. According to [2, Thm. 2.64], on a general metric space (X, d) one has $\lim_{r \to \infty} f_r(x) = \text{cl}\, f(x)$ $(x \in R^n)$ (where cl denotes lower semicontinuous hull), for any function $f : X \to \overline{R}$ satisfying the quadratic growth condition (with $\| \cdot - \cdot \|$ replaced by $d(\cdot, \cdot)$ in the definition of f_r and in (3.11)). Next, we are going to show how this result can be derived, for functions defined on R^n (or more generally, on a Hilbert space), from the result for convex functions, using the methods of generalized conjugation theory. First, we will consider the case when f does not satisfy the quadratic growth condition.

Proposition 3.2. If $f: R^n \to \bar{R}$ does not satisfy the quadratic growth condition, then $f_r \equiv -\infty$ $(r > 0)$.

Proof: If, for some $r > 0$ and $x_0 \in R^n$, we have $f_r(x_0) > -\infty$, from the definition of f_r it follows that the function $f + (1/2r) \| \cdot - x_0 \|^2$ is bounded from below, which implies the quadratic growth condition.

Adapting the definition of J. P. Vial [50] to the extended real valued case, we say that $f: R^n \to \bar{R}$ is weakly convex if there exists $\rho > 0$ such that $f + \rho \| \cdot \|^2$ is l.s.c., proper and convex (see [50, Prop. 4.3]).

Lemma 3.3. If $f: R^n \to \bar{R}$ is weakly convex, then

$$\lim_{r \to 0} f_r(x) = f(x) \qquad (x \in R^n)$$

Proof: By the assumption on f, there exists $\rho > 0$ such that the function $g = f + \rho \| \cdot \|^2$ is l.s.c, proper and convex. Let $r \in (0, 1/2\rho)$ and $s(r) = r/1-2r\rho$. Clearly, $s(r) > 0$ and $\lim_{r \to 0} s(r)=0$. Then, for any $x \in R^n$, we have:

$$f_r(x) = \inf_{y \in R^n} \{ f(y) + 1/2r \| x-y \|^2 \} = \inf_{y \in R^n} \{ g(y) - \rho \| y \|^2 + 1/2r \| x-y \|^2 \} =$$

$$= \inf_{y \in R^n} \{ g(y) - \rho (\| x-y \|^2 - \| x \|^2 + 2 \langle x, y \rangle) + 1/2r \| x-y \|^2 \} =$$

$$= \inf_{y \in R^n} \{ (g - 2\rho \langle x, \cdot \rangle)(y) + (1/2r - \rho) \| x-y \|^2 \} + \rho \| x \|^2 =$$

$$= \inf_{y \in R^n} \{ (g - 2\rho \langle x, \cdot \rangle)(y) + (1/(2s(r))) \| x-y \|^2 \} + \rho \| x \|^2 =$$

$$= (g - 2\rho \langle x, \cdot \rangle)_{s(r)}(x) + \rho \| x \|^2 \underset{r \to 0}{\longrightarrow} (g - 2\rho \langle x, \cdot \rangle)(x) + \rho \| x \|^2 =$$

$$= g(x) - \rho \| x \|^2 = f(x).$$

Proposition 3.3. If $f: R^n \to \bar{R}$ satisfies the quadratic growth condition, then

$$\lim_{r \to \infty} f_r(x) = \text{cl}\, f(x) \qquad (x \in R^n) \tag{3.13}$$

Proof: For any $\rho > 0$, let us denote by $c(\rho)$ the conjugation operator of Lemma 3.2. Let $\rho(r) = 1/2r$ $(r > 0)$. Given any $\rho_0 > 0$, for every $r \in (0, 1/2\rho_0)$ we have $f^{c(\rho(r))c(\rho(r))^*} \geq f^{c(\rho_0)c(\rho_0)^*}$ (see [25, p.205]). Hence, by (3.6) and Lemma 3.3 we have

$$\lim_{r\to 0} f_r = \lim_{r\to 0} (f^{c(\rho(r))c(\rho(r))^*})_r \geq \lim_{r\to 0} (f^{c(\rho_0)c(\rho_0)^*})_r = f^{c(\rho_0)c(\rho_0)^*}$$

Since $\rho_0 > 0$ is arbitrary, this implies that

$$\lim_{r\to 0} f_r \geq \sup_{\rho\to 0} f^{c(\rho)c(\rho)^*} = f^{cc^*} = \mathrm{cl}\ f, \tag{3.14}$$

where c is the conjugation operator corresponding to the coupling function φ of (3.10) (for the last equality see [10, p.286]). On the other hand, since the infimum of a function coincides with that of its l.s.c. hull and $\mathrm{cl}(f+g) = \mathrm{cl}f + g$ for any continuous finite functions g, we have $f_r = (\mathrm{cl}\ f)_r$ $(r>0)$, whence

$$\lim_{r\to 0} f_r = \lim_{r\to 0} (\mathrm{cl}\ f)_r \leq \mathrm{cl}\ f.$$

This, together with (3.14), gives (3.13).

Since $\lim_{r\to\infty} f_{[r]} = \sup_{r>0} f_{[r]}$, from [25, Cor. 3.6 and 3.2] it follows that Propositions3.2 and3.3 remain valid when one replaces f_r by $f_{[r]}$ in their statements.

Finally, let us mention that the regulazation operators $f \overset{r}{\longmapsto} f_r$ and $f \overset{[r]}{\longmapsto} f_{[r]}$ are characterized by the following properties:

$$\inf_{i\in I} f^i \overset{r}{\longmapsto} \inf_{i\in I} f^i{}_r, \quad \inf_{i\in I} f^i \overset{[r]}{\longmapsto} \inf_{i\in I} f^i{}_{[r]} \qquad (f^i: R^n \to \overline{R},\ i \in I)$$

$$f \dotplus d \overset{r}{\longmapsto} f_r \dotplus d, \quad f \dotplus d \overset{[r]}{\longmapsto} f_{[r]} \dotplus d \qquad (f: R^n \to \overline{R},\ d \in \overline{R})$$

$$\chi_{(x_0)} \overset{r}{\longmapsto} 1/2r \,\|\cdot - x_0\|^2, \quad \chi_{(x_0)} \overset{[r]}{\longmapsto} 1/r \,\|\cdot - x_0\| \qquad (x_0 \in R^n)$$

This follows from Theorem 3.1 of [41], using that $-f_r$ and $-f_{[r]}$ can be interpreted as certain conjugates.

4. The Φ-Conjugate of the difference of two functions

Let Φ be a nonempty family of real functions defined on an arbitrary set X, satisfying the condition $\Phi + \Phi \subset \Phi$ (i.e., being closed under addition). The conjugation operator (see Section 2) $c : f \in \overline{R}^X \to f^c \in \overline{R}^\Phi$ corresponding to the natural coupling function $\varphi: X \times \Phi \to R$, given by $\varphi(x, w) = w(x)$ $(x \in X, w \in \Phi)$, satisfies:

$$f^c(w) = \sup_{x \in X} \{ w(x) - f(x) \} \qquad (f \in \overline{R}^X, w \in \Phi),$$

$$g^{c^*}(w) = \sup_{w \in \Phi} \{ w(x) - g(x) \} \qquad (g \in \overline{R}^\Phi, x \in X).$$

This type of generalized conjugation was first introduced and studied by S. Dolecki and S. Kurcyusz [10]. In this Section, we will analyze the Φ-conjugate of the difference of two functions, generalizing the formula obtained by B.N. Pshenichnyi [34] (see also [11, 18]) for the convex conjugate. This formula was proved by J.B. Hiriart-Urruty [18] to be equivalent to Toland-Singer duality [47, 38, 48, 49, 39]. The same situation holds in the context of Φ-conjugation, under the assumption we have made that $\Phi + \Phi \subset \Phi$. The close connection existing between Toland-Singer type duality results (Section 3) and the formula on the Φ-conjugate of the difference of two functions is based on the assertions stated in the following lemma, whose proof is straightforward:

Lemma 4.1. Let $f, h : X \to \overline{R}$ and $w, v \in \Phi$. Then

a) $(f \dot{-} h)^c (w) = - \inf_{x \in X} \{ (f - w) \dot{-} h(x) \}$

b) $(f - w)^c (v) = f^c(w + v)$

c) $(f + w)^c (w) = - \inf_{x \in X} f(x).$

Statements a) and c) allow us to relate the Φ-conjugate of a difference to the infimum of a difference (and conversely). The interest of statement b) lies in that it permits to compute the Φ-conjugate of f-w, for any $w \in \Phi$, with the only information of f^c. Combining Lemma 4.1 with the results of Section 3, we can state the following properties of the Φ-conjugate of the difference of two functions (similarly, we could derive those results from the ones of this Section, using again Lemma 4.1):

Lemma 4.2. Let $f, h : X \to \overline{R}$ and $w \in \Phi$. Then

$$(f - h^{cc^*})^c (w) = \sup_{v \in \Phi} \{ f^c (w+v) - h^c (v) \}.$$

Corollary 4.1. Let $f, h : X \to \overline{R}$ and $w \in \Phi$. Then

$$(f - h)^c (w) \geq \sup_{v \in \Phi} \{ f^c (w+v) - h^c (v) \}.$$

Corollary 4.2. Let $f, h : X \to \overline{R}$. Then $(f - h^{cc^*})^c = (f^{cc^*} - h^{cc^*})^c$.

In the particular case when c is the convex conjugation operator, the preceding results have been obtained by M.Volle [53].

Lemma 4.3. Let $f, h : X \to \overline{R}$, $x_0 \in X$ and $w \in \Phi$. Then

$$(\chi_{\{x_0\}} - h)^c (w) = h(x_0) + w(x_0).$$

Proof: The supremum in

$$(\chi_{\{x_0\}} \dot{-} h)^c (w) = \sup \{ w(x) - (\chi_{\{x_0\}} (x) \dot{-} h(x)) \}$$

is attained at x_0.

Theorem 4.1. Let $h : X \to \overline{R}$. The following statements are equivalent:

1° $h = h^{cc^*}$.

2° $(f \dot{-} h)^c = (f \dot{-} h^{cc^*})^c$ $\quad (f \in \overline{R}^X)$.

3° $(f \dot{-} h)^c (w) = \sup_{v \in \Phi} \{ f^c (w + v) \dot{-} h^c (v) \}$ $\quad (f \in \overline{R}^X, w \in \Phi)$.

4° $(f \dot{-} h)^c = (f^{cc^*} \dot{-} h^{cc^*})^c$ $\quad (f \in \overline{R}^X)$.

These statements imply

5° There exists a formula expressing $(f - h)^c$ in terms of f^c

6° $(f \dot{-} h)^c = (f^{cc^*} \dot{-} h)^c$ $\quad (f \in \overline{R}^X)$.

Moreover, 5° and 6° are equivalent and, when the constant function $0 \in R^X$ satisfies the condition $0 = 0^{cc^*}$, they are equivalent also to 1°- 4°.

Proof: The implications 1°$\Rightarrow$ 2° $\Rightarrow$ .... $\Rightarrow$ 6° can be proved as the corresponding ones in Theorem 3.1. The implication 4°$\Rightarrow$ 2° is an immediate consequence of Corollary 4.2. To prove 2°$\Rightarrow$ 1°, take $f = \chi_{\{x_0\}}$ in 4°, for any $x_0 \in X$. Then, taking any $w \in \Phi$, by Lemma 4.3 and 2°, we have

$$h(x_0) + w(x_0) = (\chi_{\{x_0\}} \dot{-} h)^c (w) = (\chi_{\{x_0\}} \dot{-} h^{cc^*})(w) = h^{cc^*} (x_0) + w(x_0),$$

whence, since x_0 is arbitrary, $h = h^{cc^*}$.

The implication $6° \Rightarrow 5°$ is obvious.
Let us now prove the implication $6° \Rightarrow 1°$, under the assumption $0 = 0^{cc^*}$. Applying the operator c^* to the equality in $6°$, with $f=h$, we obtain

$$h^{cc^*} \dot{-} h \geq (h^{cc^*} \dot{-} h)^{cc^*} = (h \dot{-} h)^{cc^*} \geq 0^{cc^*} = 0,$$

whence $h^{cc^*} \geq h$, which gives $1°$.

Remark 4.1. In fact, the equivalence between statements $1°$- $4°$ and $5°$- $6°$ holds for any h if and only if $0 = 0^{cc^*}$, since $h = 0$ satisfies $6°$.

Among the conjugation operators we have studied in the preceding section, we will now examine that corresponding to the coupling function φ of (3.10) since, by a slight modification, it may be converted into a Φ-conjugation operator, with Φ satisfying the condition $\Phi + \Phi \subset \Phi$. Indeed, let $\Phi = \{ -\rho \| \cdot - y \|^2 + d \mid \rho > 0, y \in R^n, d \in R \}$.
Then, it is easy to check that, for any $\rho_1, \rho_2 > 0$, $y_1, y_2 \in R^n$ and $d_1, d_2 \in R$, one has:

$$(-\rho_1 \| \cdot - y_1 \|^2 + d_1) + (-\rho_2 \| \cdot - y_2 \|^2 + d_2) =$$

$$= -(\rho_1 + \rho_2) \| \cdot - (1/(\rho_1+\rho_2))(\rho_1 y_1 + \rho_2 y_2 \|^2 + (d_1 + d_2 - ((\rho_1 \rho_2)/(\rho_1+\rho_2)) \| y_1 - y_2 \|^2) \in \Phi.$$

By identifying any $-\rho \| \cdot - y \| + d \in \Phi$ with (ρ, y, d), we have the following formula for the Φ-conjugate of $f : R^n \to \overline{R}$:

$$f^c(\rho, y, d) = \sup_{y \in R^n} \{ -\rho \| x-y \|^2 + d - f(x) \} = d - f_r(y) \qquad (\rho > 0, y \in R^n, d \in R),$$

where $r = 1/2\rho$. It is also easy to show that f^{cc^*} coincides with the second conjugate corresponding to the coupling function of (3.10). Hence, the results in this Section admit an interpretation in terms of Moreau-Yosida approximates and lower semicontinuous hulls. In particular, the formula of $3°$ in Theorem 4.1 yields a relation between $(f - h)_r$ and the Moreau-Yosida approximates of f and h, under the assumption that h is l.s.c. and satisfies the quadratic growth condition.

5. Volle's duality

Throughout this Section, we will consider a special type of conjugation operator. Given two sets X, W and $G \subset X \times W$, the coupling function $\varphi : X \times W \to \overline{R}$ will be $\varphi = -\chi_G$. The corresponding conjugation operators $c : f \in \overline{R}^X \to f^c \in \overline{R}^W$ and $c^* : g \in \overline{R}^W \to g^{c^*} \in \overline{R}^X$ are thus given by

(5.1) $$f^c(w) = -\inf\{f(x) \mid x \in \Gamma^{-1}(w)\} \qquad (f \in \overline{R}^X, w \in W)$$

and $$g^{c^*}(x) = -\inf\{g(w) \mid w \in \Gamma(x)\} \qquad (g \in \overline{R}^W, x \in X),$$

where $\Gamma : X \to 2W$ is the multifunction, with inverse $\Gamma^{-1}: W \to 2^X$, whose graph is G, i.e.,

$$\Gamma(x) = \{w \in W \mid (x, w) \in G\} \qquad (x \in X),$$
$$\Gamma^{-1}(w) = \{x \in X \mid (x, w) \in G\} \qquad (w \in W).$$

We will also use the notations

$$\Gamma(A) = \bigcup_{x \in A} \Gamma(x), \quad \Gamma^{-1}(B) = \bigcup_{w \in B} \Gamma^{-1}(w) \qquad (A \subset X, B \subset W).$$

The second conjugate of any function $f : X \to \overline{R}$ satisfies:

(5.2) $$f^{cc^*}(x_0) = \sup_{w \in \Gamma(x_0)} \inf_{w \in \Gamma^{-1}(w)} f(x) \qquad (x_0 \in X).$$

Without loss of generality, we will assume that $\Gamma(x) \neq \emptyset$ $(x \in X)$ or, in other words, that $\Gamma^{-1}(W) = X$. Indeed, formula (5.1) shows that the only values of f which are relevant for the calculation of f^c are those which it takes on $\Gamma^{-1}(W)$. On the other hand, by (5.2) one has $f^{cc^*}(x_0) = -\infty$ whenever $\Gamma(x_0) = \emptyset$, for any $f : X \to \overline{R}$, whence this c-conjugation does not provide any useful information on f outside $\Gamma^{-1}(W)$.
This kind of conjugation operators, which are useful, e.g., in quasiconvex analysis, have been studied in [24, 40, 41, 42, 51, 52]. In [52], M. Volle obtained the following analogue to Toland-Singer duality formula, valid for any conjugation of type (5.1):

(5.3) $$\inf_{x \in X} \max\{f(x), -h(x)\} = \inf_{w \in W} \max\{h^c(w), -f^c(w)\},$$

where $f, h : X \to \overline{R}$ and the level sets of h, $S_\lambda(h) = h^{-1}([-\infty, \lambda]) = \{x \in X \mid h(x) \leq \lambda\}$ $(\lambda \in R)$ are assumed to be intersections of sets of the form $X \setminus \Gamma^{-1}(w)$ with $w \in W$, that is, $S_\lambda(h) = X \setminus \Gamma^{-1}(W \setminus \Gamma(S_\lambda(h)))$ $(\lambda \in R)$ (observe that, for any $A \subset X$, the smallest set containing A which can be expressed as an intersection of sets of the form $X \setminus \Gamma^{-1}(w)$ is just $X \setminus \Gamma^{-1}(W \setminus \Gamma(A))$). This property is equivalent to the equality $h = h^{cc^*}$ [24]. Without this assumption, the inequality $\leq$ still holds in (5.3) [52]. In this Section, we will show that Volle's duality formula (5.3) admits a treatment similar to that we have applied to Toland-Singer duality result in Section 3. In fact, M. Volle has proved [52, p. 127] that (5.3) implies Toland-Singer duality formula, since one can conveniently associate to any conjugation operator a new conjugation operator of type (5.1).

Lemma 5.1. [52, p.118]. Let $f, h : X \to \overline{R}$. Then

$$\inf_{x \in X} \max \{ f(x), -h^{cc^*}(x) \} = \inf_{w \in W} \max \{ h^c(w), -f^c(w) \}.$$

Corollary 5.1. [52, p.118]. Let $f, h : X \to \overline{R}$. Then

$$\inf_{x \in X} \max \{ f(x), -h(x) \} \leq \inf_{w \in W} \max \{ h^c(w), -f^c(w) \}.$$

Corollary 5.2. [52, p.119]. Let $f, h : X \to \overline{R}$. Then

$$\inf_{x \in X} \max \{ f(x), -h^{cc^*}(x) \} = \inf_{x \in X} \max \{ f^{cc^*}(x), -h^{cc^*}(x) \}.$$

In [52], M. Volle has used the following representation function $\eta_A : X \to \overline{R}$ of a set $A \subset X$:

$$\eta_A(x) = \begin{cases} -\infty & \text{if } x \in A \\ +\infty & \text{if } x \notin A \end{cases} \qquad (x \in X).$$

Letting $f = \eta_{\{x_0\}}$, for any $x_0 \in X$, in (5.3), one obtains the equality $-h(x_0) = -h^{cc^*}(x_0)$ (since $\eta^c_{\{x_0\}}(w) = -\eta_{\Gamma^{-1}(w)}(x_0)$), which proves that (5.3) holds for any $f : X \to \overline{R}$ if and only if $h = h^{cc^*}$. A different method of proof is employed in the next theorem, using the following lemma:

Lemma 5.2. [52, p. 118]. Let $A \subset X$. Then $\eta^{cc^*}{}_A = \eta_{X \setminus \Gamma^{-1}(W \setminus \Gamma(A))}$.

Theorem 5.1. Let $h : X \to \overline{R}$. The following statements are equivalent:

1° $h = h^{cc^*}$.

2° $\inf_{x \in X} \max \{ f(x), -h(x) \} = \inf_{x \in X} \max \{ f(x), -h^{cc^*}(x) \} \qquad (f \in \overline{R}^X).$

3° $\inf_{x \in X} \max \{ f(x), -h(x) \} = \inf_{w \in W} \max \{ h^c(w), -f^c(w) \} \qquad (f \in \overline{R}^X).$

4° $\inf_{x \in X} \max \{ f(x), -h(x) \} = \inf_{x \in X} \max \{ f^{cc^*}(x), -h^{cc^*}(x) \} \qquad (f \in \overline{R}^X).$

5° There exists a formula expressing $\inf_{x \in X} \max \{ f(x), -h(x) \}$ in terms of f^c.

6° $\inf_{x \in X} \max \{ f(x), -h(x) \} = \inf_{x \in X} \max \{ f^{cc^*}(x), - h(x) \} \qquad (f \in \overline{R}^X).$

Proof: Similarly to Theorem 4.1, each of the implications $1° \Rightarrow 2° \Rightarrow \ldots\ldots \Rightarrow 6°$ is either evident or an immediate consequence of the preceding results of this Section. To prove the implication $6° \Rightarrow 1°$, let $\lambda \in R$ and take $f = \eta_{S_\lambda(h)}$ in 6°. The left hand side gives

$$(5.4) \qquad \inf_{x\in X} \max \{ \eta_{S_\lambda(h)}(x), -h(x) \} = \inf_{x\in S_\lambda(h)} \{ -h(x) \} = - \sup_{x\in S_\lambda(h)} h(x) \geq -\lambda,$$

while the right hand side is, by Lemma 5.2,

$$(5.5) \qquad \inf_{x\in X} \max \{ \eta^{cc^*}{}_{S_\lambda(h)}(x), -h(x) \} = \inf_{x\in X} \max \{ \eta_{X\setminus \Gamma^{-1}(W\setminus \Gamma(S_\lambda(h)))}(x), -h(x) \} =$$

$$= - \sup \{ h(x) \mid x \in X \setminus \Gamma^{-1}(W \setminus \Gamma(S_\lambda(h))) \} .$$

Let $x \in X \setminus \Gamma^{-1}(W \setminus \Gamma(S_\lambda(h)))$. Then, by (5.5),6° and (5.4), we have $-h(x) \geq -\lambda$, i.e., $x \in S_\lambda(h)$. This proves the equality $S_\lambda(h) = X \setminus \Gamma^{-1}(W \setminus S_\lambda(h))$ $(\lambda \in R)$ (since the inclusion $\subset$ always holds), which is equivalent to 1°.

At this point, a natural question is whether there is some formula for the c-conjugate of $\max\{f, -h\}$ which is related to Volle's duality. In order to give a positive answer, we will assume in the sequel that the following condition holds: there exists an internal operation $\vee$ on W, such that

$$\Gamma^{-1}(w \vee v) = \Gamma^{-1}(w) \cap \Gamma^{-1}(v) \qquad (w, v \in W).$$

This assumption is the analogue to the condition $\Phi + \Phi \subset \Phi$ that we have imposed in Section 4. It holds, for example, when $(W, \geq)$ is a complete lattice and $\Gamma^{-1}: (W, \geq) \to (2^X, \supseteq)$ is a duality (see Section 1). The proof of the following lemma is easy.

Lemma 5.3. Let $f, h : X \to \overline{R}$ and $w, v \in W$. Then

a) $\max\{ f, -h\}^c (w) = - \inf_{x\in X} \max \{ \max \{ f(x), \eta_{\Gamma^{-1}(w)}(x) \}, -h(x) \}.$

b) $\max \{ f, \eta_{\Gamma^{-1}(w)} \}^c (v) = f^c(w \vee v).$

From Lemmas 5.3. and 5.1, we obtain:

Lemma 5.4. Let $f, h : X \to \overline{R}$ and $w \in W$. Then

$$\max \{ f, -h^{cc^*} \}^c (w) = \sup_{v\in W} \min \{ f^c(w \vee v), -h^c(v) \}.$$

Corollary 5.3. Let $f, h : X \to \overline{R}$ and $w \in \Phi$. Then

$$\max\{f, -h\}^c(w) \geq \sup_{v\in W} \min\{f^c(w \vee v), -h^c(v)\}.$$

Corollary 5.4. Let $f, h : X \to \overline{R}$. Then

$$\max\{f, -h^{cc^*}\}^c = \max\{f^{cc^*}, -h^{cc^*}\}^c$$

Using these results, we get the following analogue to Theorem 4.1:

Theorem 5.2. Let $h : X \to \overline{R}$. The following statements are equivalent:

1° $h = h^{cc^*}$

2° $\max\{f, -h\}^c = \max\{f, -h^{cc^*}\}^c \qquad (f \in \overline{R}^X).$

3° $\max\{f, -h\}^c(w) = \sup_{v\in W} \min\{f^c(w \vee v), h^c(v)\} \qquad (f \in \overline{R}^X, w \in W).$

4° $\max\{f, -h\}^c = \max\{f^{cc^*}, -h^{cc^*}\}^c \qquad (f \in \overline{R}^X).$

5° There exists a formula expressing $\max\{f, -h\}^c$ in terms of f^c.

6° $\max\{f, -h\}^c = \max\{f^{cc^*}, -h\}^c \qquad (f \in \overline{R}^X).$

Proof: Each of the implications $1° \Rightarrow 2° \Rightarrow \ldots\ldots \Rightarrow 6°$ is either obvious or an easy consequence of Lemma 5.4 and Corollary 5.4. To prove the implication $6° \Rightarrow 1°$, let us apply both sides of 6° to $w \in W$. In this way, after a change of signs, we get:

$$\inf_{x\in\Gamma^{-1}(w)} \max\{f(x), -h(x)\} = \inf_{x\in\Gamma^{-1}(w)} \max\{f^{cc^*}(x), -h(x)\} \qquad (f \in \overline{R}^X, w \in W).$$

Then, taking $\inf_{w\in W}$ of both sides of this equality, by $\Gamma^{-1}(W) = X$ we arrive at 6° of Theorem 5.1, which terminates the proof.

Conclusions

After giving a brief description of an abstract approach to generalized conjugation theory, we have examined Toland-Singer type duality results, relating them to Moreau-Yosida and Lipschitz regularizations of functions. This new approach to regularizations clarifies several aspects and permits to give easier proofs of old results, as well as to obtain some new properties. We have also studied the Φ-conjugate of the difference of two functions and an analogue to it for level set conjugations, which is related to a duality formula due to M. Volle.

These developments suggest that the generalized conjugation theory may serve as an useful tool in some branches of nonconvex analysis, where it could bring a new insight.

Acknowledgement: We are grateful to Mr. J. Gómez Urgellés for many useful discussions on Moreau-Yosida approximates, which stimulated our interest on this subject, and to Professor J. B. Hiriart-Urruty for several interesting remarks.

References

/1/ **Anderson J.R., Duffin W.N., Trapp G.E.**, Parallel subtraction of matrices, Proceedings of the National Academy of Sciences of the U.S.A. 69, 1972, pp. 2530-2531.

/2/ **Attouch H.**, Variational convergence for functions and operators, Pitman, London, 1984.

/3/ **Attouch H., Azé D., Wets R.**, On continuity properties of the partial Legendre-Fenchel transform: convergence of sequences of augmented Lagrangian functions, Moreau-Yosida approximates and subdifferential operators, in: J.B. Hiriart-Urruty(Ed.), Fermat Days 85: Mathematics for Optimization, North-Holland, Amsterdam - New York-Oxford-Tokyo, 1986, pp.1-42.

/4/ **Attouch H., Wets R.J.B.**, Isometries for the Legendre-Fenchel transform, Transactions of the American Mathematical Society 296, 1986, pp. 33-60.

/5/ **Balder E.J.**, An extension of duality-stability relations to non convex problems, SIAM Journal on Control and Optimization 15, 1977, pp.329-343.

/6/ **Ben-Tal A., Ben-Israel A.**, F-convex functions: properties and applications, in: S. Schaible, W.T. Ziemba (Eds), Generalized Concavity in Optimization and Economics, Academic Press, New York, 1981, pp. 301-334.

/7/ **Birkhoff G.**, Lattice Theory, American Mathematical Society, New York, 1948.

/8/ **Brézis H.**, Opérateurs maximaux monotones et semi-groupes de contraction dans les espaces de Hilbert, North-Holland, Amsterdam-New York-Oxford-Tokyo, 1973.

/9/ **Dolecki S., Kurcyusz S.**, Convexité généralisée et optimization, Comptes Rendus de Séances de l'Acedemie des Sciences de Paris 283, 1976, pp.91-94.

/10/ **Dolecki S., Kurcyusz S.**, On Φ-convexity in extremal problems, SIAM Journal on Control and Optimization 16, 1978, pp.277-300.

/11/ **Ellaia R., Hiriart-Urruty J.B.**, The conjugate of the difference of convex functions, Journal of Optimization Theory and Applications 49, 1986, pp. 493-498.

/12/ **Evers J.J., Van Maaren H.**, Duality principles in mathematics and their relation to conjugate functions, Nieuw Archiv voor Wiskunde 3, 1985, pp. 23-68.

/13/ **Franchetti C., Singer I.**, Deviation and farthest points in normed linear spaces, Revue roumaine de mathématiques pures et appliquées 24, 1979, pp. 373-381.

/14/ **Fujishige S.**, Theory of submodular programs: a Fenchel-type minmax theorem and subgradients of submodular functions, Mathematical Programming 29, 1984, pp. 142-155.

[15] **Gabay D.**, Minimizing the difference of two convex functions: Part I: Algorithms based on exact regularization, INRIA, working paper, 1982.

[16] **Hiriart-Urruty J.B.**, Extension of Lipschitz functions, Journal of Mathematical Analysis and Applications 77, 1980, pp. 539-554.

[17] **Hiriart-Urruty J.B.**, Lipschitz r-continuity of the approximate subdifferential of a convex function, Mathematica Scandinavica 47, 1980, pp. 123-134.

[18] **Hiriart-Urruty J.B.**, Generalized differentiability, duality and optimization for problems dealing with differences of convex functions, in: J.Ponstein (Ed.), Convexity and Duality in Optimization, Springer Verlag, Berlin-Heidelberg-New York-Tokyo, 1985, pp. 37-70.

[19] **Hiriart-Urruty J.B.**, A general formula on the conjugate of the difference of functions, Canadian Mathematical Bulletin 29, 1986, pp.482-485.

[20] **Hiriart-Urruty J.B.,Mazure M.L.**, Formulations variationelles de l'addition parallèle et la soustraction parallèle d'opérateurs semi-définis positifs, Comptes Rendus des Séances de l'Academie des Sciences de Paris 302, 1986, pp.527-530.

[21] **Hiriart-Urruty J.B., Plazanet Ph.**, Moreau's decomposition theorem revisited, Preprint, 1987.

[22] **Hörmander L.**, Sur la fonction d'appui des ensembles convexes dans un espace localement convexe, Arkiv für Mathematik 3, 1954, pp. 181-186.

[23] **Laurent P.J.**, Approximation et optimisation, Hermann, Paris, 1972.

[24] **Martínez-Legaz J.E.**, Conjugación asociada a un grafo, in: Actas IX Jornadas Matemáticas Hispano-Lusas, Universidad de Salamanca, Salamanca, 1982, pp.837-839.

[25] **Martínez-Legaz J.E.**, On lower subdifferentiable functions, in: K.-H. Hoffmann, J.B. Hiriart-Urruty, C. Lemarechal, J. Zowe (Eds.), Trends in Mathematical Optimization. 4th French-German Conference on Optimization, Birkhäuser Verlag, Basel-Boston, 1988, pp.197-232.

[26] **Martínez-Legaz J.E.**, Quasiconvex duality theory by generalized conjugation methods, Optimization, to appear.

[27] **Martínez-Legaz J.E., Singer I.**, Dualities between complete lattices, INCREST, Preprint Series in Mathematics, n° 19/1987, (submitted).

[28] **Mazure M.L.**, La soustraction parallèle d'opérateurs interprétée comme déconvolution de formes quadratiques convexes, Optimization 18, 1987, pp. 465-484.

[29] **Mc Shane E.J.**, Extension of range of functions, Bulletin of the American Mathematical Society 40, 1934, pp.837-842.

[30] **Moreau J.J.**, Fonctionelles convexes, Séminaire sur les Equation aux Dérivées Partielles II, Collège de France, Paris, 1966-1967.

[31] **Moreau J.J.**, Inf-convolution, sous-additivité, convexité des fonctions numériques, Journal de Mathématiques Pures et Appliquées 49, 1970, pp. 109-154.

[32] **Penot J.P., Volle M.**, On strongly convex and paraconvex dualities, this volume.

[33] **Poljak B.T.**, Existence theorems and convergence of minimizing sequences in extremum problems with restrictions, Soviet Mathematics Doklady 166, 1966, pp. 72-75.

/34/ **Pshenichnyi B.N.**, Lecons sur les Jeux Différentiels, Cahiers de l'IRIA, vol. 4, 1971.

/35/ **Pshenichnyi B.N., Daniline Y.**, Méthodes Numériques dans les Problèmes d'Extremum, Mir, Moscou, 1977.

/36/ **Rockafellar R.T.**, Convex analysis, Princeton University Press, Princeton 1970.

/37/ **Rockafellar R.T.**, Augmented Lagrange multiplier functions and duality in nonconvex programming, SIAM Journal on Control and Optimization 12, 1974, pp. 268-285.

/38/ **Singer I.**, A Fenchel-Rochafellar type duality theorem for maximization, Bulletin of the Australian Mathematical Society 29, 1979, pp.193-198.

/39/ **Singer I.**, Maximization of lower semi-continuous convex functionals on bounded subsets of locally convex spaces. II: Quasi-Lagrangian duality theorems, Resultate der Mathematik 3, 1980, pp. 235-248.

/40/ **Singer I.**, Generalized convexity, functional hulls and applications to conjugate duality in optimization, in: G. Hammer, D. Pallaschke (Eds.), Selected Topics in Operations Research and Mathematical Economics, Springer Verlag, Berlin-Heidelberg-New York, 1984, pp. 49-79.

/41/ **Singer I.**, Conjugation operators, ibidem, pp. 80-97.

/42/ **Singer I.**, Some relations between dualities, polarities, coupling functionals and conjugations, Journal of Mathematical Analysis and Applications 116, 1986, pp.1-33.

/43/ **Singer I.**, A general theory of dual optimization problems, Journal of Mathematical Analysis and Applications 116, 1986, pp.77-131.

/44/ **Singer I.**, Generalizations of convex supremization duality, in: B.L. Lin, S. Simons (Eds.). Nonlinear and Convex Analysis, Marcel Dekker, New York, 1987, pp. 253-270.

/45/ **Singer I.**, Infimal generators and dualities between complete lattices, Annali di Matematica Pura ed Applicata 148, 1987, pp. 289-352.

/46/ **Stoer J., Witzgall C.**, Convexity and optimization in finite dimensions. I, Springer Verlag, Berlin-Heidelberg-New York, 1970.

/47/ **Toland J.F.**, Duality in nonconvex optimization, Journal of Mathematical Analysis and Applications 66, 1978, pp. 399-415.

/48/ **Toland J.F.**, A duality principle for nonconvex optimization and the calculus of variations, Archive for Rational Mechanics and Analysis 71, 1979, pp. 41-61.

/49/ **Toland J.F.**, On subdifferential calculus and duality in nonconvex optimization, Bulletin de la Societé Mathematique de France 60, 1979, pp. 177-183.

/50/ **Vial J.P.**, Strong and weak convexity of sets and functions, Mathematics of Operations Research 8, 1983, pp. 279-311.

/51/ **Volle M.**, Conjugaison par tranches, Annali di Matematica Pura ed Applicata 139, 1985, pp. 279-311.

/52/ **Volle M.**, Contributions a la dualité en optimisation et a l'epi-convergence, Université de Pau et des Pays de l'Adour, Thèse, 1986.

/53/ **Volle M.**, Concave duality: Application to problems dealing with difference of functions, Mathematical Programming 41, 1988, pp. 261-278.

/54/ **Whitney H.**, Analytic extensions of differentiable functions defined in closed sets, Transactions of the American Mathematical Society 36, 1934, pp. 63-89.

ON STRONGLY CONVEX AND PARACONVEX DUALITIES

J. P. Penot - M. Volle

Abstract

In this paper the notions of strongly convex functions and paraconvex functions are examined under the light of generalized dualities; exact generalized conjugations for strongly convex functions and paraconvex functions are established and a condition ensuring the paraconvexity of the value function of a perturbed constrained minimization problem is given. Furthermore a dual problem with respect to generalized conjugation is introduced and a condition ensuring the absence of duality gap is given.

Introduction

Among the generalized concepts of convexity, the notions of strongly convex functions and paraconvex functions take an important place (see [3], [7], [16], [27], [29], [31], [35] for instance). It is the purpose of this paper to examine these notions under the light of generalized dualities. Let us first recall some basic facts about these notions.

Given an Hilbert space Z and a real number r, an extended real-valued function f defined on Z is said to be r-convex if it fulfills the following inequality

(1) $$f(z_t) \geq (1-t)\, f(z_0) + tf(z_1) - (r/2\ t)(1-t) \| z_0 - z_1 \|^2$$

whenever

$$f(z_0), f(z_1) \in \mathrm{IR},\ t \in]0,1[,\ z_t = (1-t)\, z_0 + tz_1.$$

If the real number is positive f is more than convex and, following the terminology of J.P. Vial [28], [29] we call it r-strongly convex.

When the real number r = -s is negative, f may be less than convex. So we call it s-paraconvex, accordingly with the terminology of S. Rolewicz [27].

If -f is r-strongly convex (resp. r-paraconvex) we shall say that f is r-strongly concave (resp. r-paraconcave).

Due to the Hilbert space structure, it is well known that inequality (1) amounts to the convexity of the function $x \to f(x) - (r/2) \| x \|^2$ Let us give some examples.

1) The quadratic functional

$$q_A(z) = 1/2 \langle Az, z \rangle$$

associated to the continuous linear operator $A : Z \to Z$ is r-paraconvex with $r = \| A \|$.
If A is r-elliptic (r>0), that is for any $z \in Z$,

$$\langle Az, z \rangle \geq r \| z \|^2$$

then q_A is r-strongly convex. In particular, if the dimension of Z is finite, A is positive definite and r is the lowest eigenvalue of A, then q_A is r-strongly convex.
2) If S is any subset of Z and

$$d_S(z) = \inf \{ \| x - z \| : x \in S \}$$

the distance function to S then $1/2\ d_S^2$ is 1-paraconcave whatever S may be (Asplund [1]).
The paper is organized as follows. In § I we characterize the usual Fenchel conjugate of a strongly convex function. This enables us to point out some classes of continuously differentiable functions which are invariant by the Fenchel transformation.
After a review of the use of generalized conjugation in optimization (§ II), exact generalized conjugations for strongly convex functions and paraconvex functions are presented in § III. Applications are given in quadratic programming and for solving infimal convolution equations.
In a more general setting, a condition ensuring the paraconvenxity of the value function of a perturbed constrained minimization problem is given in terms of paraconvexity of the constraint multifunction in the sense of Rolewicz [26]. We present some methods of construction of such multifunctions, in particular in the case of inequality constraints. When the value function is paraconvex it is natural to consider the dual problem with respect to our generalized conjugation and to give a condition ensuring the absence of duality gap and a condition for the existence of optimal dual solutions. This is what is done in § IV.

I. Fenchel conjugation, strong convexity and paraconvexity

The purpose of this section is to show how, in some sense, the Fenchel conjugation exchanges the concepts of strong convexity and paraconvexity. So we construct some interesting classes of continuously differentiable convex functions which are stable by taking Fenchel conjugates.
Let us recall the expression of the Fenchel conjugate f^* of f at any point $y \in Z$:

$$f^*(y) = \sup_{z \in Z} \{ \langle y, z \rangle - f(z) \}.$$

We shall need the following notion due to Moreau [18] (see [21] Def. 1.4 for the related notion of index of convexity), concerning functions from Z to $]-\infty, +\infty]$: a convex function φ_1 is <u>more convex</u> than a convex function φ_2 (or, equivalently, φ_2 is <u>less convex</u> than φ_1) if there exists a convex function γ such that $\varphi_1 = \varphi_2 + \gamma$. Then, given a convex proper lower semicontinuous function φ , the following assertions are equivalent ([18] p 289):

a) φ is less convex than $1/2 \|\ \|^2$
b) φ^* is more convex than $1/2 \|\ \|^2$

Let us give a short proof of this assertion using the Pshenichnyi-Hiriart-Urruty's formula

$$(g - h)^*(z) = \sup_{w \in Z} [g^*(w + z) \dot{-} h^*(w)] \quad \forall z \in Z$$

for $g, h : Z \to \mathbb{R} \cup \{+\infty\}$ with h convex proper and l.s.c., where $r \dot{-} s$ the extension of the subtraction given by $r \dot{-} s = +\infty$ as soon as $r = +\infty$ or $s = -\infty$ and $r \dot{-} s = -(s \dot{-} r)$. If a) holds we can write $g := 1/2 \|\,.\,\|^2 = \varphi + h$ for some convex function h which necessarily finitely valued and u.s.c. hence continuous. Thus $\varphi = g - h$ and

$$\varphi^*(z) = \sup [\, 1/2 \|w+z\|^2 - h^*(w) : w \in Z \,] = 1/2 \|z\|^2 + (h^* - 1/2 \|\ \|^2)^* (z)$$

and b) holds. Conversely if $\varphi^* = 1/2 \|\ \|^2 + k$ with k convex, as $1/2 \|\ \|^2$ has domain Z and is continuous we have

$$\varphi(z) = \varphi^{**}(z) = \inf [\, 1/2 \|z-w\|^2 + k^*(w) : w \in Z \,] = 1/2 \|z\|^2 - (k^* + 1/2 \|\ \|^2)^* (z)$$

hence $\varphi + h = 1/2 \|\ \|^2$ with $h = (k^* + 1/2 \|\ \|^2)^*$ convex and finitely valued.
Now we can characterize strongly convex functions in terms of their conjugates. The characterization we give is a special case of [3] Prop. 1.8 which in turn enters into the general duality between rotundity and smoothness (Asplund and Rockafellar [2]).

Theorem 1. Let f be a proper lower semicontinuous convex function and let $r > 0$. The following properties are equivalent:

i) f is r-strongly convex
ii) f^* is $1/r$ paraconcave and finitely valued.

Proof: i) $\Rightarrow$ ii) By setting $g := f - r/2 \|\ \|^2$ we obtain a convex function and we have $1/r\, f = 1/2 \|\ \|^2 + 1/r\, g$; hence $r^{-1} f$ is more convex than $1/2 \|\ \|^2$ and, consequently, $(r^{-1} f)^*$ is less convex than $1/2 \|\ \|^2$. Then there exists a convex function k such that $1/2 \|\ \|^2 = (1/r\, f)^* + k$. It follows that $(1/(2r)) \|\ \|^2 = f^* + h$ where $h(y) := r\, k(y/r)$ is a finitely valued convex function, as well as f^*. Then $h - (1/(2r)) \|\ \|^2 = -f^*$ and f^* is $1/r$ -paraconcave.

ii) $\Rightarrow$ i) By assumption $h := -f^* + (1/(2r)) \| \|^2$ is convex and finite valued. In this way $r f^*$ is less convex than $1/2 \| \|^2$ so that $(r f^*)^*$ is more convex than $1/2 \| \|^2$. We have then $(r f^*)^* = 1/2 \| \|^2 + k$ where k is a convex function. It follows that $f = f^{**} = r/2 \| \|^2 + g$ where $g(x) := 1/r\, h(rx)$ is convex. Hence f is r-strongly convex. □

An easy consequence of Theorem 1 is that, for each positive real number r, the class F_r of finitely valued l.s.c. functions that are both r-strongly convex and $1/r$ -paraconcave is stable with respect to the Fenchel transformation.

For describing these classes $(F_r)_{r>0}$ some basic facts about the notion of inf-convolution will be usefull.

Let us recall that for $f, g : Z \to]-\infty, +\infty]$,

$$(f \,\square\, g)(z) = \inf_{x \in Z} \{ f(z - x) + g(x) \}$$

denotes the infimal convlution of f and g on the space Z. Denoting by Γ_0 the set of l.s.c. proper convex functions defined on Z, it is well known that each function that belongs to the class

$$D = 1/2 \| \|^2 \,\square\, \Gamma_0 := \{1/2 \| \|^2 \,\square\, f : f \in \Gamma_0\}$$

is convex, finitely-valued and continuously differentiable on Z [18]. Moreover, the decomposition Theorem of Moreau says that for any $f \in \Gamma_0$,

$$f \,\square\, 1/2 \| \|^2 + f^* \,\square\, 1/2 \| \|^2 = 1/2 \| \|^2.$$

The converse property has recently been established [12]:

Theorem 2. Let g, h be convex functions such that $g + h = 1/2 \| \|^2$. Then, there exists a unique $f \in \Gamma_0$ such that $g = 1/2 \| \|^2 \,\square\, f$, $h = f^* \,\square\, 1/2 \| \|^2$.

Now we are in a position to describe the classes F_r for $r>0$ introduced above.

Proposition 3. Denoting by A the class of continuous affine functionals and setting $D = 1/2 \| \|^2 \,\square\, \Gamma_0$ we have

$$F_r = \begin{cases} \emptyset & \text{if } r > 1 \\ 1/2 \| \|^2 + A & \text{if } r = 1 \\ r/2 \| \|^2 + (1/r - r) D = (1/(2r)) \| \|^2 - (1/r - r) D & \text{if } 0 < r < 1 \end{cases}$$

Proof: If $f \in F_r$ there exist two convex functions g, h such that

$$(2) \qquad f - r/2 \| \|^2 = g$$

(3) $$-f + (1/(2r)) \| \ \|^2 = h.$$

As f is finitely valued we obtain

(4) $$1/2\,(1/r - r) \| \ \|^2 = g + h.$$

Now the convexity of $g + h$ implies $1/r - r \geq 0$, i.e. $0 < r \leq 1$. Hence $F_r = \emptyset$ for $r > 1$.
Let us study the case $r = 1$.
By (4) we have $g + h = 0$ from what it follows that g and h are both convex and concave. To obtain the continuity of g note that the lower semicontinuity of f and the relation (2) entail the lower semicontinuity of g. Moreover g is finitely valued. Now it is well known that a convex l.s.c. finitely valued function is continuous (e.g. [8]). Consequently g and h are affine continuous and $F_1 \subset A + 1/2 \| \ \|^2$. Conversely, if $f \in A + 1/2 \| \ \|^2$ then $f - 1/2 \| \ \|^2$ and $-f + 1/2 \| \ \|^2$ belong to A from what it follows that f is finitely and both 1-strongly convex and 1-paraconcave : $f \in F_1$.
We conclude the proof with the case $0 < r < 1$. For $f \in F_r$, setting $c = (1/r - r)$ we have by (4) $(1/c)\,g + (1/c)\,h = 1/2 \| \ \|^2$. From Theorem 2 there exists $k \in \Gamma_0$ such that $(1/c)\,g = k \,\square\, (1/2) \| \ \|^2 1/c\,h = k^* \,\square\, 1/2 \| \ \|^2$. By using (2) and (3) we have then

(5) $$f \in r/2 \| \ \|^2 + (1/r - r)\,D$$

(6) $$f \in (1/(2r)) \| \ \|^2 - (1/r - r)\,D.$$

Conversely, let us suppose that (5) holds. We see immediately that f is r-strongly convex and finitely valued. It remains to show that f is $1/r$ -paraconcave. But by (5) there exists $j \in \Gamma_0$ such that $-f + (1/(2r)) \| \ \|^2 = c\,(1/2 \| \ \|^2 - j \,\square\, 1/2 \| \ \|^2) = c\,(j^* \,\square\, 1/2 \| \ \|^2)$, by Moreau's decomposition Theorem. The fact that (6) entails $f \in F_r$ can be proved in the same way.

□

Making use of the concepts of strongly convex functions and paraconvex functions, we have pointed out a decreasing family $(F_r)_{0<r\leq 1}$ of classes of continuously differentiable convex functions stable with respect to the Fenchel transformation. Before studyng more specific dualities for strongly convex and paraconvex functions, we recall some basic facts about generalized conjugation.

II. A review of the use of generalized conjugation in optimization

The usefulness of generalized conjugation in duality theory has been first observed by J.J.Moreau [18] and developed by many authors [4] [7] [9] [13] [19] [21] [23] [25] [28] [33]...

Let us recall the main notions related to this approach. Given any sets Y, Z and a coupling functional $c : Y \times Z \to \overline{IR}$ the classes of c-affine functions and c-regular functions on Z are respectively defined by

$$A_c = \{ c(y;.) + s : y \in Y, s \in R \}$$

$$\Gamma_c = \{ \sup_{h \in H} h \ : H \subset A_c \}$$

while the c-conjugate of $q : Z \to \overline{IR}$ is given by

$$q^c(y) = \sup_{z \in Z} \{c(y,z)-q(z)\}.$$

A similar formula is used to define the c-conjugate of any function from Y to $\overline{IR}$ as a function from Z to $\overline{IR}$. The c-biconjugate $q^{cc} := (q^c)^c$ is always c-regular. More precisely

$$q^{cc} = \sup \{ h \in A_c : h \leq q \}$$

so that $q=q^{cc}$ if and only if q is c-regular.
Beside the usual framework where Y and Z are locally convex vector spaces (l.c.v.s.) in separated duality with respect to the bilinear mapping $c(y,z)= \langle y, z \rangle$ and where $A_c = A$ and $\Gamma_c = \Gamma_0 \cup \{+ \infty\}^Z \cup \{- \infty\}^Z$, let us give some other examples.

a) <u>Quasiconvex duality</u>
Let Y, Z be l.c.v.s. in duality. A mapping f from Z to $\overline{IR}$ is called quasiconvex if all its sublevel sets

$$\{f \leq r\} := \{z \in Z : f(z) \leq r\} \ r \in IR$$

are convex; f is said to be quasiconcave if - f is quasiconvex. There exist several approaches to quasiconvex duality theory [5] [14] [20] [23] [25] [28]. In [25] we take advantage of the fact that for any nondecreasing function $h : IR \to \overline{IR}$ and $y \in Y$, the mapping $h \circ y := h \circ \langle y,\cdot \rangle$ is both quasiconvex and quasiconcave: it is quasi-affine. Then we consider the coupling functional

$$c(h \circ y, z) := h(\langle y, z \rangle)$$

and show that f is c-regular iff f is evenly quasiconvex, that is to say $\{f \leq r\}$ is an intersection of open half spaces for each $r \in IR$ ([10]). When the nondecreasing functions h are supposed to be l.s.c. then the class of c-regular functions coincide with the class of l.s.c. quasiconvex functions. More generally it is shown in [32] that the generalized Moreau-Fenchel conjugation is versatile enough to characterize functions whose sublevel sets are closed with respect to a given Moore closure operator (convex hull, topological closure, decreasing hull,......).

b) Exact duality for α-Hölder functions

In [15] Martinez-Legaz considers the case where $Y = Z = \mathbb{R}^n$, $\alpha \in]0, 1]$, $M > 0$, and

$$c(y, z) = -M \| y - z \|^{\alpha}.$$

Then the class of c-regular functions is characterized as the class of α-Hölder functions f with constant M:

$$| f(x) - f(z) | \leq M \| x - z \|^{\alpha}.$$

In fact this is still valid in a mere metric space if one takes $c(y, z) = -M\, d^{\alpha}(y, z)$, d being the distance function of the space. □

The main interest of the generalized Moreau-Fenchel conjugation lies in the pertubational approach to a minimization problem

$$(\mathbf{P}) \quad \min \{ f(x) : x \in \Gamma(b) \}$$

where X is a set, $f : X \to \mathbb{R}$ is the objective function, $\Gamma : Z \to X$ is the constraint multifunction and $b \in Z$. The use of generalized conjugation will reveal a dual problem, provide sensitivity results, and yield an estimate of the value inf(**P**) of (**P**). We shall see later that it may also make easier the search of optimal solutions to (**P**). To obtain the dual problem one considers the value function

$$v(z) = \inf \{ f(x) : x \in \Gamma(z) \}.$$

The c-dual problem of (**P**) is then defined as

$$(\mathbf{Q}_c) \quad \max \{ c(y, b) \dot{-} v^c(y) : y \in Y \}$$

To avoid the use of the operation $\dot{-}$ we suppose that the coupling functional c is finitely valued, for this will be the case in the application we have in view here. The values of (**P**) and $(\mathbf{Q}_c)$ are related by

$$\inf(\mathbf{P}) = v(b) \geq v^{cc}(b) = \sup(\mathbf{Q}_c).$$

To describe the set $S(Q_c)$ of optimal solutions of (Q_c), recall that the c-subdifferential of $q : Z \to IR \cup \{+\infty\}$ at a point $a \in Z$ where q is finite is defined by

$$\partial_c q(a) = \{ y \in Y : \forall z \in Z : q(z) - q(a) \geq c(y, z) - c(y, a) \}.$$

Then if the value of (Q_c) is finite we have

$$S(Q_c) = \partial_c v^{cc}(b).$$

In particular, if $\partial_c v(b) \neq \emptyset$ then the duality gap between (P) and (Q_c) is zero and $S(Q_c) = \partial_c v(b)$.

We represent below several kinds of c-affine minorants. These figures show that the duality gap can be reduced by taking an appropriate choice of the coupling functional.

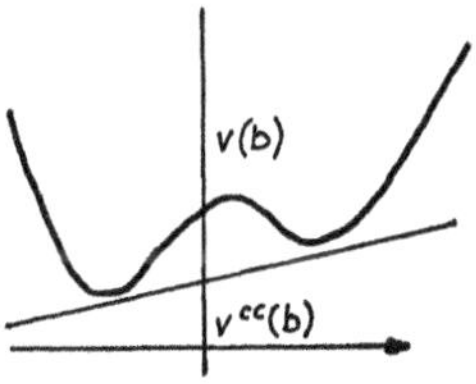

Usual affine minorant

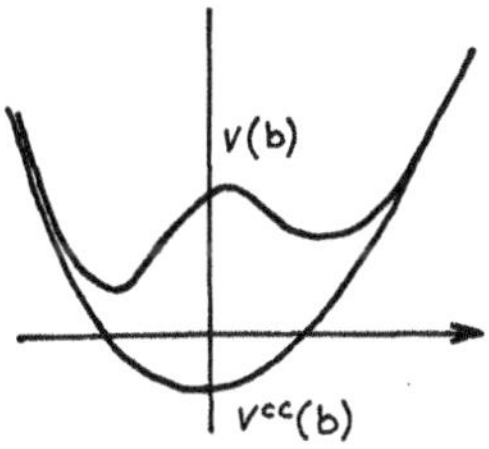

Convex c-affine minorant

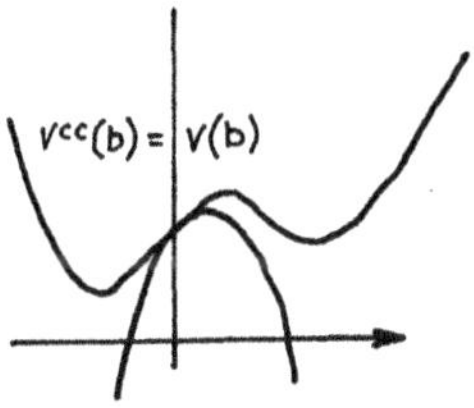

Concave c-affine minorant

III. Exact dualities for strongly convex functions and paraconvex functions

Our purpose is to characterize strongly convex functions and paraconvex functions as regular functions with respect to a coupling functional. Given a real number r and a Hilbert space $Y = Z$ we first consider

$$\gamma : Y \times Z \to IR \quad \gamma(y, z) = \langle y, z \rangle - r/2 \, \| z \|^2$$

We have then for any extended real-valued function q (resp.p) defined on Z (resp. Y)

$$q^\gamma(y) = \sup_{z \in Z} \{ \langle y, z \rangle - r/2 \, \| z \|^2 - q(z) \}$$

$$(\text{resp. } p^{\gamma}(z) = \sup_{y \in Y} \{ \langle y, z \rangle - r/2 \| z \|^2 - p(z) \})$$

so that

$$q^{\gamma} = (q + r/2 \ \| \ \|^2)^*$$

$$(\text{resp. } p^{\gamma} = p^* - r/2 \| \ \|^2)$$

and finally

(7) $$q^{\gamma\gamma} = (q + r/2 \| \ \|^2)^{**} - r/2 \| \ \|^2$$

Another adequate coupling functional is

$$c : Y \times Z \to \overline{IR} \quad c(y, z) = -r/2 \| y - z \|^2.$$

Here we have

$$q^c(y) = \sup_{z \in Z} \{ - r/2 \| y - z \|^2 - q(z) \} = -r/2 \| y \|^2 + \sup_{z \in Z} \{ r \langle y, z \rangle - r/2 \| z \|^2 - q(z) \}$$

so that

$$q^c(y) = -r/2 \| y \|^2 + (q + r/2 \| \ \|^2)^* (ry)$$

and, by an easy computation, we get for $r \neq 0$,

(8) $$q^{cc} = (q + r/2 \| \ \|^2)^{**} - r/2 \| \ \|^2 = q^{\gamma\gamma}$$

This last formula announced in [19] has also been obtained by Martinez-Legaz [16]. See also [7] for a slightly different approach). Hence we have proved.

Proposition 4. For any $r \in IR$ and any proper function q, the following properties are equivalent:

a) $q + r/2 \| \ \|^2 \in \Gamma_0$
b) q is γ-regular
c) q is c-regular.

and, for $r \neq 0$,
The choice of c may be justified by the fact that c is symmetric. On the other hand, using

γ we can include the classical Fenchel duality by taking $r = 0$.
Let us observe that in both cases the c-affine functions

$$z \to -r/2 \| y - z \|^2 + s$$

and the corresponding γ-affine functions satisfy the equality in (1): an interesting analogy with the class of usual affine functionals.

III.1 - Application to quadratic minimization

The next example shows that solving the generalized dual problem (Q_γ) makes easier the search of optimal solutions of the primal problem.
Let D be a symmetric n x n matrix, $l \in IR^n$, $b \in IR^m$, $A : IR^n \to IR^m$ linear and onto. We consider the quadratic program

$$(\mathbf{P}) : \mathrm{Min}\ \{1/2 \langle Dx, x \rangle - \langle l, x \rangle : Ax = b\}$$

and we assume the existence of a real number r such that

$$D + r\, A^T A \text{ is positive definite (p.d.).}$$

This condition holds in each of the following two cases:

(i) with $r = 0$ or r negative and $|r|$ small enough if D is p.d.
(ii) with $r > 0$ large enough if D is p.d. on the kernel of A ([6] p 296)

So, the results below are valid for $r > 0$ large enough if D is p.d. on ker A, and for $r < 0$ with $| r |$ small enough if D is p.d..
The γ-conjugate of the value function

$$v(z) = \inf \{ q_D(x) - \langle l, x \rangle : Ax = z \}$$

associated to $(\mathbf{P})$ is given by

$$\begin{aligned} v^\gamma(y) &= \sup_{z \in IR^m} \left\{ \langle y, z \rangle - r/2 \| z \|^2 + \sup \{ \langle l, x \rangle - q_D(x) : Ax = z \} \right\} \\ &= \sup_{z \in IR^m} \{ \langle l, x \rangle - q_D(x) + \langle y, Ax \rangle - r/2 \| Ax \|^2 \} \\ &= \sup_{z \in IR^m} \{ \langle l + A^T y, x \rangle - q_{D+rA^TA}(x) \} \\ &= (q_{D+rA^TA})^* (l + A^T y) \end{aligned}$$

As $D + rA^TA$ is p.d. we have then (e.g.[25] p. 208)

$$v^{\gamma}(y) = q_{(D+rA^TA)^{-1}}(1 + A^Ty).$$

Consequently the dual problem (Q_γ) consists in maximizing over the whole space $\mathbb{R}^m$ the concave differentiable function

$$y \to \langle y, b \rangle - r/2 \, \| b \|^2 - q_{(D+rA^TA)^{-1}}(1 + A^Ty).$$

the gradient of which vanishes at $\bar{y}$ iff

(9)
$$b = A\,(D + rA^TA)^{-1}(1 + A^T\bar{y}).$$

As A is onto and as $(D+rA^TA)^{-1}$ is definite positive, such a $\bar{y}$ does exist and is unique so that (Q_γ) has a unique optimal solution $\bar{y}$ completely determined by (9). Moreover, taking $\bar{x} = (D+r\;A^TA)^{-1}\;(1 + A^T\bar{y}\;)$ we obtain a feasible solution for (**P**) and a straightforward computation shows that $q_D(\bar{x}) - \langle 1, \bar{x} \rangle = \sup\,(Q_\gamma)$. Hence

(10)
$$\sup\,(Q_\gamma) \le \inf\,(P) \le q_D(\bar{x}) - \langle 1, \bar{x} \rangle = \sup\,(Q_\gamma)$$

so that $\bar{x}$ is an optimal solution of (**P**).

III.2 - Application to infimal convolution equations

Recently ([11]) the Fenchel conjugation has been used for solving the infimal convolution equation

(E)
$$f \,\square\, \xi = g$$

where f and g are given and convex. Here we consider the case $f = r/2 \,\|\;\|^2$, $r \in \mathbb{R}$, $r \neq 0$. Hence f may be concave and we do not make any assumption on g (we only suppose that -g is a proper function defined on the Hilbert space Z). By setting

$$c\,(y, z) = -r/2 \,\| y - z \|^2$$

we rewrite (E) as

$$\xi^c = -\,g.$$

By taking the c-biconjugate in both sides in the above formula we get

$$\xi^c = \xi^{ccc} = (-g)^{cc}$$

so that

$$-g = (-g)^{cc},$$

a necessary condition for the set of solutions S(E) of E to be nonempty. It can be written using Proposition 4 as

$$g \in r/2 \| \ \|^2 - \Gamma_0.$$

In fact the c-regularity of $-g$ suffices for having $S(E) \neq \emptyset$ as for $\xi_0 := (-g)^c$ we have then $(\xi_0)^c = -g$.
Any other solution ξ of (E) satisfies

$$\xi^c = -g \Rightarrow \xi \geq \xi^{cc} = \xi_0.$$

Thus when $\xi \in \Gamma_0 - r/2 \| \ \|^2$ we have $\xi = \xi^{cc} = \xi_0$. Let us give the expression of ξ_0:

$$\xi_0(z) = (-g)^c(z) = \sup_{x \in Z} \{g(y) - r/2 \| y - z \|^2\}$$

that is the deconvolution of g by $r/2 \| \ \|^2$, $g \,\Box\, r/2 \| \ \|^2$ introduced in the convex case in [11] (see also [34]). Hence we have obtained (see also [16] Proposition 3.1 for a different approach obtained independently).

Theorem 5. Given a non zero real number r and a function g defined on an Hilbert space, with $g \neq -\infty$, $g(z) < +\infty$ for each z in Z, the equation

$$r/2 \| \ \|^2 \Box\, \xi = g$$

has a solution iff $g \in r/2 \| \ \|^2 - \Gamma_0$ (i.e. iff g is r-concave and u.s.c.).
In this case the deconvolution

$$\xi_0 = g \,\Box\, r/2 \| \ \|^2 := -((-g) \,\Box\, r/2 \| \ \|^2).$$

is the smallest solution and the unique solution belonging to $\Gamma_0 - r/2 \| \ \|^2$.

We must pay a particular attention to the case $r=1$ and $\xi \in \Gamma_0$ for then $g \in 1/2 \| \ \|^2 \Box\, \Gamma_0 = D$, a class we have used intensively before. We know that D is a class of continuously differentiable convex functions. Let us present now a characterization of the class D.

Theorem 6 . D is the class of proper convex continuous functions which are 1-paraconcave:

$$1/2 \| \|^2 \,\Box\, \Gamma_0 = 1/2 \| \|^2 - \Gamma_0) \cap \Gamma_0.$$

Proof: If $g \in D$ we must show that g is 1-paraconcave. Now there exists f such that $-g = f^c$, where $c(y, z) = -1/2 \| y - z \|^2$ so that $-g$ is c-regular, i.e. $-g + 1/2 \| \|^2 \in \Gamma_0$. Conversely, if $g \in (1/2 \| \|^2 - \Gamma_0) \cap \Gamma_0$ take $f = -((-g) \,\Box\, 1/2 \| \|^2)$. Then, by Theorem 5, $1/2 \| \|^2 \Box f = g$. The lower semicontinuity and the convexity of g entail the lower semicontinuity and the convexity of f for $f(z) = \sup_{u \in Z} \{ g(z + u) - 1/2 \| u \|^2 \}$ is then a suprema of such functions. The properness of f follows easily from the properness of g(use $1/2 \| \|^2 \Box f = g$). □

Corollary 7 - For any $r > 0$ the class $D_r := r/2 \| \|^2 \,\Box\, \Gamma_0$ is the class of proper convex continuous functions which are r-paraconcave.

Proof: This follows from the fact that Γ_0 is a convex cone:

$$D_r = r\,[1/2 \| \|^2 \Box \Gamma_0] = r\,[\Gamma_0 \cap (1/2 \| \|^2 - \Gamma_0)] = \Gamma_0 \cap (r/2 \| \|^2 - \Gamma_0). \ \Box$$

IV . Paraconvex minimization problems

IV.1 On Paraconvex multifunctions

Following the terminology of Rolewicz [27] up to a slight modification we say shall that the multifunction Γ from the Hilbert space Z to the normed space X is L-paraconvex ($L \in \mathbb{R}_+$) if

(11) $$(1-t)\,\Gamma(z_0) \,\bar{+}\, t\,\Gamma(z_1) \subset \Gamma(z_t) \,\bar{+}\, L/2\, t(1-t) \| z_0 - z_1 \|^2 B_X$$

for any $t \in [0, 1]$, $z_0, z_1 \in Z$, $z_t = (1-t) z_0 + t z_1$, B_X being the closed unit ball of X and $A \,\bar{+}\, B$ denoting the closure of the sum A+B of two subsets A, B of X.
This property is weaker than the convexity of the graph of Γ and stronger than the convexity of the values of Γ. A stronger property is the following one. A multifunction $\Gamma : Z \to X$ is said to be L-strongly convex if for any t in [0, 1], z_0, z_1 in Z

(12) $$(1-t)\,\Gamma(z_0) + t\,\Gamma(z_1) + L/2\, t(1-t) \| z_0 - z_1 \|^2 B_X \subset \Gamma((1-t) z_0 + t z_1).$$

When Γ is an epigraphical relation, that is when $X = \mathbb{R}$ and there exists $h : Z \to \overline{\mathbb{R}}$ such that $\Gamma z = \mathbb{R} \cap [\, h(z), +\infty\, [$ then Γ is L-paraconvex (resp. L-strongly convex) iff h is L-paraconvex (L-strongly convex).

Other examples will be presented below.

The following characterization follows from the properties of the support function $A \to h(A,y) = \sup \{ \langle y, x \rangle : x \in A \}$ to a subset A of X when y is a given element of the dual of X.

Proposition 8 . The multifunction $\Gamma : Z \to X$ is L-paraconvex (resp. L-strongly convex) iff for each y in the unit sphere of the dual space of X the function $z \to p(y, z) = -h(\Gamma(z), -y)$ is L-paraconvex (resp. L-strongly convex).

Another characterization similar to characterizations of convex or pseudo-lipschitzian multifunctions can be given by using the following definition.

Definition 9 . An extended real-valued function $f : X \times Y \to \overline{\mathbb{R}}$ on a product of normed linear spaces is said to be convex-L-paraconvex (resp. positively convex-L-strongly convex) if for any $(x_0, y_0), (x_1, y_1)$ in $X \times Y$, $t \in [0, 1]$ one has with $x_t = (1-t)\, x_0 + t x_1$, $y_t = (1-t)\, y_0 + t y_1$

$$f(x_t, y_t) \leq (1-t) f(x_0, y_0) + t f(x_1, y_1) + L/2\, t\, (1-t)\, \| y_0 - y_1 \|^2$$

$$(\text{resp. } f(x_t, y_t) \leq \max (0, (1-t) f(x_0, y_0) + t f(x_1, y_1) - L/2\, t\, (1-t)\, \| y_0 - y_1 \|^2))$$

Proposition 10 . A closed-valued multifunction $\Gamma : Z \to X$ is L-paraconvex (resp. L-strongly convex) iff its associated distance function $d_\Gamma : (x, z) \to d(x, \Gamma(z))$ is convex-L-paraconvex (resp. positively convex-L-strongly convex).

Proof: Suppose Γ is L-paraconvex and let $(x_0, z_0), (x_1, z_1)$ in $X \times Z$, $t \in [0, 1]$, $x_t = (1-t)x_0 + t x_1$, $z_t = (1-t)z_0 + t z_1$. For any $y_0 \in \Gamma(z_0)$, $y_1 \in \Gamma(z_1)$ we can find sequences (w^n) in $\Gamma(z_t)$ and (u^n) in B_X with

$$(1-t)\, y_0 + t y_1 = \lim (w^n + L/2\, t(1-t) \| z_0 - z_1 \|^2 u^n).$$

Therefore

$$d(x_t, \Gamma(z_t)) \leq \limsup_{n \to +\infty} \| x_t - w^n \| \leq (1-t) \| x_0 - y_0 \| + t\, \| x_1 - y_1 \| + L/2\, t\, (1-t) \| z_0 - z_1 \|^2$$

Taking the infimum over $(y_0, y_1) \in \Gamma(z_0) \times \Gamma(z_1)$ we get that d_Γ is convex-L-paraconvex. Conversely, given x_i, z_i $(i = 0, t, 1)$ as above and such that $x_0 \in \Gamma(z_0)$, $x_1 \in \Gamma(z_1)$, and supposing d_Γ is convex-L-paraconvex we have

$$d(x_t, \Gamma(z_t) \leq L/2\, t(1-t) \| z_o - z_1 \|^2.$$

Since for any $r > 0$ and any nonempty subset A of X we have $d(x, A) \leq r$ iff $x \in A + rB$ we get inclusion (11).
Now suppose Γ is L-strongly convex. Given x_i, z_i (i=0, t, 1) as above we deduce from (12) and the relations

$$d(.\,, A+sB_X) = \max(0, d(.\,, A)-s)$$

$$d(x_t, (1-t)\, C+tD) = (1-t)\, d(x_o, C) + td(x_1, D)$$

for $A, C, D \subset X$, $s \in IR$, $0 \leq t \leq 1$, that

$$d(x_t, \Gamma(z_t) \leq \max(0, (1-t)\, d(x_o, \Gamma(z_o)) + td(x_1, \Gamma(z_1)) - L/2\, t(1-t) \| z_o - z_1 \|^2).$$

Conversely, when this relation holds, taking u arbitrary in $L/2\, t(1-t) \| z_o - z_1 \|^2 B_X$ and x_i in $u + \Gamma(z_i)$ for $i=0, 1$, $x_t = (1-t)x_o + tx_1$ we get $d(x_t, \Gamma(z_t)) = 0$ hence (12). □

The notions of strongly convex and paraconvex multifunctions can be used simultaneously as in the following result.

Proposition 11 . Suppose Γ is r-paraconvex and Φ is s-strongly convex. Then for $\Psi(z) = \Gamma(z) \dot{-} \Phi(z)$

(a) if $r > s$ Ψ is (r-s)-paraconvex
(b) if $r < s$ Ψ is (s-r)-strongly convex
(c) if $r = s$ Ψ is convex.

Proof : Let $w_t = (1-t)w_o + tw_1$, $t \in [0, 1]$, $w_o, w_1 \in W$. Then

$$(1-t)\,\Gamma(w_o) + t\,\Gamma(w_1) \subset \Gamma(w_t) \dot{+} r/2\, t(1-t) \| w_1 - w_o \|^2 B_X$$

$$(1-t)\,\Phi(w_o) + t\,\Phi(w_1) + s/2\, t(1-t) \| w_1 - w_o \|^2 B_X \subset \Phi(w_t)$$

hence,

$$(1-t)\,\Psi(w_o) + t\,\Psi(w_1) + s/2\, t(1-t) \| w_1 - w_o \|^2 B_X \subset \Psi(w_t) \dot{+} r/2\, t(1-t) \| w_1 - w_o \|^2 B_X.$$

Since $\Psi(w_t)$ is closed (and $\Psi(w_t) \dot{+} (r-s)/2\, t(1-t) \| w_1 - w_o \|^2 B_X$ is closed for $r > s$) the Rädstrom's cancellation rule gives result. □

For what concerns duality theory, an important question is to know when the value function associated to the problem

$$(\mathbf{P}) \qquad \min \{ f(x) : x \in \Gamma(0) \}$$

is paraconvex. In this direction the next result shows that the notion of paraconvex relation can be used to answer this question. It can be deduced from a more general theorem about composition of Lipschitz relations with paraconvex relations ([26] Theorem 1); it can also be given a simple direct proof. Here we deduce it from Proposition 10.

Theorem 12. Assume that f is convex and K-Lipschitz on X and Γ is L-paraconvex. Then the value function $v(z) = \inf \{ f(x) : x \in \Gamma(z) \}$ is a KL-paraconvex function of z.

Proof: It is easy to see that for each $z \in Z$ we have

$$v(z) = \inf \{ f(x) + Kd(x, \Gamma(z)) : x \in X \}.$$

In fact if there exists $x \in X$ with $v(z) > f(x) + Kd(x, \Gamma(z))$ we can find $y \in \Gamma(z)$ with $Kd(x, y) < v(z) - f(x) \le f(y) - f(x) \le Kd(x, y)$, a contradiction. Then the result follows from Proposition 10 and the following lemma, taking $F(x, z) = f(x) + Kd(x, \Gamma(z))$ and $M = KL$.

Lemma 13 (compare with [17] Proposition 4.1) Let $F : X \times Z \to \mathbb{R}$ be convex-M-paraconvex. Then $v(z) = \inf \{ F(x, z) : x \in X \}$ is M-paraconvex.

Proof: It suffices to observe that

$$v(z) + M/2 \| z \|^2 = \inf \{ F(x, z) + M/2 \| z \|^2 : x \in X \}$$

and that $(x, z) \to F(x, z) + M/2 \| z \|^2$ is convex. □

The preceding result does not cover the case of quadratic minimization under linear constraints. This case is encompassed by the following result. Here we define the gap between two subsets A, B of a n.v.s. X by

$$\text{gap}(A, B) = \inf \{ d(a, b) : a \in A, b \in B \}.$$

with the usual convention $\inf \emptyset = +\infty$.

Theorem 14. Suppose f is L-paraconvex, Γ is convex and for some positive number K the solution multifunction $S : Z \to X$ of (P) satisfies

$$\text{gap}(S(z), S(z')) \le K\, d(z, z').$$

Then the value function v of (P) is LK^2-paraconvex.

Proof: Given $t \in [0, 1]$, z_0, z_1 in Z, $z_t = (1-t)\, z_0 + tz_1$ for any $x_0 \in S(z_0)$, $x_1 \in S(z_1)$ we have $x_t := (1-t)x_0 + tx_1 \in \Gamma(z_t)$ and

$$v(z_t) \le f(x_t) \le (1-t)\, f(x_0) + tf(x_1) + 1/2\, Lt(1-t)\, \| x_0 - x_1 \|^2.$$

Taking the infimum over $(x_0, x_1) \in S(z_0) \times S(z_1)$ and using the assumption about S we get the result. □

We present now two ways of obtaining paraconvex multifunctions.

Proposition 15. Let θ be a real-valued function defined on the Hilbert space Z, let B_X be the closed unit ball of the normed vector space X and let Ω be given by

$$\Omega(z) = \theta(z)\, B_X.$$

Then Ω is paraconvex iff $|\theta|$ is paraconcave (with the same constant).

Proof: The L-paraconvexity of Ω can be expressed in the following equivalent ways:

$$t\theta(z_0)\, B_X + (1-t)\, \theta(z_1)\, B_X \subset \theta(z_t)\, B_X + L/2\; t(1-t)\, \| z_0 - z_1 \|^2\, B_X$$

$$(t|\theta(z_0)| + (1-t)\, |\theta(z_1)|)\, B_X \subset (|\theta(z_t)| + L/2\; t(1-t)\, \| z_0 - z_1 \|^2)\, B_X$$

$$t|\theta(z_0)| + (1-t)\, |\theta(z_1)| \le |\theta(z_t)| + L/2\; t(1-t)\, \| z_0 - z_1 \|^2$$

that is exactly the L-paraconvexity of $|\theta|$. □

Proposition 16. Let Φ and Γ be two multifunctions from Z to X and let Φ + Γ be given by

$$(\Phi + \Gamma)\, (z) = \Phi(z) + \Gamma(z).$$

Assume that Φ is K-paraconvex and Γ is L-paraconvex. Then Φ + Γ is (K+L)-paraconvex. In particular if Φ is convex and Γ is L-paraconvex Φ + Γ is L-paraconvex.
The proof consists in a routine calculation.

Remark : It is easy to see that $\Psi = \mathrm{cl}\, (\Gamma + \Phi)$ is also K L-paraconvex since a multifunction Ω given by $(\mathrm{cl}\, \Omega)(z) = \mathrm{cl}(\Omega(z))$ is M-paraconvex.

Corollary 17 . Assume that Φ is a convex multifunction from Z to X and θ is a real-valued function defined on Z such that $|\theta|$ is r-paraconcave. Then the enlargement Γ of Φ given by

$$\Gamma(z) = \Phi(z) + \theta(z) B_X$$

is a r-paraconvex multifunction.

Since the inverse of a paraconvex multifunction Φ is not paraconvex in general (take for Φ the epigraph of $k : \mathbb{R} \to \mathbb{R}$ given by $k(t) = -t^2$) it is not easy to decide whether a multifunction of the type

$$\Gamma(z) = \{ g \le z \}$$

with $g : X \to \mathbb{R}$, is a paraconvex multifunction. Since this question is important when dealing with perturbed minimization problems as **(P)** let us give an instance in which Corollary 17 brings a positive answer.

Let us take in Corollary 17 $\Phi(z) = \{ k \le z \}$ where $k : X \to \mathbb{R}$ is convex and let us suppose $\theta : \mathbb{R} \to \mathbb{R}$ is paraconcave, nondecreasing and upper semicontinuous. Then let us show that $\Gamma(z) = \Phi(z) + \theta(z) B_X$ is of the form $\{ g \le z \}$ for some function $g : X \to \mathbb{R}$. Introducing the quasi-inverse ([24]) Ψ of θ given by $\Psi(s) = \inf \{ r \in \mathbb{R} : \theta(r) \ge s \}$ we obtain

$$\| x \| \le \theta(z) \Leftrightarrow \Psi(\| x \|) \le z,$$

so that we have $\{ h \le z \} = \theta(z) B_X$ for $h := \psi^{o} \| \ \|$. Now let g be the quasi-inf-convolution $h \nabla k$ of h and k (see [32]), given by

$$(h \nabla k)(x) = \inf_{u \in X} \max (k(x-u), h(u)).$$

Then it is easy to see ([25] p. 40) that for each $z \in \mathbb{R}$

$$\{ g < z \} = \{ h < z \} + \{ k < z \}.$$

When X is finite dimensional, and more generally when X is reflexive, we obtain the announced relation

$$\{ g \le z \} = \{ h \le z \} + \{ k \le z \}$$

since k is then weakly lower semicontinuous and $h \in \psi^{o}\| \ \|$ has weakly compact sublevel sets.

So,

$$z \to \{ g \le z \} = \{ k \le z \} + \theta(z) B_X$$

is another example of paraconvex multifunction.

IV.2 - Duality results in paraconvex optimization

Let us turn to the problem

$$(\mathbf{P}) : \min \{ f(x) : x \in \Gamma(0) \}.$$

Let us suppose its value is finite, f is bounded on bounded sets and the value function v is r-paraconvex. Conditions ensuring this property have been given in Theorems 12 and 14. Then there exists a convex function φ such that

$$v + r/2 \| \|^2 = \varphi.$$

The dual of $(\mathbf{P})$ with respect to the coupling functional $c(y, z) = -r/2 \| y - z \|^2$ is easily seen to be

$$(\mathbf{Q}_c) : \max \inf_{x \in X} \{ f(x) + r/2 \, (d^2(y, \Gamma^{-1}(x)) - \| y \|^2) \} \text{ for } y \in Z$$

where $z \in \Gamma^{-1}(x)$ iff $x \in \Gamma(z)$. It is the concave problem of maximizing $y \to -\varphi^*(ry)$ on Z. Note that in the case of equality constraints, $\Gamma(z) = A^{-1}(z)$, we have $d^2(y, \Gamma^{-1}(x)) = \| y - Ax \|^2$ and $(\mathbf{Q}_c)$ takes the form

$$\text{maximize } -(f + q_{rA^TA})^* \, (rA^Ty) : y \in Z,$$

which is equivalent to the dual problem, computed in section III.1 when f is quadratic. To obtain a zero duality gap between $(\mathbf{P})$ and $(\mathbf{Q}_c)$ and to get optimal solutions for $(\mathbf{Q}_c)$ we have to express that the set of c-subgradients of v at the origin is nonempty. Using the definition of this set we get easily

$$y \in \partial_c v(0) \Leftrightarrow r\, y \in \partial\varphi(0)$$

where $\partial\varphi(0)$ denotes the classical convex subdifferential of φ at the origin. As φ is convex, $\partial_c v(0)$ is nonempty if φ is majorized on a neighborhood of the origin ([8] prop. 5.2). As f is bounded on bounded sets, it is easy to see that this occurs if

(13) $$\exists\alpha > 0, \exists\beta > 0 : \forall z \in \alpha B_Z : \Gamma z \cap \beta B_X \neq \emptyset.$$

Let us observe that condition (13) is weaker than the lower semicontinuity of the multifunction Γ at zero and that, in the case of inequality constraints, it holds if the usual Slater's condition holds, or more generally if for some $\beta > 0$ the set $g(\beta B_X) + C$ is a

neighborhood of 0 in Z, with $\Gamma(z) = g^{-1}(z-C)$, C being the ordering cone of Z.
We conclude with the following result which summarizes the preceding observations.

Theorem 18 . Assume that f is bounded on bounded sets, the value function v is paraconvex and condition (13) holds. Then $\inf(\mathbf{P}) = \sup(\mathbf{Q}_c)$ and $(\mathbf{Q}_c)$ has optimal solutions.

References

[1] **Asplund E.**, Differentiability of the metric projection in finite-dimensional Enclidean space, Proc. Amer. Math. Soc. 38 (1973), 218-219.

[2] **Asplund E. and Rockafellar R.T.**, Gradients of convex functions, Trans. Amer. Math. Soc. 139 (1969), 443-467.

[3] **Aze D. and Penot J.-P.**, Uniformly convex functions and uniformly smooth convex functions, Preprint (1988).

[4] **Balder E.J.**, An extension of duality-stability relations to nonconvex problems, SIAM Journal on Control and Optimization, 15 (1977), 329-343.

[5] **Crouzeix J.-P.**, Contribution à l'étude des fonctions quasiconvexes, Thèse d'Etat, Université de Clermont II, 1977.

[6] **Debreu G.**, Definite and semidefinite quadratic forms, Econometrica 20 (1952), 295-300.

[7] **Dolecky S., Kurcyusz S.**, On Φ-convexity in extremal problems, SIAM Journal on Control and Optimization 16 (1978) 277-300.

[8] **Ekeland I., Temam R.**, Analyse convexe et problèmes variationnels, Dunod, Paris, 1974.

[9] **Elster K.H., Nehse R.**, Zur theorie der Polarkunktionale, Meth. Oper. forsh 5(1974), 3-21.

[10] **Fenchel W.**, A remark on convex sets and polarity, Medd. Lunds Univ. Mat. Sem. (Suppl. 1952) 82-89.

[11] **Hiriart-Urruty J.-B., Mazure M.-L.**, Formulations variationelles de l'addition parallèle et de la soustraction parallèle d'opérateurs semi-définis positifs, C.R.A.S. Paris, 302, Serie I, n.15 (1986), 527-530.

[12] **Hiriart-Urruty J.-B., Plazanet P.**, Moreau's decomposition Theorem revisited, to appear.

[13] **Lindberg P.O.**, A generalization of Fenchel conjugation, Proc. IX Int. Symp. on Math. Prog. Vol.2, Akad.Kiado Budapest 1979.

[14] **Martinez-Legaz J.E.**, Un concepto genelizado de conjugacion, application a las funciones quasiconvexas, Thesis, Barcelona, 1981.

[15] **Martinez-Legaz J.E.**, On lower subdifferentiable functions, in: Trends in Mathematical optimization, Hoffmann K.-H ed al ed. Birkhauser Inter. Series of Num. Math. Vol. 84 (1988), 197-232.

[16] **Martinez-Legaz J.E.**, Generalized conjugation and related topics, Inter.Worshop on Generalized Concavity, Fractional Programming and Economic Applications, Pisa May-June 1988.

[17] **Montrucchio L.**, Lipschitz continuous policy functions for strongly concave optimization problems, J. Math. Eco. 16 (1987) 259-273.

[18] **Moreau J.J.**, Proximité et dualité dans un esapce Hilbertien, Bull. Soc. Math. France 93 (1965), 273-299.

[19] **Moreau J.J.**, Inf-convolution, sous additivité, convexité des fonctions numeriques, J. Math. pures ed appliquées 49 (1970) 109-154.

[20] **Passy U., Prisman E.Z.**, Conjugacy in quasiconvex programming, Math. Prog. 36, 121-146.

[21] **Penot J.-P.**, Modified and augumented Lagrangian theory revisited and augmented, Fermat Days (1985) Toulouse (un published lecture).

[22] **Penot J.-P. and Bourgeard M.L.**, Approximation and decomposition properties of some classes of locally d.c. functions, Math. Progr. 41 (2)(1988), 195-228.

[23] **Penot J.-P., Volle M.**, Dualité de Fenchel et quasi-convexité, C.R.A.S. Paris t 304, Series I, n. 13 (1987), 371-374.

[24] **Penot J.-P., Volle M.**, Inversion of real valued functions and applications, To appear in Zeits, für Oper. Research.

[25] **Penot J.-P., Volle M.**, On quasi-convex duality, to appear in Math. Oper. Research.

[26] **Rockafellar R.T.**, Convex analysis, Princeton Univ. Press 1970.

[27] **Rolewicz, S.**, On paraconvex multifunctions, Proceedings of III Symposium über Operations Research, Mannheim, Sept. 1978, 539-546.

[28] **Singer I.**, Surrogate conjugate functionals and surrogate convexity, Applicable Anal. 16 (1983), 291-327.

[29] **Vial J. P.**, Strong convexity of sets and functions, Journal of Math. Economics 9 (1982), 187-205.

[30] **Vial J. P.**, Strong and weak convexity of sets and functions, Math. of Oper. Research 8 n. 2 (1983), 231-259.

[31] **Vladimirov A.A., Nesterov Yu.E. and Chekanov Yu.N.**, On uniformly convex functionals, Vestnik Moskov. Univ. Ser. XV Vycisl. Mat. K bernet. n. 3 (1978) 12-13.

[32] **Volle M.**, Conjugaison par tranches, Ann. di. Mat. pura ed applicata IV, Vol. 139(1985), 279-312.

[33] **Volle M.**, Contribuitions à la dualité en optimisation et à l'épiconvergence, Thèse d'Etat, Univ. of Pau, 1986.

[34] **Volle M.**, Concave duality: Application to problems dealing with difference of functions, Math. Prog. 41 (1988), 261-278.

[35] **Zalinescu C.**, On uniformly convex functions, J. Math. Anal. and Appl. 95 (2)(1983) 344-374.

GENERALIZED CONVEXITY AND FRACTIONAL OPTIMIZATION

K.H. Elster - A. Wolf

Besides Lagrange duality, the use of conjugate functions was an important tool to establish an approach to duality theory in nonlinear optimization, denoted as Fenchel duality.
In the paper, a general concept of conjugation is introduced, based on polarities and hence called P-conjugation. Several well-known conjugations turn out to be special cases of P-conjugation, p.e. level-set-conjugation, conjugation by coupling functionales, Φ-conjugation and Fenchel conjugation.
The concept of P-conjugation can be connected with a generalized convexity of sets and functions (P-convexity), expressed by a corresponding representation theorem.
A special case of the P-conjugation, which is denoted as M-conjugation, is used for treating fractional optimization problems.
Some duality theorems for such problems can be given.

1. Introduction

The notion of duality is closely connected with the theory of nonlinear optimization. Besides Lagrange duality, conjugate functions establish another approach to duality theory, well-known as Fenchel duality.
Fenchel's classical paper [6], published 1953, gave rise to numerous results devoted to duality in nonlinear optimization and extending Fenchel's theory in different directions ([4]).
In this present paper, the general concept of P-conjugation presented in [4], which is based on polarities, will be connected with generalized convexity of sets and functions. Moreover, a special case of P-conjugation is used for the investigation of fractional optimization problems. Complementary to earlier results about such problems ([7]), several duality assertions will be given in the following.

2. Some Results Concerning Polarities

In the following we will summarize some notions and assertions concerning polarities (cf. [5], [4]).

Definition 2.1. Let X, Y be nonempty sets. A mapping $\Delta : 2^{X \times R} \to 2^{Y \times R}$ is called polarity, if there exists a functional $p : X \times R \times Y \times R \to R$ such that for each subset

$A \subseteq X \times R$ holds

$$\Delta(A) = \bigcap_{(x,k)\in A} \{ (y,l) \in Y \times R \mid p(x,k,y,l) \geq 0 \}.$$

The functional p is said to be a generating functional of the polarity Δ.
The dual polarity Δ^* of a polarity Δ is defined by $\Delta^* : 2^{Y\times R} \to 2^{X\times R}$ with

$$\Delta^*(B) := \bigcap_{(y,l)\in B} \{ (x,k) \in X \times R \mid p(x,k,y,l) \geq 0 \}, \quad B \subseteq Y \times R.$$

Several properties of polarities are given in [4] and [5].
Example. If X, Y are nonempty sets and $\varphi : X \times Y \to R$ a so-called coupling functional, then the mapping $\Delta : 2^{X\times R} \to 2^{Y\times R}$, where

$$\Delta(A) := \bigcap_{(x,k)\in A} \{ (y,l) \in Y \times R \mid -\varphi(x,y) + k + l \geq 0 \}, \quad A \subseteq X \times R,$$

is a polarity. This polarity is useful to describe the generalized Fenchel-conjugation with the coupling functional φ as a special P-conjugation.

Definition 2.2. Let $p : X \times R \times Y \times R \to R$ be a functional.
(i) p is said to be epigraphical, if

$$\forall (x_0, k_0, y_0) \in X \times R \times Y \quad \exists l_* \in R : \ l_* := \min \{ l \in R \mid p(x_0, k_0, y_0, l) \geq 0 \}$$

and

$$\forall (x_0, k_0, y_0, l_0) \in X \times R \times Y \times R : \ p(x_0, k_0, y_0, l_0) \geq 0 \Rightarrow (p(x_0, k_0, y_0, l) \geq 0 \ \forall l \geq l_0)$$

(ii) p is said to be hypographical, if

$$\forall (x_0, k_0, y_0) \in X \times R \times Y \quad \exists l^* \in R : l^* := \max \{ l \in R \mid p(x_0, k_0, y_0, l) \geq 0 \}$$

and

$$\forall (x_0, k_0, y_0, l_0) \in X \times R \times Y \times R : \ p(x_0, k_0, y_0, l_0) \geq 0 \Rightarrow (p(x_0, k_0, y_0, l) \geq 0 \ \forall l \leq l_0).$$

If p is an epigraphical functional, then the set

$$\{ (x,k,y,l) \in X \times R \times Y \times R \mid p(x,k,y,l) \geq 0 \}$$

can be considered as the epigraph of a functional $e : X \times R \times Y \to R$.

If p is a hypographical functional, then there exists a corresponding functional $h : X \times R \times Y \to R$.

Consequently we say, a polarity Δ is epigraphical (resp. hypographical) if there exists an epigraphical (resp. a hypographical) generating functional of Δ.

Now we will give some assertions.

Theorem 2.1 Let $\Delta : 2^{X \times R} \to 2^{Y \times R}$ be a polarity.

(i) Δ is epigraphical iff there exists a generating functional p of Δ such that (on $X \times R \times Y \times R$) holds $p(x, k, y, l) = -e(x, k, y) + l$, where $e : X \times R \times Y \to R$.

(ii) Δ is hypographical iff there exists a generating functional p of Δ such that on ($X \times R \times Y \times R$ holds $p(x, k, y, l) = h(x, k, y) - l$, where $h : X \times R \times Y \to R$.

(iii) The functionals e, h in (i) resp. in (ii) are uniquely determined.

Proof. We consider the case of an epigraphical functional.Obviously, the functional $-e+l$ is epigraphical by Definition 2.2. Conversely, let $\Delta : 2^{X \times R} \to 2^{Y \times R}$
be an epigraphical polarity with the generating functional p and let

$$e(x, k, y) := \min \{ l \in R \mid p(x, k, y, l) \geq 0 \}.$$

Here the minimum exists on $X \times R \times Y$ by Definition 2.2, (i).

So we have to prove that $\forall (x, k, y, l) \in X \times R \times Y \times R : p(x, k, y, l) \geq 0 \Leftrightarrow -e(x, k, y) + l \geq 0$, since in that case the functionals p and $-e+l$ generate the same polarity.Using the definition of e we have

$$\begin{aligned} -e(x, k, y) + l \geq 0 &\Leftrightarrow -\min \{ l \in R \mid p(x, k, y, l) \geq 0 \} + l \geq 0 \\ &\Leftrightarrow l \geq \min \{ l \in R \mid p(x, k, y, l) \geq 0 \} \\ &\Leftrightarrow p(x, k, y, l) \geq 0 , \end{aligned}$$

where the last equivalence is true because of Definition 2.2. (ii).

To prove the uniqueness we assume that there exists two functionals $e_1, e_2 : X \times R \times Y \to R$ such that

$$(2.1) \qquad -e_1(x, k, y) + l \geq 0 \Leftrightarrow -e_2(x, k, y) + l \geq 0$$

and there is at least one $(x_0, k_0, y_0) \in X \times R \times Y$ such that

$$e_1(x_0, k_0, y_0) = e_2(x_0, k_0, y_0) + d, \quad d \neq 0.$$

For $\qquad x = x_0, \ y = y_0, \ k = k_0, \ l_0 = e_1(x_0, k_0, y_0) - d/2$

we obtain by (2.1) immediately $-e_1(x_0, k_0, y_0) + e_1(x_0, k_0, y_0) - d/2 \geq 0 \Leftrightarrow -e_2(x_0, k_0, y_0) + e_1(x_0, k_0, y_0) - d/2 \geq 0$

and hence $-d/2 \geq 0 \Leftrightarrow d/2 \geq 0$ which contradicts to $d \neq 0$.

A similar approach as in Definition 2.2 with respect to the variable k instead of l leads to *-epigraphical and *-hypographical functionals and corresponding polarities (cf. [4]). So it is easy to prove the following theorem

Theorem 2.2 A polarity $\Delta : 2^{X\times R} \to 2^{Y\times R}$ is *-epigraphical (resp. *-hypographical) iff its dual polarity $\Delta^* : 2^{Y\times R} \to 2^{X\times R}$ is epigraphical (resp. hypographical).

In connection with generalized conjugate functions introduced in the following section we use the notion of symmetrical polarity.

Definition 2.3. Let $\Delta_1, \Delta_2 : 2^{X\times R} \to 2^{Y\times R}$ be polarities. Δ_1 is said to be symmetrical to Δ_2 if there exist a generating functional p_1 of Δ_1 and a generating functional p_2 of Δ_2 such that on $(X \times R \times Y \times R)$ holds $p_1 = -p_2$.

Now we can prove the following properties.

Theorem 2.2
(i) If $\Delta : 2^{X\times R} \to 2^{Y\times R}$ is a polarity, then there exists one and only one polarity Δ symmetrical to Δ.
(ii) Let $\Delta_1, \Delta_2 : 2^{X\times R} \to 2^{Y\times R}$ be polarities. Then holds: Δ_1 symmetrical to $\Delta_2 \Leftrightarrow \Delta_2$ symmetrical to Δ_1.
(iii) If $\Delta : 2^{X\times R} \to 2^{Y\times R}$ is a polarity and Δ is symmetrical to Δ, then Δ epigraphical $\Leftrightarrow \Delta$ hypographical.

Proof. The proof of (i) and (ii) is an easy consequence of Definition 2.3 and the fact that two functionals $p_1, p_2 : X \times R \times Y \times R \to R$ generate the same polarity Δ, if

$$\forall (x, k, y, 1) \in X \times R \times Y \times R : \quad p_1(x, k, y, 1) \geq 0 \Leftrightarrow p_2(x, k, y, 1) \geq 0.$$

To prove (iii) we assume Δ to be epigraphical. Because of Definition 2.2, the generating functional of Δ is epigraphical, too. The functional p satisfies consequently condition (i) of Definition 2.2. By Definition 2.3 the symmetrical polarity Δ has the generating functional $-p$, which satisfies condition (ii) of Definition 2.2 and hence it is hypographical. Consequently, the polarity Δ ist hypographical. The proof of the other implication is analogously.

3. P-Conjugation and P-Convexity

Using the notion introduced in Section 2 we will define generalized conjugate functions.

Definition 3.1 Let $\Delta_1, \Delta_2 : 2^{X\times R} \to 2^{Y\times R}$ be polarities and $f : X \to R$ a function. Then $f^{\Delta_1} : Y \to R$, where $f^{\Delta_1}(y) := \inf \{ 1 \in R \mid (y, 1) \in \Delta_1 (\text{graph } f) \}$,

is called the upper P-conjugate function of f (with respect to the polarity Δ_1) and $f_{\Delta_2}(y)$ $= Y \to R$, where $f_{\Delta_2}(y) := \sup \{ 1 \in R \mid (y, 1) \in \Delta_2 (\text{graph } f) \}$,
is called the lower P-conjugate function of f (with respect to the polarity Δ_2).

For most of the conjugations known from the literature the polarities Δ_1 and Δ_2 according Definitions 3.1. are symmetrical. But this isn't necessary; an appropriate counter-example is the M-conjugation (cf. [4]).
If $\Delta_1 : 2^{X \times R} \to 2^{Y \times R}$ is an epigraphical polarity with the generating functional p_1, then for each function $f : X \to R$ holds

$$(3.1) \qquad p_1(x, f(x), y, f^{\Delta_1}(y)) \geq 0$$

Correspondingly, if Δ_2 is a hypographical polarity with the generating functional p_2, then for each function $f : X \to R$ holds

$$(3.2) \qquad p_2(x, f(x), y, f_{\Delta_2}(y)) \geq 0.$$

The inequalities (3.1) and (3.2) are generalizations of the well-knowm Fenchel-Young-inequality.
In the case of epigraphical or hypographical polarities we can conclude from the identity of two (either upper or lower) P-conjugate functions to the identity of the used polarities. In the general case this proposition doesn't hold ([4]).
If we assume epigraphical (resp. hypographical) polarities it is possible to give an appropriate representation of P-conjugate functions and properties of P-biconjugate functions.

Theorem 3.1. Let $\Delta_1 : 2^{X \times R} \to 2^{Y \times R}$ be an epigraphical, $\Delta_2 : 2^{X \times R} \to 2^{Y \times R}$ be a hypographical polarity and $f : X \to R$.
(i) There exist uniquely determined functionals $e : X \times R \times Y \to R$ and $h : X \times R \times Y \to R$ such that $f^{\Delta_1}(y) = \sup_{x \in X} e(x, f(x), y)$, $\quad f_{\Delta_2}(y) = \inf_{x \in X} h(x, f(x), y)$,

(ii) $f^{\Delta_1 \Delta_1^*}(x) = \inf \{ k \mid (x, k) \in \Delta_1^* (\Delta_1(\text{graph } f)) \}$,
$f_{\Delta_2 \Delta_2^*}(x) = \sup \{ k \mid (x, k) \in \Delta_2^* (\Delta_2(\text{graph } f)) \}$.

(iii) $\forall x \in X : f(x) \geq f^{\Delta_1 \Delta_1^*}(x)$,
$\forall x \in X : f(x) \geq f_{\Delta_2 \Delta_2^*}(x)$.

Proof. (i) : By Theorem 2.1 exist uniquely determined functionals e, h on $(X \times R \times Y \times R)$ such that $-e+1$ is a generating functional of Δ_1 and $h-1$ is a generating functional of Δ_2.

Hence $\quad f^{\Delta_1}(y) = \inf\{\, l \in R \mid (y, l) \in \Delta_1(\text{graph } f)\,\}$

$= \inf\{\, l \in R \mid \forall (x,k) \in \text{graph } f : -e(x,k,y) + l \geq 0\,\}$

$= \sup\{\, e(x,k,y) \mid (x,k) \in \text{graph } f\,\}$

$= \sup_{x \in X} e(x, f(x), y)$.

The second assertion of **(i)** can be shown analogously.

(ii) : Because of Definition 3.1 we have for the P-biconjugate functions of f with respect to the polarities Δ_1, Δ_2 the representation

$$f^{\Delta_1\Delta_1^*}(x) = \inf\{\, k \mid (x,k) \in \Delta_1^*(\text{graph } f^{\Delta_1})\,\},$$

$$f_{\Delta_2\Delta_2^*}(x) = \sup\{\, k \mid (x,k) \in \Delta_2^*(\text{graph } f_{\Delta_2})\,\}.$$

Since Δ_1 is epigraphical we obtain from Definition 2.2: graph $f^{\Delta_1} \subseteq \Delta_1(\text{graph } f)$.

Because of a well-known property of polarities (cf. [5]) we conclude

$$\Delta_1^*(\text{graph } f^{\Delta_1}) \supseteq \Delta_1^*(\Delta_1(\text{graph } f)).$$

Now let $(x_0, k_0) \in \Delta_1^*(\text{graph } f^{\Delta_1})$. If p_1 is a generating functional of Δ_1, then we obtain because of Definition 2.1 $\quad \forall (y, l) \in \text{graph } f^{\Delta_1} : p_1(x_0, k_0, y, l) \geq 0$.

Since Δ_1 is epigraphical, we have by Definition 2.2: $\forall (y, l) \in \text{epi } f^{\Delta_1} : p_1(x_0, k_0, y, l) \geq 0$.

From that definition follows furthermore epi $f^{\Delta_1} = \Delta_1(\text{graph } f)$, hence

$$\forall (y, l) \in \Delta_1(\text{graph } f) : p_1(x_0, k_0, y, l) \geq 0 \quad \text{and}$$

$$(x_0, k_0) \in \Delta_1^*(\Delta_1 \text{graph } f) \quad \text{or finally}$$

$$\Delta_1^*(\text{graph } f^{\Delta_1}) \subseteq \Delta_1^*(\Delta_1 \text{graph } f).$$

We obtain in such a way $\Delta_1^*(\text{graph } f^{\Delta_1}) = \Delta_1^*(\Delta_1 \text{graph } f)$.

By this result we conclude the first assertion of (ii). The second assertion can be proved similarly.

(iii) : To show this assertion we only have to use the well-known hull property of polarities (cf. [5]): $\quad \Delta^*(\Delta(A)) \supseteq A$

for all polarities $\Delta : 2^{X\times R} \to 2^{Y\times R}$ and all subsets $A \subseteq X\times R$, in connection with the propositions

$$\Delta_1^*(\text{graph } f^{\Delta_1}) = \Delta_1^*(\Delta_1 \text{graph } f),$$

$$\Delta_2^*(\text{graph } f_{\Delta_2}) = \Delta_2^*(\Delta_2 \text{graph } f)$$

for epigraphical (resp. hypographical) polarities Δ_1 (resp. Δ_2) and functions $f \in 2^{X\times R}$. Those propositions were proved in (ii).

Remark. In the proof of Theorem 3.1, assertion (i), was shown that for polarities Δ_1 with

$$f^{\Delta_1}(y) = \sup_{x \in X} e(x, f(x), y), \quad f \in R^X$$

the functional $-e+1$ is a generating functional on $(X \times R \times Y \times R)$. Obviously, that functional is an epigraphical one and hence Δ_1 is an epigraphical polarity. Analogously we can show that polarities Δ_2 with

$$f_{\Delta_2}(y) = \inf_{x \in X} h(x, f(x), y), \quad f \in R^X$$

are hypographical polarities.

In the following we introduce the notion of P-convexity concerning sets $A \subseteq X \times R$ and functions $f: X \to R$.

Definition 3.2. Let $\Delta : 2^{X \times R} \to 2^{Y \times R}$ be a polarity.

(i) The set $A \subseteq X \times R$ is called P-convex with respect to Δ, if $A = \Delta^*(\Delta(A))$.

(ii) The function $f: X \to R$ is called P-convex with respect to Δ, if on X is

$$f(x) = \inf \{ k \mid (x, k) \in \Delta^* (\Delta(\text{graph } f)) \}.$$

The function f is called P-concave with respect to Δ, if on X is

$$f(x) = \sup \{ k \mid (x, k) \in \Delta^* (\Delta(\text{graph } f)) \}.$$

P-convex functions can be characterized by the

Theorem 3.2. Let $\Delta_1 : 2^{X \times R} \to 2^{Y \times R}$ be an epigraphical polarity, $\Delta_2 : 2^{X \times R} \to 2^{Y \times R}$ be a hypographical polarity and $f: X \to R$. Then

(i) f P-convex with respect to $\Delta_1 \Leftrightarrow f = f^{\Delta_1 \Delta_1^*}$.

(ii) f P-concave with respect to $\Delta_2 \Leftrightarrow f = f_{\Delta_2 \Delta_2^*}$.

(iii) f^{Δ_1} is P-convex with respect to Δ_1^*.

(iv) f_{Δ_2} is P-concave with respect to Δ_2^*.

Proof. The assertion (i) and (ii) are direct consequences of Definition 3.2 and Theorem 3.1.
(iii) : By Definition 3.1 is $f^{\Delta_1}(y) = \inf \{ 1 \mid (y, 1) \in \Delta_1 (\text{graph } f) \}$.
Because of $\Delta_1 = \Delta_1 \Delta_1^* \Delta_1$ (cf. [5]) we obtain

$$f^{\Delta_1}(y) = \inf \{ 1 \mid (y, 1) \in \Delta_1 (\Delta_1^*(\Delta_1 (\text{graph } f))) \}. \tag{3.3}$$

In the proof of Theorem 3.1, (ii) was shown, that for each epigraphical polarity Δ_1 holds:

$$\Delta_1^* (\text{graph } f^{\Delta_1}) = \Delta_1^* (\text{graph } f)).$$

Hence $\Delta_1 (\Delta_1^*(\text{graph } f^{\Delta_1})) = \Delta_1(\Delta_1^*(\Delta_1 (\text{graph } f))).$

We conclude $\quad f^{\Delta_1}(y) = \inf \{ 1 \mid (y, 1) \in \Delta_1 (\Delta_1^*(\Delta_1 (\text{graph } f^{\Delta_1})) \}.$

Hence the assertion (iii) follows by Definition 3.2, (ii).

Assertion (iv) can be proved in a similar way.

A general property of P-convexity is given by

Theorem 3.3. (i) For a each set $A \subseteq X \times R$ and each set $Y \neq \emptyset$ there exists a polarity $\Delta : 2^{X\times R} \to 2^{Y\times R}$ such that A is P-convex with respect to Δ.

(ii) For each function $f : X \to R$ and each set $Y \neq \emptyset$ there exists a polarity $\Delta : 2^{X\times R} \to 2^{Y\times R}$ such that f is P-convex with respect to Δ.

Proof. (i): If $p : X \times R \times Y \times R \to R$ with

$$p (x, k, y, 1) := \begin{cases} 0 & \text{if} \quad (x, k) \in A, \\ -1 & \text{if} \quad (x, k) \notin A \end{cases}$$

is considered as a generating functional of a polarity Δ, then we have

$$\begin{aligned} \Delta(A) &= \{ (y, 1) \mid \forall (x, k) \in A : p(x, k, y, 1) \geq 0 \} \\ &= Y \times R . \end{aligned}$$

Hence

$$\begin{aligned} \Delta^*(\Delta(A)) &= \{ (x, k) \mid \forall (y, 1) \in \Delta(A) : p(x, k, y, 1) \geq 0 \} \\ &= \{ (x, k) \mid \forall (y, 1) \in Y \times R : p(x, k, y, 1) \geq 0 \} \\ &= A . \end{aligned}$$

Subsequently, A is P-convex with respect to Δ.

(ii) : Assertion (ii) follows from (i) because of Definition 3.2, (ii) by $A = \text{graph } f$.

Examples. 1. Let $X = R^n, Y = R^{n+1}, \langle \cdot , \cdot \rangle : R^n \times R^n \to R$ the common scalar product. Then the polarity $\Delta : 2^{X\times R} \to 2^{Y\times R}$ described by the generating functional

$$p(x, k, (y, y_0), 1) := - \langle x, y \rangle - k1+y_0, \quad x \in X \ , \ (y, y_0) \in Y,$$

has the representation

$$\Delta(A) = \bigcap_{(x,k)\in A} \{ ((y, yo), 1) \mid - \langle x, y \rangle - k1+y_0 \geq 0 \}, \quad A \subseteq X \times R.$$

For the dual polarity $\Delta^* : 2^{Y\times R} \rightarrow 2^{X\times R}$ holds

$$\Delta^*(B) = \bigcap_{((y,y_0),1)\in B} \{ (x,k) \mid -\langle x,y\rangle - k1 + y_0 \geq 0 \},\ B \subseteq Y\times R.$$

If $((y,y_0),1)$ is fixed, then $\Delta^*((y,y_0),1)$ is a closed halfspace in $X\times R$.
By tha Hahn-Banach-Theorem we conclude that $\Delta^*\Delta(A)$ is the closed convex hull of A (cf. [5], p. 11).

2. Let $X = Y$ be a nonempty set and let $\Delta : 2^{X\times R} \rightarrow 2^{Y\times R}$ be the complement operator of sets. Then this mapping is a polarity by De Morgan's rule.
Obviously is $\Delta^* = \Delta$, i.e. the polarity coincides with its dual polarity. Moreover we have $\Delta^*\Delta = I$, where I is the identity, that means $\Delta^*\Delta(A) = A$, $A \subseteq X\times R$.
Hence the P-convex hull of a set coincides always with the set itself.

4. M-Conjugation and Fractional Optimization

Applications of duality theory, p.e. numerical methods, are often related to such conjugation concepts which are based on linear separation. That means, they are finally connected with the Fenchel conjugation. But, on the other hand, we have also the corresponding disadvantages: to omit duality gaps, we need the assumption of convexity with respect to the involved functions. Subsequently it is of interest to construct such a conjugation, where both linear separation is used and which is not restricted to Fenchel-conjugation. Such concepts were given in [3] (Φ-conjugation) and [4], [9] (M-conjugation).
In the following M-conjugation will be handled, since a special case of this conjugation can be used for the investigation of fractional optimization problems.
Let f and g be real-valued functions with $f, g > 0$ and $D(f), D(g) \subseteq R^n$. We introduce the functions, f^M, g_M : $R^{n+1} \rightarrow R$ where

$$f^M(y,r) := \sup_{x\in D(f)} (\langle x,y\rangle + r) / f(x),$$

$$D(f^M) := \{ (y,r) \in R^{n+1} \mid \sup_{x\in D(f)} (\langle x,y\rangle + r) / f(x) < +\infty \}$$

and

$$g_M(y,r) := \inf_{x\in D(g)} (\langle x,y\rangle + r) / g(x),$$

$$D(g_M) := \{ (y,r) \in R^{n+1} \mid \inf_{x\in D(g)} (\langle x,y\rangle + r) / g(x) < -\infty \}$$

Remark (cf [4], pp. 36-38): f^M (resp. g_M) can be considered as a special upper (resp. lower) M-conjugate function of f (resp. g). Moreover, the M-conjugation is a special case of the P-conjugation.
The supremum (resp. the infimum) in the definition of f^M (resp. g_M) instead of the infimum (resp. the supremum) occuring in Definition 3.1 is a consequence of Theorem 3.1.
The conjugation introduced above is of importance for treating dual fractional optimization problems (cf. [1], [2], [3]):

$$(P) \quad f(x) / g(x) \to \min, \; x \in D(f) \cap D(g) \;,$$

$$(D) \quad g_M(y,r) / f^M(y,r) \to \max, \; (y,r) \in D(f^M) \cap D(g_M) \,.$$

For the problems (P), (D) we formulate duality theorems.

Theorem 4.1 (Weak Duality Theorem). Let

$$D(f) \cap D(g) \neq \emptyset \;,\; D(f^M) \cap D(g_M) \neq \emptyset \;,\; f^M > 0 \,.$$

Then
$$\inf_{x \in D(f) \cap D(g)} f(x)/g(x) \geq \sup_{(y,r) \in D(f^M) \cap D(g_M)} g_M(y,r) / f^M(y,r)$$

Proof. By the definition of M-conjugate functions we have

$$f(x) f^M(y,r) \geq \langle x, y \rangle + r \qquad \forall x \in D(f), \qquad \forall (y,r) \in D(f^M),$$
$$g(x) g_M(y,r) \leq \langle x, y \rangle + r \qquad \forall x \in D(f), \qquad \forall (y,r) \in D(g_M).$$

Combining these inequalities we obtain

$$f(x) f^M(y,r) \geq g(x) g_M(y,r) \; \forall x \in D(f) \cap D(g), \; \forall (y,r) \in D(f^M) \cap D(g_M)$$

and hence $f(x)/g(x) \geq g_M(y,r)/f^M(y,r) \quad \forall x \in D(f) \cap D(g), \; \forall (y,r) \in D(f^M) \cap D(g_M)$.
The assertion of the theorem is an easy consequence.

Theorem 4.2 (Strong Duality Theorem I). Let

$$D(f) \cap D(g) \neq \emptyset \,,\; f^M > 0 \,,$$
$$\mathrm{ri}\; D(\alpha g) \cap \mathrm{ri}\; D(f) \neq \emptyset \,,$$

where
$$\alpha := \inf_{x \in D(f) \cap D(g)} f(x)/g(x) > -\infty \,.$$

If the function f is convex and the function αg is concave, then there exists a $(y_0, r_0) \in R^{n+1}$ such that

$$\inf_{x \in D(f) \cap D(g)} f(x)/g(x) = \max_{(y,r) \in D(f^M) \cap D(g_M)} g_M(y,r)/f^M(y,r) = g_M(y_0, r_0)/f^M(y_0, r_0)$$

Proof. By the definition of α we have

(4.1) $$\alpha \le f(x)/g(x) \quad \forall x \in D(f) \cap D(g)$$

or $$\alpha g(x) \le f(x) \quad \forall x \in D(f) \cap D(g)$$

According the assumptions, the intersection of ri D(f) and ri D(α g) is nonempty, the function f is convex, the function αg is concave. Hence, by an appropriate separation theorem there exists an $(y_o, r_o) \in R^{n+1}$ such that $\forall x \in D(f) \cap D(g)$:

(4.2) $$\alpha g(x) \le \langle x, y_o \rangle + r_o \le f(x)$$

From the left-hand inequality of (4.2) we obtain

$$\alpha \le (\langle x, y_o \rangle + r_o)/g(x) \qquad \forall x \in D(f) \cap D(g)$$

and thus $$\alpha \le \inf_{x \in D(f) \cap D(g)} (\langle x, y_o \rangle + r_o)/g(x)$$

Because of $\alpha > -\infty$ and the definition of the M-conjugate functions we have

(4.3) $$\alpha \le g_M(y_o, r_o).$$

From the right-hand inequality of (4.2) we obtain

$$(\langle x, y_o \rangle + r_o)/f(x) \le 1 \qquad \forall x \in D(f) \cap D(g)$$

and thus $$\sup_{x \in D(f) \cap D(g)} (\langle x, y_o \rangle + r_o)/f(x) \le 1.$$

By the definition of M-conjugate functions follows $f^M(y_o, r_o) \le 1$;
by (4.3) we conclude $\alpha \le g_M(y_o, r_o)/f^M(y_o, r_o)$.
By Theorem 4.1 we obtain the assertion of the theorem.

In Theorem 4.2, the assumptions concerning f and g are not symmetrical. Thus it is obvious that there will be another duality theorem, where f and g "change their rolls".

Theorem 4.3 (Strong Duality Theorem II). Let

$$D(f) \cap D(g) \ne \emptyset, \quad f^M > 0,$$
$$\text{ri } D(\alpha^{-1} f) \cap \text{ri } D(g) \ne \emptyset,$$

where $$\alpha := \inf_{x \in D(f) \cap D(g)} f(x)/g(x) > -\infty, \ \alpha \ne 0.$$

If the function g is concave and the function $\alpha^{-1} f$ is convex, then there exists a $(y_0, r_0) \in R^{n+1}$ such that

$$\inf_{x \in D(f) \cap D(g)} f(x)/g(x) = \max_{(y,r) \in D(f^M) \cap D(g_M)} g_M(y,r)/f^M(y,r) = g_M(y_0,r_0)/f^M(y_0,r_0).$$

Besides direct duality theorems we can also derive inverse duality theorems.

Theorem 4.4 (Inverse Duality Theorem). Let

$$D(f^M) \cap D(g_M) \neq \emptyset,$$

$$\forall x \in D(f) \cap D(g) : f(x) = f^{MM}(x), \; g(x) = g_{MM}(x); \quad f^M, g_M > 0,$$

$$\beta := \sup_{(y,r) \in D(f^M) \cap D(g_M)} g_M(y,r)/f^M(y,r) < \infty$$

If there is an $x_0 \in D(f) \cap D(g)$ such that

$$\beta f^M(y,r) \geq \langle x_0, y \rangle + r \geq g_M(y,r) \qquad \forall (y,r) \in D(f^M) \cap D(g_M) \tag{4.4}$$

then holds $\displaystyle \sup_{(y,r) \in D(f^M) \cap D(g_M)} g_M(y,r)/f^M(y,r) \quad \min_{x \in D(f) \cap D(g)} f(x)/g(x) = f(x_0)/g(x_0).$

Proof. By the left-hand side of (4.4) we have $\beta f^M(y,r) \geq \langle x_0, y \rangle + r$ and also $\beta \geq \langle x_0, y \rangle + r/f^M(y,r)$.

It follows $\displaystyle \beta \geq \sup_{(y,r) \in D(f^M)} (\langle x_0, y \rangle + r)/f^M(y,r) = f^{MM}(x_0).$

Because of $\beta < \infty$ we have $x_0 \in D(f^{MM})$ and consequently with $f^{MM}(x_0) = f(x_0)$

$$\beta \geq f(x_0). \tag{4.5}$$

Because of the right-hand side of (4.4) we obtain $\langle x_0, y \rangle + r \geq g_M(y,r)$.

This leads to $\displaystyle g_{MM}(x_0) = \inf_{(y,r) \in D(g_M)} (\langle x_0, y \rangle + r)/g_M(y,r) \geq 1$

and because of the assumption $g(x_0) = g_{MM}(x_0)$ it follows

$$g(x_0) \geq 1. \tag{4.6}$$

From (4.5) and (4.6) we obtain $(f(x_0)/g(x_0)) \geq 1$.
Because of the weak duality assertion (Theorem 4.1) the assertion of our theorem follows.

Analogously to the direct duality theorems, we can obtain another ("symmetrical") inverse duality theorem, when the assumption (4.4) is replaced by

$$(4.4') \qquad \beta^{-1} g_M(y,r) \leq \langle x_0, y \rangle + r \leq f^M(y,r) \quad \forall (y,r) \in D(f^M) \cap D(g_M).$$

References

/1/ **Beoni C.**, A generalization of Fenchel duality theory. JOTA 49 (1986) 3, 375-388.
/2/ **Cambini A., Martein L.**, On the Fenchel-like and Lagrangian duality in fractional programming. In Beeckmann, M.J. et al. (eds.): X Symp. on Operations Research München 1985. Methods of Operations Research Vol. 53 (1986), 21-32.
/3/ **Deumlich R., Elster K.-H.**, Fractional programming in view of generalized conjugate functions. In: Brucker P., Pauly R., (eds.): IX Symp. on Operations Research Osnabrück 1984. Methods on Operations Research Vol. 49 (1985), 3-16.
/4/ **Elster K.-H., Wolf A.**, On a general concept of conjugate functions as an approach to nonconvex optimization problems. Univ. Pisa, Dipartimento di Matematica, Sezione di Matematica Applicata. Gruppo di Ottimizzazione e Ricerca Operativa. Paper 149, 1988, 54 pp.
/5/ **Evers J.J.M., Van Maaren H.**, Duality principles in mathematics and their relations to conjugate functions. Twente Univ. of Technology. Mimeo 336, 1981. Nieuw Archiv voor Wiskunde 3 (1985), 23-68.
/6/ **Fenchel W.**, Convex cones, sets and functions. Lecture Notes Dep. of Mathematics, Princeton University, Princeton 1953.
/7/ **Schaible S., Ziemba W.T.**, Generalized Concavity in Optimization and Economics. Academic Press, New York 1981, 782 pp.
/8/ **Volle M.**, Conjugaison par tranches. Annali di Mathematica pura e Applicata. 89 (1985) 4, 279-312.
/9/ **Wolf A.**, Konjugationen und ihre Anwendungen in der nichtkonvexen Optimierung. Diss. A. TH Ilmenau, 1988.

DUALITY IN MULTIOBJECTIVE FRACTIONAL PROGRAMMING

C.R. Bector - S. Chandra - C. Singh

Abstract

A class of multiobjective fractional programming problems is considered and duality results are established through a parametric approach. These results give a multiobjective analog of some of the well known results of a single ratio programming problem.

1. Introduction

Duality for multiobjective nonlinear programming problems has been of much interest in the recent past and many contributions have been made to its development (Bitran [3], Brumelle [4], Craven [5], Kawasaki [9], Singh [15, 16], Tanino and Sawaragi [18] and Weir [19]). These studies differe in their approaches as well as the sense in wich "optimality" is defined for the multiobjective programming problem.

Duality in fractional programming involving the optimization of a single ratio has been studied extensively (e.g. see [14]) in the past but there have been very few studies in multiobjective fractional programming duality. Certain special nonlinear fractional cases have been dealt with in Singh [17], Weir [20] and Lin [10] while duality for pseudolinear multiobjective programming problems has been studied by Bector and Chandra [1].

The purpose of the present paper is to consider the duality of a general multiobjective fractional programming problem by relating the primal problem to a parametric multiobjective programming problem. This development is in the spirit of Jagannathan [8] and Bector et al. [2] and depends heavily on the approach of Singh [16].

2. Preliminaries

For $x, y \in R^n$, by $x \leq y$ we mean $x_i \leq y_i$ for all i, $x \leq y$ means $x_i \leq y_i$ for all $i \neq s$, $1 \leq i \leq n$ and $x_s < y_s$ for at least one s, $1 \leq s \leq n$. By $x < y$ we mean $x_i < y_i$ for all i, $1 \leq i \leq n$. All vectors will be taken as column vectors.

Let $S \subsetneq R^n$ be open. Let f, g and h be vector valued functions defined on S and taking values in R^p, R^p and R^m respectively. We assume that all these functions are differentiable and the symbol ∇_x denotes the usual gradient operator with recpect to x. The symbol $()^T$ will stand for the transpose.

We now state the multiobjective programming problem to be studied here:

(P) $$\text{V-minimize} \left(\frac{f_1(x)}{g_1(x)}, \ldots\ldots\ldots, \frac{f_p(x)}{g_p(x)} \right)$$

subject to

(1) $$h_j(x) \le 0 \quad (j = 1, 2, \ldots\ldots, m)$$

(2) $$x \in S$$

Hence the symbol "V-minimize" stands for vector minimization and minimality is taken in terms of "efficient points" or 'Pareto optimal solutions" as defined below. We assume that $g_i(x) > 0$ and $f_i(x) \ge 0$ whenever $g_i(x)$ is not linear for all $i = 1, 2, \ldots, p$, $x \in S$. Let X denote the set of all feasible solutions of P and

$$\Phi(x) = (\Phi_1(x), \Phi_2(x), \ldots\ldots, \Phi_p(x)), \text{ where } \Phi(x) = f_i(x)/\, g_i(x), \; i = 1, 2, \ldots, p.$$

Definition 1 ([12]). A point $\bar{x} \in X$ is said to be an efficient solution of (P) if there does not exist any $x \in X$ such that $\Phi(x) \le \Phi(\bar{x})$.
For a vector maximization problem an efficient solution is defined similarly.

Definition 2 ([12]). let $Y \subseteq R^t$. The vector $q \in R^t$ is called a convergence vector for Y at $y^o \in Y$ if there exists a sequence $\{y_k\}$ in Y and a sequence $\{\alpha_k\}$ of positive real numbers such that

$$\lim_{k \to \infty} y_k = y^o, \quad \lim_{k \to \infty} \alpha_k = 0, \quad \lim_{k \to \infty} \frac{y_k - y^o}{\alpha_k} = q$$

Following Marùsciac [12], let $C(X, \bar{x})$ be the set of all convergence vectors for X at $\bar{x}$. Further let

$$D = \{ d \in R^n : \nabla h_j(\bar{x})\, d \le 0 \}$$

where J is the set of all active constraints at $\bar{x}$ i.e. $J = \{ j : h_j(\bar{x}) = 0, j = 1, 2, \ldots\ldots, n \}$.

Definition 3 ([15]). The constraints functions h will be said to satisfy a constraint qualification at $\bar{x}$ if $D \subseteq C(X, \bar{x})$.

3. Dual problem: formulation and motivation

Taking motivation from Dinkelbach [6] and Jagannathan [7], we consider the following parametric vector optimization problem (P_v) for each $v \in R^p_+$, where R^p_+ denotes the nonnegative orthant of R^p.

$$(P_v) \qquad \text{V-minimize} \quad (f_1(x) - v_1 g_1(x), \ldots\ldots, f_p(x) - v_p g_p(x))$$

subject to

$$h_j(x) \le 0, \quad (j = 1, 2, \ldots\ldots, m)$$

$$x \in S$$

We now prove the following

Lemma 1. Let $\bar{x}$ be efficient to (P). Then there exists $\bar{v} \in R^p_+$ such that $\bar{x}$ is also efficient to $(P_{\bar{v}})$.

Proof. For $i = 1, 2, \ldots, p$ let $\bar{v}_i = \dfrac{f_i(\bar{x})}{g_i(\bar{x})}$ and $\bar{v} = (\bar{v}_1, \bar{v}_2, \ldots\ldots, \bar{v}_p)$.

Let $\bar{x}$ be not efficient for $(P_{\bar{v}})$. This implies that there is an x feasible for $(P_{\bar{v}})$ such that

$$f_i(x) - \bar{v}_i g_i(x) \le f_i(\bar{x}) - \bar{v}_i g_i(\bar{x}) \qquad \text{for all } i = 1, 2, \ldots\ldots, p$$

and

$$f_i(x) - \bar{v}_i g_i(x) < f_i(\bar{x}) - \bar{v}_i g_i(\bar{x}) \qquad \text{for at least one } i \text{ (say } i = i_0).$$

Since $\bar{v}_i = (f_i(\bar{x})/g_i(\bar{x}))$, therefore, $f_i(\bar{x}) - \bar{v}_i g_i(\bar{x}) = 0$ for $i = 1, 2, \ldots\ldots, p$.
Hence from above we have,

$$\frac{f_i(x)}{g_i(x)} \le \bar{v}_i = \frac{f_i(\bar{x})}{g_i(\bar{x})} \qquad \text{for all } i$$

and

$$\frac{f_{i_0}(x)}{g_{i_0}(x)} < \bar{v}_{i_0} = \frac{f_{i_0}(\bar{x})}{g_{i_0}(\bar{x})}$$

This contradicts that $\bar{x}$ is efficient to (P). Hence the result follows.

Remark 1.1. : The converse of Lemma 1.1 also holds provided we assume

$$\bar{v}_i = f_i(\bar{x})/g_i(\bar{x}) \text{ i.e. } f_i(\bar{x}) - \bar{v}_i\ g_i(\bar{x}) = 0 \text{ for } i = 1, 2, \ldots, p.$$

Now in view of Lemma 1 above and in analogy with the traditions in fractional programming duality (Jagannathan [8], Bector, Chandra and Bector [2]) we introduce the following problem (D) as the dual of (P):

(D) $$\text{V-maximize } (v_1, v_2, \ldots, v_p)$$

subject to

(3) $$\nabla\ [\ \lambda^T f(u) - \sum_{i=1}^{p} \lambda_i v_i g_i(u) + \mu^T h(u)\ ] = 0$$

(4) $$f_i(u) - v_i g_i(u) \geq 0 \qquad (i = 1, 2, \ldots, p)$$

(5) $$\mu^T h(u) \geq 0$$

(6) $$v \in R^p_+,\ \mu \in R^m_+,\ \lambda \in R^p \text{ and } \lambda \geq 0$$

(7) $$u \in S$$

where $f(u) = (f_1(u), \ldots, f_p(u))$ and $h(u) = (h_1(u), \ldots, h_m(u))$.
In the analysis to follow we shall assure that S is a convex set. Other assumptions will be stated at appropriate places.
Let Z be the set of feasible solutions of (D). As in [16], let $C(Z; (\bar{u}, \bar{\lambda}, \bar{\mu}, \bar{v}))$ be the set of all convergence vectors for Z at $\bar{z} = (\bar{u}, \bar{\lambda}, \bar{\mu}, \bar{v})$ and let $E = \{ d \in R^n:\ \nabla_z H(\bar{z})\ d = 0, \nabla_z L_1(\bar{z})\ d \leq 0\}$ where $z = (u, \lambda, \mu, v)$, $H(z) = \nabla\ [\ \lambda^T f(u) - \sum_{i=1}^{p} \lambda_i v_i g_i(u) + \mu^T h(u)\ ]$ and $L_1(\bar{z})$ are other dual constraints ((4) and (5)) which are active at $\bar{z}$ after they have been expressed in the form of primal constraints.

Definition 4. ([12, 16]) Dual constraints will be said to satisfy a constraint qualification at $z = (\bar{u}, \bar{\lambda}, \bar{\mu}, \bar{v})$ if $E \subseteq C(Z, \bar{z})$.

4. Duality

We shall now be proving various duality theorems between (P) and (D)

Theorem 1 (Weak Duality). Let x be feasible to (P) and (u, λ, μ, v) be feasible to (D). At u, let $\sum_{i=1}^{p} \lambda_i(f_i - v_i g_i)$ be strictly pseudoconvex for all $v_i \in R_+$ and $\mu^T h$ be quasiconvex. Then

$$f(x)/\, g(x) \;\nleq\; v$$

Proof. Suppose $\dfrac{f(x)}{g(x)} \leq v$. This implies that

$$\frac{f_i(x)}{g_i(x)} \leq v_i \qquad \text{for all } i$$

and
$$\frac{f_i(x)}{g_i(x)} < v_i \qquad \text{for at least one } i, \text{ say } i = i_0$$

i.e.,
$$f_i(x) - v_i g_i(x) \leq 0 \qquad \text{for all } i$$

$$f_{i_0}(x) - v_{i_0}\, g_{i_0}(x) < 0$$

In view of (4) and $\lambda \geq 0$ we have

$$(8) \qquad \sum_{i=1}^{p} \lambda_i(f_i(x) - v_i g_i(x)) < 0 \leq \sum_{i=1}^{p} \lambda_i(f_i(u) - v_i g_i(u))$$

But $\sum_{i=1}^{p} \lambda_i(f_i - v_i g_i)$ is strictly pseudo convex at u for $v_i \in R_+$ and hence now (8) gives

$$(9) \qquad (x - u)^T \nabla\Big(\sum_{i=1}^{p} \lambda_i(f_i(u) - v_i g_i(u)) \Big) < 0$$

Further $\mu^T h(x) \leq 0 \leq \mu^T h(u)$ and hence due to the quasiconvexity of $\mu^T h$ at u we have

$$(10) \qquad (x - u)^T \nabla(\mu^T h(u)) \leq 0$$

Adding (9) and (10) we get

$$(x - u)^T \nabla[\lambda^T f(u) - \sum_{i=1}^{p} \lambda_i v_i g_i(u) + \mu^T h(u)] < 0,$$

which contradicts (3). This proves the theorem.

Corollary 1. Let $\bar{x}$ be feasible to (P) and $(\bar{u}, \bar{\lambda}, \bar{\mu}, \bar{v})$ be feasible to (D) with $(f(\bar{x})/g(\bar{x})) = \bar{v}$. Let hypotheses of Theorem 1 hold. Then $\bar{x}$ and $(\bar{u}, \bar{\lambda}, \bar{\mu}, \bar{v})$ are efficient to (P) and (D) respectively.

Proof. Let $\bar{x}$ be not efficient to (P). Then there exists an x^0 feasible for (P) such that

$$\frac{f(x^0)}{g(x^0)} \leq \frac{f(\bar{x})}{g(\bar{x})} = \bar{v}$$

which contradicts the conclusion of Theorem 1. Hence $\bar{x}$ is efficient. The efficiency of $(\bar{u}, \bar{\lambda}, \bar{\mu}, \bar{v})$ follows similarly.

Theorem 2 (Strong Duality). Let $\bar{x}$ be efficient to (P) and the primal constraint qualification (Definition 3) hold at $\bar{x}$. Then there exists $\bar{\lambda} \in R^p_+$, $\bar{\mu} \in R^m_+$, $\bar{v} \in R^p_+$ such that $(\bar{x}, \bar{\lambda}, \bar{\mu}, \bar{v})$ is feasible to (D). If in addition the weak duality theorem holds then $(\bar{x}, \bar{\lambda}, \bar{\mu}, \bar{v})$ is efficient to (D).

Proof. Since $\bar{x}$ is efficient to (P), by Lemma 1, $\bar{x}$ is efficient to (P_v) for $\bar{v}_i = (f_i(\bar{x}) / g_i(\bar{x}))$. Hence using necessary optimality conditions for (P_v) (Singh [15]), there exists $\bar{\lambda} \in R^p_+$ $(\bar{\lambda} \geq 0)$, $\mu \in R^m_+$ such that

$$\nabla [\lambda^T f(x) - \sum_{i=1}^{p} \bar{\lambda}_i v_i g_i(\bar{x}) + \mu^T h(\bar{x})] = 0,$$

$$\bar{\mu}^T h(\bar{x}) = 0$$

Further for the above mentioned choice of v_i, $f_i(\bar{x}) - v_i g_i(\bar{x}) = 0$ for $i = 1, 2, \ldots, p$. Hence $(\bar{x}, \bar{\lambda}, \bar{\mu}, \bar{v})$ is feasible to (D). The rest of the theorem follows by observing that $(f(\bar{x})/g(\bar{x})) = \bar{v}$ and using Corollary 1.

Theorem 3 (Converse Duality). Let $(\bar{u}, \bar{\lambda}, \bar{\mu}, \bar{v})$ be efficient to (D) at which the dual constraint qualification of Section 3(Definition 4) holds. If,

(i) $\bar{\lambda} > 0$

(ii) $\{\nabla f_i(u) - \bar{v}_i g_i(\bar{u}))$, $i = 1, 2, \ldots, p\}$ are linearly independent, and

(iii) the matrix $\nabla^2(\lambda^T f(u) - \sum \lambda_i v_i g_i(u) + \mu^T h(u))$ is positive or negative definite at $(\bar{u}, \bar{\lambda}, \bar{\mu}, \bar{v})$.
then $\bar{u}$ is feasible to (P). If in addition,

(iv) the weak duality theorem holds,
then $\bar{u}$ is efficient to (P).

Proof. Since $(\bar{u}, \bar{\lambda}, \bar{\mu}, \bar{v})$ is efficient to (D) by rewriting (D) in the form of (P) and using necessary optimality conditions of [15], we observe that there exists $\xi \in R^p_+$, $\xi \geq 0$, $\eta \in R^n$, $\delta \in R^p_+$, $\rho \in R$, $w \in R^m_+$, $\theta \in R^p_+$, $\sigma \in R^p_+$ such that

$$(11) \qquad \nabla^2 [\lambda^T f(\bar{u}) - \sum_{i=1}^{p} \bar{\lambda}_i \bar{v}_i g_i(\bar{u}) + \bar{\mu}^T h(\bar{u})] \eta + \sum_{i=1}^{p} \delta_i \bar{v}_i \nabla g_i(\bar{u}) - \nabla [\sum_{i=1}^{p} \delta_i f_i(\bar{u}) - \rho\mu^T h(\bar{u})] = 0$$

$$(12) \qquad \eta^T(\nabla(f_i(\bar{u}) - \bar{v}_i g_i(\bar{u})) - \sigma_i = 0 \qquad (i = 1, 2, \ldots, p),$$

$$(13) \qquad \eta^T(\nabla(h_j(\bar{u}) - \rho h_j(\bar{u})) - \omega_j = 0 \qquad (j = 1, 2, \ldots, m),$$

$$(14) \qquad -\xi_i - \lambda_i \eta^T(\nabla(g_i(\bar{u}) + \delta_i g_i(\bar{u}) - \theta_i = 0 \qquad (i = 1, 2, \ldots, p),$$

$$(15) \qquad \delta_i(f_i(\bar{u}) - \bar{v}_i g_i(\bar{u})) = 0 \qquad (i = 1, 2, \ldots, p),$$

$$(16) \qquad \rho(\bar{\mu}^T h(\bar{u})) = 0$$

$$(17) \qquad \omega^T \bar{\mu} = 0, \ \theta^T \bar{v} = 0, \ \sigma^T \bar{\lambda} = 0$$

$\bar{\lambda} > 0$ and (17) yield, $\sigma = 0$. Therefore, (12) gives

$$(18) \qquad \eta^T(\nabla(f_i(\bar{u}) - \bar{v}_i g_i(\bar{u})) = 0 \qquad (i = 1, 2, \ldots, p).$$

Multiplying (13) by $\bar{\mu}_j$ and summing over j and using (17) we get

$$(19) \qquad \eta^T(\nabla \bar{\mu}^T h(\bar{u})) - \rho(\bar{\mu}^T h(\bar{u})) = 0$$

i.e., $\eta^T(\nabla \bar{\mu}^T h(\bar{u})) = 0$ as the second term of (19) is zero due to (16).
Pre-multiplying (11) by η^T and using (18) and (19) we get

$$\eta^T \nabla^2 [\bar{\lambda}^T f(\bar{u}) - \sum \bar{\lambda}_i \bar{v}_i g_i(\bar{u})) + \bar{\mu}^T h(\bar{u})] \eta = 0,$$

which by hypotesis (iii) of the theorem implies $\eta = 0$.
Next, from (3) we substitute for $\nabla(\bar{\mu}^T h(\bar{u}))$ in (11) to get

$$\sum_{i=1}^{p} (\delta_i - \rho)\, \nabla(f_i(\bar{u}) - \bar{v}_i g_i(\bar{u})) = 0$$

Since, by hypothesis (ii) of the theorem, the vectors $\{ \nabla(f_i(\bar{u}) - \bar{v}_i g_i(\bar{u})),\ i = 1, 2, \ldots, p \}$ are linearly independent, therefore, the above equations give $\delta_i = \rho$ for all i.
We now claim that $\rho > 0$. If $\rho = 0$ then this means that $\delta_i = 0$ for all i which because of (14) gives that $\xi_i = \theta_i = 0$ for all $i = 1, 2, \ldots, p$.
But $\xi \geq 0$ and hence ξ_i cannot be zero for all i. Thus $\rho > 0$ as claimed.
Therefore, (13) gives $\rho h_j(\bar{u}) - \omega_j = 0$ for all $j = 1, 2, \ldots, m$ i.e. $h_j(\bar{u}) = -(\omega_j / \rho) \leq 0$ for all $j = 1, 2, \ldots, m$. Thus $\bar{u}$ is feasible to (P). Also, as $\delta_i = \rho > 0$ for all i, (15) implies that $(f(\bar{u})/g(\bar{u})) = \bar{v}$. The rest of the theorem is now a consequence of Corollary 1.

Theorem 4 (Strict Converse-Like Duality). Let $\bar{x}$ be feasible to (P) and $(\bar{u}, \bar{\lambda}, \bar{\mu}, \bar{v})$ be feasible to (D) such that $\Phi(\bar{x}) = \bar{v}$. At $\bar{u}$ let $\sum_{i=1}^{p} \bar{\lambda}_i (f_i - v_i g_i)$ be quasiconvex and $\bar{\mu}^T h$ be strictly pseudoconvex. Then, $\bar{x} = \bar{u}$.

Proof. Let $\bar{x} \neq \bar{u}$. Since $(\bar{u}, \bar{\lambda}, \bar{\mu}, \bar{v})$ is feasible to (D) we have $\bar{\mu}^T h(\bar{u}) \geq 0$ and

$$\nabla[\lambda^T f(\bar{u})) - \sum_{i=1}^{p} \bar{\lambda}_i \bar{v}_i g_i(\bar{u}) + \bar{\mu}^T h(\bar{u})] = 0 \tag{20}$$

Also $\bar{\mu} \geq 0$ and $h(\bar{x}) \leq 0$. Therefore $\bar{\mu}^T h(\bar{x}) \leq 0$. Hence $\bar{\mu}^T h(\bar{x}) \leq \bar{\mu}^T h(\bar{u})$. But $\bar{\mu}^T h$ is assumed to be strictly pseudoconvex at $\bar{u}$, therefore

$$(\bar{x} - \bar{u})^T \nabla(\bar{\mu}^T h(\bar{u})) < 0 \tag{21}$$

Now (20) and (21) imply

$$(\bar{x} - \bar{u})^T \nabla \sum_{i=1}^{p} \bar{\lambda}_i (f_i - \bar{v}_i g_i(u)) > 0$$

This along with quasiconvexity of $\sum_{i=1}^{p} \bar{\lambda}_i (f_i - \bar{v}_i g_i)$ yields,

$$\sum_{i=1}^{p} \bar{\lambda}_i (f_i(\bar{x}) - \bar{v}_i g_i(\bar{x}) > \sum_{i=1}^{p} \bar{\lambda}_i (f_i(\bar{u}) - \bar{v}_i g_i(\bar{u})). \tag{22}$$

Now the hypothesis $\Phi(\bar{x}) = \bar{v}$ makes left hand side of (22) equal to zero contradicting (4) since $(\bar{u}, \bar{\lambda}, \bar{\mu}, \bar{v})$ is feasible to (D). Hence $\bar{x} = \bar{u}$.

Remark 1. In the statement of Theorem 4, the strictly pseudo convexity of $\bar{\mu}^T h \bar{\mu} \neq 0$. This will happen for example if we assume that $\bar{\lambda} > 0$ and vectors $\{ \nabla(f_i(\bar{u}) - \bar{v}_i g_i(\bar{u})), i = 1, 2, \ldots, p \}$ are linearly independent.

Remark 2. We se that for $p = 1$, the form of the dual (D) reduces to the usual fractional dual (Jagannathan [8]) where duality theorems holds under more general conditions of pseudo convexity of $f(x)-vg(x)$ and quasi convexity of $\mu^T h$ (Mond and Weir [13]).

References

[1] **Bector C.R. and Chandra S.**, Duality for Pseudolinear Multiobjective Programs, Research Report 5-87, Faculty of Management, University of Manitoba, (Canada)

[2] **Bector C.R., Chandra S. and Bector M.K.**, Generalized Fractional Programming Duality: A Parametric Approach, Journal of Optimization Theory and Applications (to appear)

[3] **Bitran G.R.**, Duality for Nonlinear Multicriterion Optimization Problems, Journal of Optimization Theory and Applications, 35 (1981) pp.367-401.

[4] **Brumelle S.**, Duality for Multiple Objective Convex Programs, Mathematics of Operations Research, 6 (1981) pp. 159-172.

[5] **Craven B.D.**, Strong Vector Minimization and Duality, Zeitschrift fur Angewandte Mathematik and Mechanik, 60 (1980) pp.1-5.

[6] **Dinkelbach W.**, On Nonlinear Fractional Programming, Management Science, 13 (1967) pp. 492-498.

[7] **Jagannathan R.**, On some Properties of Programming Problems in Parametric Form Pertaining to Fractional Programming, Management Science, 12 (1966) pp.609-615.

[8] **Jagannathann R.**, Duality for Nonlinear Fractional Programs, Zeitschrift fur Operations Research, 17 (1973) pp. 1-3.

[9] **Kawasaki H.**, A Duality Theorem in Multiobjective Nonlinear Programming, Mathematics of Operations Research, 7 (1982) pp. 95-110.

[10] **Lin C.Y.**, Duality Theory for Multiobjective Fractional Programming, Numerical Methods Journal, Chinese University, 4 (1982) pp.289-299.

[11] **Luc D.T.**, On Duality Theory in Multiobjective Programming, Journal of Optimization Theory and Applications, 43 (1984) pp.457-482.

[12] **Mar'usciac I.**, On Fritz-John Type of Optimality Criterion in Multiobjective Optimization, L'Analise Numerique et la Theorie de L'Approximation, 11 (1982) pp.109-114.

[13] **Mond B. and Weir T.**, Generalized Concavity and Duality, Generalized Concavity in Optimization and Economics, S.Schaible and W.T.Ziemba (ed.), Academic Press, New York (1981)

[14] **Schaible S.**, Bibliography in Fractional Programming, Zeitschrift fur Operations Research, 26 (1982) pp.211-241.

[15] **Singh C.**, Optimality Conditions in Multiobjective Differentiable Programming, Journal of Optimization Theory and Applications, 53 (1987) pp.115-123.

[16] **Singh C.**, Duality Theory in Multiobjective Differentiable Programming, Journal of Information and Optimization Sciences (to appear)

[17] **Singh C.**, A Class of Multiple-Criteria Fractional Programming Problems, Journal of Mathematical Analysis and Applications, 115 (1986) pp.202-213.

[18] **Tanino T. and Sawaragi Y.**, Duality in Multiobjective Programming, Journal of Optimization Theory and Applications 27 (1979) pp. 509-529.

[19] **Weir T.**, Proper Efficiency and Duality for Vector Valued Optimization Problems, Research Report 12-85, Department of Mathematics, Royal Military College, Duntroon (Australia)

[20] **Weir T.**, A Dual for a Multiple Objective Fractional Programming Problem, Journal of Information and Optimization Sciences, 7 (1986) pp. 261-299.

AN APPROACH TO LAGRANGIAN DUALITY IN VECTOR OPTIMIZATION

L. Martein

Abstract

For a vector extremum problem, optimality conditions, regularity and duality can be studied by means of vector-valued weak separation functions; in this paper duality theorems are established for a wide class of multiobjective problems, by choosing a suitable class of vector-valued weak separation functions which is strictly related to a generalized vector-valued Lagrangian function. The suggested approach allows us to obtain some regulatiry conditions.

Introduction

In recent papers [3, 6, 10] by means of theorems of the alternative, weak and strong separation functions have been introduced in order to obtain a general framework within which optimality conditions and related topics can be studied. More exactly, sufficient optimality conditions, exterior penality method and duality have been deduced and/or interpreted by means of weak separation functions while penality interior scheme, and necessary optimality conditions can be deduced and/or interpreted by using strong separation functions.

Recently [4, 5, 11, 12, 13], for a vector extremum problem, optimality conditions and regularity have been studied by means of vector-valued weak separation functions; as a natural extension of the studies we will suggest, in this paper, a general approach to Lagrangian duality which allows us to obtain known results and new ones. More exactly duality will be established for a wide classe of multiobjective problems by choosing, in a sutitable way, a class of vector-valued weak separation functions which is strictly related to a generalized vector-valued Lagrangian function.

Furtermore the given approach points out that the role played by regularity in vector optimization has not yet been sufficiently investigated.

1. Statement of the Problem

Recently (see [3, 6]), theorems of the alternative for generalized systems have been studied and their use in scalar optimization exploited.

As a consequence of this analysis, the concept of image of a constrained extremum problem has been developed and recognized as a powerful tool in developing some basic topics in optimization such as optimality conditions, regularity and duality.

Since the aim of this, paper is to extend such an approach to study duality in vector optimization, we will briefly summarize some fundamental ideas.
Consider the scalar constrained problem

$$P_1 \quad : \qquad \min \Phi(x)\,;\, x \in R_S = \{\, x \in X : g(x) \geq 0 \,\}$$

where $\Phi : IR^n \to IR$ and $g : IR^n \to IR^m$.
Let $\overline{x}$ be a feasible point for P_1 and set:

$$f(x) \triangleq \Phi(\overline{x}) - \Phi(x),\ F(x) = (f(x), g(x)),\ \mathcal{K} = F(X)\,,\ \mathcal{H} = (IR_+ \setminus \{0\}\,) \times IR_+^m.$$

It is easy to prove (see [3, 6]) that $\overline{x}$ is a minimum point for P_1 iff

(1.1) $$\mathcal{K} \cap \mathcal{H} = \emptyset,$$

so that any condition which ensures the validity of (1.1) becomes a sufficient optimality condition.
The disjunction between $\mathcal{K}$ and $\mathcal{H}$ can be studied by means of suitable weak separation functions; we say that $w : IR \times IR^m \to IR$ is a weak separation function iff

(1.2) $$w(u\,,v) \geq 0 \qquad \forall (u\,,v) \in \mathcal{H}$$

Furthermore, if w guarantees alternative, that is the following condition is verified:

(1.3) $$w(F(x)) \leq 0 \qquad \forall\, x \in X,$$

then (1.1) holds; such a property points out the role of weak separation functions in studying the disjunction between $\mathcal{K}$ and $\mathcal{H}$ and related topics.
A general class of weak separation functions is as follows [3, 4, 5, 6, 12]:

$$W_1 = \{\, w : IR \times IR^m \to IR,\ w(u, v; \omega) = u + \gamma(v; \omega),\ \omega \in \Omega \,\}$$

where the set Ω of parameters ω and $\gamma : IR^m \to IR$ are such that

(1.4a) $$\mathrm{lev}_{\geq 0}\ \gamma \supseteq IR_+^m \qquad \forall\, \omega \in \Omega$$

(1.4b) $$\bigcap_{\omega \in \Omega} \mathrm{lev}_{\geq 0}\ \gamma = IR_+^m$$

(1.4c) $$\gamma(\cdot\,;\overline{\omega}) > 0 \Rightarrow \exists\, \hat{\omega} \in \Omega \quad \text{such that} \quad \gamma(\cdot\,;\hat{\omega}) < \gamma(\cdot\,;\overline{\omega})$$

where $\mathrm{lev}_{\geq 0}\ \gamma = \{\, v : \gamma(v) \geq 0 \,\}$.

Now we will give the scheme of the general approach to duality proposed in [3, 6].
Let W_1 be a class of weak separation functions and suppose that $\bar{x}$ is an optimal solution for problem P_1; condition (1.4) implies that $w(f(\bar{x}), g(\bar{x}); \omega) \geq 0 \quad \forall \omega \in \Omega$, so that:

$$(1.5) \qquad \max_{x \in X} \; w(f(x), g(x); \omega) \geq 0 \qquad \forall \omega \in \Omega .$$

Here and in the sequel, for the sake of simplicity, we will suppose that any maximum or minimum is attained.
Taking into account that

$$w(f(x), g(x); \omega) = \phi(\bar{x}) - \phi(x) + \gamma(g(x); \omega)$$

condition (1.5) implies

$$(1.6) \qquad \phi(\bar{x}) \geq \max_{\omega \in \Omega} \; \min_{x \in X} \; (\phi(x) - \gamma(g(x); \omega)) .$$

Let us introduce $L(x ; \omega) \overset{\Delta}{=} \phi(x) - \gamma(g(x); \omega)$ the generalized Lagrangian function associated with P_1; relation (1.6) brings us to a natural definition of the problem:

$$P_1^* : \max_{\omega \in \Omega} \; \min_{x \in X} L(x ; \omega)$$

which is called the dual of problem P_1.
With respect to the primal problem P_1 and the dual problem P_1^*, (1.6) is referred to as the weak duality theorem.
Let us note that when, in the given class W_1 of weak separation functions, there exists an element which guarantees alternative, that is $\exists \bar{\omega} \in \Omega : w(f(x), g(x); \bar{\omega}) \leq 0 \quad \forall x \in X$, then (1.6) holds as an equality.
As a consequence, the duality gap is zero iff in a given class W_1 there exists an element which guarantees alternative (besides the weak one). When problem P_1 is convex, it is natural to choose $\gamma(v ; \omega)$ as the linear function $\langle \omega ; v \rangle$; in [9] a class of exponential weak separation functions is proposed for some classes of nonconvex problems.
In the next section, we will see how it is possible to extend such general approach to duality in vector optimization; this study can also be viewed as an effort to complete some previous research related to optimality conditions and regularity for a vector extremum problem.

2. Lagrange Duality in Vector Optimization

Assume they are given the positive integer ℓ, m, n, the set $X \subseteq IR^n$, the non-empty closed convex cones $U \subset IR^\ell$ and $V \subset IR^m$ with apex at the origin and the functions $\phi : X \to IR^\ell$ and $g : X \to IR^m$. Suppose that U is pointed (i.e. $-x \notin U$ when $x \neq 0$ and $x \in U$) and int $U \neq \emptyset$.

For a set A, int A, and cl A denote interior A and closure A, respectively .
Consider the following vector optimization problem:

$$P : \min_U \phi(x) ; x \in R = \{ x \in X : g(x) \in V \}.$$

We say that $\bar{x} \in R$ is a vector minimum point for problem P with respect to cone U iff the following system is impossible:

(2.1) $$f(x) \triangleq \phi(\bar{x}) - \phi(x) \in U \setminus \{0\}, \ g(x) \in V, \ x \in X.$$

Set $F(x) = (f(x), g(x))$, $\mathcal{K} = F(X)$, $\mathcal{H} = (U \setminus \{0\}) \times V$, $\mathcal{E} = \mathcal{K} - \mathrm{cl}\, \mathcal{H}$; the impossibility of system (2.1) is equivalent to the disjunction between the sets $\mathcal{K}$ and $\mathcal{H}$. As in the scalar case, such a disjunction can be studied by means of vector weak separation functions defined in the following way: $w : IR^\ell \times IR^m \to IR^\ell$ is called a vector weak separation function iff

(2.2) $$w(u , v) \in U \setminus \{0\}, \quad \forall (u , v) \in \mathcal{H}.$$

In addition we say that w(u , v) guarantees alternative iff

(2.3) $$w(F(x)) \notin U \setminus \{0\}, \quad \forall x \in X.$$

Let us note that, in the scalar case (namely $\ell = 1$, $U = IR_+$) (2.2) and (2.3) collapse to (1.2) and (1.3) respectively.
In order to suggest an approach to duality similar to that given for the scalar case, we will consider the following class of vector functions:

$$W = \{ w : IR^\ell \times IR^m \to IR^\ell : w(u , v ; \omega) = u + \gamma(v ; \omega), \ \omega \in \Omega \}$$

where the set Ω of parameters ω and $\gamma : IR^m \to IR^\ell$ are such that:

(2.4a) $$\gamma(v ; \omega) \in U \qquad \forall v \in V, \quad \forall \omega \in \Omega$$

(2.4b) $$\exists \omega^\circ \in \Omega : \gamma(\cdot ; \omega^\circ) \equiv 0$$

(2.4c) $$\forall v \notin V, \ \ \forall \omega \in \Omega, \ \mathrm{F}\, \bar{\omega} \in \Omega : \gamma(v ; \omega) \in \gamma(v ; \bar{\omega}) + (U \setminus \{0\})$$

Let us note that any element of W is a weak separation function; furthermore, as shown in Lemma 2.1, W contains the subclass of linear functions $w(u, v; \Lambda) = u + \Lambda v$, $\Lambda \in V_U^*$ where V_U^* is the vector polar of V with respect to U. For such a class Λ plays the role of a parametr and $\Omega = V_U^*$. The vector polar of a cone is defined in the following way : let $C \subseteq IR^k$, $D \subseteq IR^s$ be cones with apex at the origin; the vector polar of C with respect to D is the set $C_D^* = \{ \Gamma \in IR^{s,k} : \Gamma c \in D, \forall c \in C \}$ where $IR^{s,k}$ denotes the set of all s by k real matrices. When $D = IR_+$, C_D^* collapses to the usual positive polar.

Lemma 2.1. The class of linear functions:

$$\gamma(v; \Lambda) = \Lambda v, \quad \Lambda \in V_U^*$$

satisfies property (2.4).

Proof. (2.4a) follows from the definition of vector polar; (2.4b) is obvious since $\Lambda = 0 \in V_U^*$. As regards to (2.4c) it is sufficient to consider $\overline{\Lambda} = \Lambda + u\overline{v}^T$, where $\Lambda \in V_U^*$, $u \in U \setminus \{0\}$ and $\overline{v} \in V$ is such that $\langle \overline{v}, v \rangle < 0$.

Let $\overline{x}$ be a feasible point for the vector extremum problem P, so that $w(f(\overline{x}), g(\overline{x}); \omega) = \gamma(g(\overline{x}); \omega) \in U \quad \forall \omega \in \Omega$.

As a consequence the maximal vectors[1] of the problem $\max_U_{x \in X} w(f(x), g(x); \omega)$ do not belong to $\gamma(g(\overline{x}); \omega) - (U \setminus \{0\}) \supset -(U \setminus \{0\})$[2] that is $[\max_U_{x \in X} w(f(x), g(x); \omega)] \cap (-U \setminus \{0\}) = \emptyset$ and this is equivalent to stating that

$$\phi(\overline{x}) \notin [\min_U_{x \in X} (\phi(x) - \gamma(g(x); \omega))] \cap (-U \setminus \{0\}), \quad \forall \omega \in \Omega \tag{2.5}$$

Let $L(x; \omega) \triangleq \phi(x) - \gamma(g(x); \omega)$ be the generalized vector-valued Lagrangian function associated with problem P and set:

$$\Phi(\omega) \triangleq \min_U_{x \in X} [\phi(x) - \gamma(g(x); \omega)];$$

---

[1]Let A be a subset of IR^k. A vector $x \in A$ is said to be a maximal vector of A with respect to the cone C iff $y \notin x + (C \setminus \{0\})$, $\forall y \in A$; a maximal vector of the problem $\max_C G(x)$, $x \in X$ is a maximal vector of G(X) with respect to the cone C where $G : X \to IR^k$.

[2]Let A and B subsets of IR^ℓ and $\psi : IR^n \to IR^\ell$. The following properties hold:

i) $a + U \supset U, \forall a \in -U$; $a - U \supset -U, \forall a \in U$.

ii) if $(x - A) \cap B = \emptyset$ then $x \notin A + B$.

iii) $\max_U_{x \in X} \psi(x) = -\min_U_{x \in X} (-\psi(x))$.

(2.5) implies:

(2.6) $$\phi(\overline{x}) \notin \left(\bigcup_{\omega \in \Omega} \Phi(\omega) \right) - (U \setminus \{0\})$$

and hence:

(2.7) $$\phi(\overline{x}) \notin [\ \max_U \bigcup_{\omega \in \Omega} \Phi(\omega)\ ] - (U \setminus \{0\})\,;$$

this relation leads us to introduce the problem:

$$P^*:\ \max_U \bigcup_{\omega \in \Omega} \Phi(\omega)$$

which is called the dual of problem P.
When $\ell = 1$ and $U = IR_+$, problem P^* becomes P_1^*.

Condition (2.6) allows us to state the following theorem:

Theorem 2.1. For any $x \in R$ and $y \in \Phi(\omega)$, we have $\phi(x) \notin y - (U \setminus \{0\})$.

The following theorems, together with theorem 2.1, represent some properties of efficient solutions to primal problem P in connection with the map Φ and so may be considered as duality theorems for multiobjective optimization stated in a general form.

Theorem 2.2. Let $\overline{x}$ be a vector minimum point for problem P. If there exists $\overline{\omega} \in \Omega$ such that the vector weak separation function $w(u, v; \overline{\omega})$ guarantees alternative, then $\phi(\overline{x})$ is a maximal vector of problem P^*, that is:

(2.8) $$\phi(\overline{x}) \in \max_U \bigcup_{\omega \in \Omega} \Phi(\overline{\omega})$$

Proof. Since $w(u, v; \overline{\omega})$ is a vector weak separation function which guarantees alternative and since $\overline{x}$ is a feasible point for P, then we have $\gamma(g(\overline{x})\,;\ \overline{\omega}) = 0$ so that (2.7) can be restated as:

$$\phi(x) - \gamma(g(x); \overline{\omega}) \notin (\ \phi(\overline{x}) - \gamma(g(x); \overline{\omega})) + (U \setminus \{0\}),\ \forall x \in X.$$

This relation implies $\phi(\overline{x}) \in [\ \min_U_{x \in X} L(x\,; \overline{\omega})\ ] \triangleq \Phi(\overline{\omega})$.

The thesis follows by theorem 2.1. This completes the proof.

Theorem 2.3. Suppose that there exist $\bar{x} \in R$ and $\bar{\omega} \in \Omega$ such that $\Phi(\bar{x}) \in \phi(\bar{\omega})$. Then $\bar{x}$ is a vector minimum point for problem P and $\phi(\bar{x})$ is a maximal vector for problem P*.

Proof. If $\bar{x} \in R$ is not a vector minimum point for P, then system (2.1) is possible for $x = \hat{x} \in R$, i.e. $\phi(\bar{x}) - \phi(\hat{x}) \in U\setminus\{0\}$, $g(\hat{x}) \in V$. Since $\gamma(g(\hat{x}) ; \omega) \in U$ $\forall \omega \in \Omega$ and $U + (U\setminus\{0\}) = U\setminus\{0\}$, we have:

$$\phi(\bar{x}) - \phi(\hat{x}) + \gamma(g(\hat{x}) ; \omega) \in U\setminus\{0\}, \quad \text{so that} \quad \phi(\hat{x}) - \gamma(g(\hat{x}) ; \bar{\omega}) \in \phi(\bar{x}) - (U\setminus\{0\}),$$

which contradicts the assumption $\phi(\bar{x}) \in \Phi(\bar{\omega})$.
The fact that $\phi(\bar{x})$ is a maximal vector for P* is a direct consequence of Theorem 2.1. This completes the proof.

Let us note that the duality theorem 2.2 holds when, in a given class of vector weak separation functions, it is possible to find a function which guarantees alternative.
The existence of such a function is, obviously, related either to the class of multiobjective problems or to the choosen class of vector weak separation functions. In the following section we will go more deeply into this analysis in order to obtain known results and new ones.

Remark 2.1. The approach to duality given in this section by means of the generalized vector-valued Lagrangian function $L(x ; \omega) = \phi(x) - \gamma(g(x) ; \omega)$, $\omega \in \Omega$, allows us to obtain, by means of a suitable choice of the function $\gamma(v; \omega)$, different vector-valued Lagrangian functions which have been, recently proposed [1, 4, 5, 8, 14, 15, 16] for a nonlinear multiobjective programming problem. For instance, when $\gamma(v; \omega)$ is a U-function, continuous, positively homogeneous and monotonic $L(x ; \omega)$ reduces to the one proposed by Luc in [8]; when $\gamma(v; \omega)$ is a vector-valued linear function, we obtain the vector valued Lagrangian function proposed by several authors.
As a particular case, when $\gamma(v; \omega) = e\langle \omega, v\rangle$, where $\omega \in V^*$ and e is a fixed vector of U, $L(x ; \omega)$ collapses to the one given in [8].

3. Vector Duality in the Convex Case

In this section we will consider problem P when $F(x) = (f(x), g(x))$ is cl $\mathcal{H}$-convexlike or, equivalently, when set $\mathcal{E}$ is convex [3,6]. (We recall that a function $h: X \to IR^s$ is C-convexlike, where $C \subseteq IR^s$ is a convex cone with apex at the origin and $X \subseteq IR^n$ is a convex set, iff $\forall x, y \in X$, there exists $z \in X$ such that $h(z)-(1-\alpha)h(x)-\alpha h(y) \in C$, $\forall \alpha \in [0,1]$. As a particular case, if f is a U-function and g is a V-function, then $F(x)=(f(x),g(x)$ is cl H-convexlike [3,6]).
For such a problem, a suitable choice of the class of vector weak separation functions would be the linear one, that is:

$$W_L = \{ w : IR^{\ell} \times IR^m \to IR^{\ell} : w(u, v; \Lambda) = u + \Lambda v, \quad \Lambda \in V_U^* \}.$$

The generalized vector-valued Lagrangian function associated with the class W_L is

$$L(x ; \Lambda) = \phi(x) - \Lambda g(x), \quad x \in X, \quad \Lambda \in V_U^*.$$

Taking into account the results given in section 2, the duality theorem 2.2 holds if there exists an element of the class W_L which guarantees alternative, that is if there exists $\bar{\Lambda} \in V_U^*$ such that

$$(3.1) \qquad f(x) + \bar{\Lambda} g(x) \notin U \setminus \{0\}, \qquad \forall x \in X.$$

As in the scalar case, the validity of (3.1) is related to the validity of a constraint qualification or of a regularity condition.
In vector optimization, Slater's constraint qualification (i.e. $\exists \hat{x} \in R : g(\hat{x}) \in \text{int } V$) is the one which appears must frequently, while the concept of regularity has not yet examinated throughly.
In order to clarify what we mean by regularity, let us note, first of all, that relation (3.1) is in strict connection with the exixtence of a hyperplane which separates $\mathcal{E}$ and $\mathcal{H}$ (see [4, 5, 13]).
More exactly, if $\langle \alpha , u \rangle + \langle \beta , v \rangle = 0$, $\alpha \in U^*$, $\beta \in V^*$, is the equation of a hyperplane such that:

$$(3.2) \qquad \langle \alpha , f(x) \rangle + \langle \beta , g(x) \rangle \leq 0 \qquad \forall x \in X,$$

then, setting $\bar{\Theta} \triangleq u \alpha^T$ and $\bar{\Lambda} \triangleq u \beta^T$, where u is a fixed vector of $U \setminus \{0\}$, we have $\bar{\Theta} \in U_U^*$ and $\bar{\Lambda} \in V_U^*$ (see [12, 15]) and furthermore:

$$(3.3) \qquad \bar{\Theta} f(x) + \bar{\Lambda} g(x) \notin U \setminus \{0\}, \quad \forall x \in X.$$

Any condition which ensures the validity of (3.3) with $\bar{\Theta} = I$ (I denotes the $\ell \times \ell$ identity matrix), that is the validity of (3.1), is referred to as a regularity condition.
In the scalar case, this is equivalent to require in (3.2) $\alpha \in IR_+, \alpha \neq 0$, so that the suggested concept of regularity in vector optimization turns out to be an extension of regularity in scalar optimization.
In [13], general regularity conditions are given which allow us to establish the duality theorem 2.2 in a general form.
The following theorem holds:

Theorem 3.1. Let $\bar{x}$ be a vector minimum point for problem P, where $F(x) = (f(x), g(x))$ is cl $\mathcal{H}$-convexlike and suppose that a regularity condition holds. Then $\phi(\bar{x})$ is a maximal vector for dual problem P^*.

Proof. Since $\bar{x}$ is a vector minimum point for P, then $\mathcal{E} \cap \mathcal{H} = \emptyset$, (see [4, 5]), so that, taking into account the convexity of the sets $\mathcal{E}$ and $\mathcal{H}$, there exist $\alpha \in U^*$ and $\beta \in V^*$ such that (3.2) holds. The validity of a regularity condition for P implies the existence of $\bar{\Lambda} \in V^*_U$ satisfying (3.1). Consequently $w(u, v; \bar{\Lambda}) = u + \bar{\Lambda}v$ is a vector weak separation function which guarantees alternative. The thesis follows from theorem 2.2.

The following result given in [1, 16] is a direct consequence of theorem 3.1.

Corollary 3.1. Let $\bar{x}$ be a properly efficient solution for problem P where ϕ is a (-U)-function and g is a V-function and suppose that Slater's constraint qualification is satisfied. Then:

$$\phi(\bar{x}) \in \max_U \bigcup_{\Lambda \in V^*_U} \Phi(\Lambda).$$

Proof. According to theorem 3.1, it is sufficient to prove that the assumption of proper efficience and the validity of Slater's constraint qualification imply a regularity condition. To this aim, let us note that the above-mentioned assumptions ensure the validity of (3.2) with $\alpha \in \text{int } U^*$ and $\beta \in V^*$ (see theorem 5.2.1 in [6]) and this is equivalent to state (see corollary 3.3 and theorem 4.1 in [3]) the validity of the regularity condition $T \cap (U \times \{0\}) = \{0\}$, where T is the tangent cone to $\mathcal{E}$ at $(0, g(\bar{x}))$. The proof is complete.

Remark 3.1 As it is well known [16] there are several definitions of proper efficiency; we consider Borwein's Proper Efficiency: A point $x \in R$ is said to be a properly efficient solution for P iff: $T(\phi(R)+U, \phi(x)) \cap (-U) = \{0\}$, where T is the tangent cone to $(\phi(R)+U)$ at $\phi(x) \in (\phi(R)+U)$. The tangent cone to a set $S \subseteq IR^n$ at $\bar{s} \in \text{cl } S$ is defined as the set of $\bar{s} + s$ for which there exist a sequence $\{s^r\} \subseteq S$, such that $\lim_{r \to +\infty} s^r = \bar{s}$, and a positive sequence $\{\alpha_r\} \subset IR$, such that $\lim_{r \to +\infty} \alpha_r(s - \bar{s}) = s$.

References

/1/ **Bitran G.R.**, Duality for Nonlinear Multiple-Criteria Optimization Problems, J.O.T.A., vol. 35, (1981), pp.367-401.

/2/ **Bitran G.R., Magnanti T.L.**, The Structure of Admissible Points with Respect to Cone Dominance, J.O.T.A., vol. 29 (1979), pp. 583-614.

/3/ **Cambini A.**, Nonlinear Separation Theorems, Duality and Optimality Conditions, in: Optimization and Related Fields, Lecture Notes in Math., n° 1190, Springer Verlag (1986).

/4/ **Cambini A., Martein L.**, Separation Functions and Optimality Conditions in Vector Extremum Problems, University of Pisa, Dep. of Mathematics, Optimization and Operations Reserach Section, Paper A-120, (1984).

/5/ **Cambini A., Martein L.**, Some Optimality Conditions in Vector Optimization, Journal of Information and Optimization Sciences, vol. 10, n° 1, (1989).

/6/ **Giannessi F.**, Theorems of the Alternative and Optimality Conditions, J.O.T.A., vol. 42, (1984) pp. 331-365.

/7/ **Jahn J.**, Duality in Vector Optimization, Mathematical Programming, vol. 25 (1983), pp. 343-353.

/8/ **T. Luc D.**, On Duality Theory in Multiobjective Programming, J.O.T.A., vol. 43 (1984) pp. 557-582.

/9/ **Martein L.**, Sulla Dualità Lagrangiana Esponenziale, University of Pisa, Dep. of Mathematics, Optimization and Operations Research Section, paper A-114 (1984).

/10/ **Martein L.**, Regularity Conditions for Constrained Extremum Problems, J.O.T.A., vol. 47, (1985), pp. 217-233.

/11/ **Martein L.**, Stationary Points and Necessary Conditions in Vector Extremum Problems, Journal of Information and Optimization Sciences, vol. 10, n° 1, (1989).

/12/ **Martein L.**, Lagrange Multipliers and Generalized Differentiable Functions in Vector Extremum Problems, to appear in Journal of Optimization Theory and Applications.

/13/ **Martein L.**, Some Results on Regularity in Vector Optimization, vol. 20, n° 6, (1989), pp. 787-798.

/14/ **Nakayama H.**, Duality Theory in Vector Optimization: an overview, Research Report, Dep. of Applied Mathematics, Konon University (1979).

/15/ **Tanino T., Sawaragi Y.**, Duality Theory in Multiobjective Programming, J.O.T.A., vol. 27, (1979) pp. 509-529.

/16/ **Sawaragi Y., Nakayama H., Tanino T.**, Theory of Multiobjective Optimization, Academic Press, Inc. (1985).

This research was partly supported by the Italian-Ministery of Education.

RUBINSTEIN DUALITY SCHEME FOR VECTOR OPTIMIZATION

M. Vlach

1. Introduction

G.S.Rubinstein in his paper [1] has introduced a universal scheme for constructing a dual problem to a given optimization problem with a real valued objective function. T.Q. Chien in his dissertation [2] has extended the Rubinstein scheme for optimization problems with vector valued objective functions. In this paper the basic ideas of Rubinstein's approach are expounded and its application potency is illustrated by means of fractional programming problems with vector valued objective functions.

2. Scalar Optimization

In order to demonstrate how the original Rubinstein scheme works let us suppose that f_0, $f_1, \ldots, f_m$ are real valued functions on R^n and consider the problem

$$(1) \qquad \begin{array}{ll} \text{maximize} & f_0(x) \\ \text{subject to} & x \in X \end{array}$$

where X is the set of all nonnegative solutions of the system

$$f_1(x) \geq 0,\ f_2(x) \geq 0, \ldots, f_m(x) \geq 0.$$

For each real number t, let Q_t denote the set of points $p = (p_0, p_1, \ldots, p_m)$ in R^{m+1} satisfying

$$p_0 \geq t, \qquad p_1 = 0, \ldots, p_m = 0,$$

and let S be the set of those points p in R^{m+1} for which the system

$$f_0(x) \geq p_0,\ f_1(x) \geq p_1, \ldots, f_m(x) \geq p_m$$

has nonnegative solutions. Besides the problem (1) let us consider also the problem

$$\text{(1')} \qquad \begin{array}{ll} \text{maximize} & \mu(p) \\ \text{subject to} & p \in P \end{array}$$

where μ and P are defined as follows:

$$P := S \cap \big(\bigcup_{t \in R^1} Q_t \big)$$

$$\mu(p) := \sup \{ t \mid p \in Q_t \}.$$

It is easy to see that

$$\forall_{x \in X} \; \exists_{p \in P} : f_0(x) = \mu(p)$$

$$\forall_{p \in P} \; \exists_{x \in X} : f_0(x) \geq \mu(p)$$

Therefore the two problems under consideration are equivalent in the sense that one of them has an optimal solution if and only if the other one has. We also note that the point to set mapping $t \to Q_t$ is antitone in the sense that

$$t' < t'' \Rightarrow Q_{t'} \supset Q_{t''}.$$

In Rubinstein's scheme, a given problem of the form (1) is first converted in a problem of the form (1') and then a dual problem to (1') is constructed. Therefore, as a primal problem we consider the following generalization of (1'). Given

a) an interval

$$\Omega := (\alpha, \beta), \; -\infty \leq \alpha < \beta \leq +\infty \; ;$$

b) nonempty sets S and E such that $S \subset E$;

c) a point to set mapping $Q : \Omega \to 2^E$ such that

$$t' < t'' \Rightarrow Q(t') \supset Q(t''),$$

define

$$P := S \cap \big(\bigcup_{t \in \Omega} Q(t) \big)$$

$$\mu(p) := \sup \{ t \mid p \in Q(t) \},$$

and consider the problem

$$\text{Find } \hat{p} \in P \quad \text{such that}$$
$$\mu(\hat{p}) = \max_{p \in P} \mu(p).$$

In order to be able to construct a problem which is dual to (I), we choose a nonempty class E^* of nonempty subsets of E and define dual objects S^*, Q^*, P^*, μ^* as follows:

$$S^* := \{ H \in E^* \mid H \supset S \},$$

$$Q^*(t) := \{ H \in E^* \mid H \cap (\bigcup_{t<\tau<\beta} Q\tau) = \emptyset \},$$

$$P^* := S^* \cap (\bigcup_{t \in \Omega} Q^*(t)),$$

$$\mu^*(H) := \inf \{ t \mid H \in Q^*(t) \}.$$

The dual problem is then defined as the problem

$$(I^*) \qquad \text{Find } \hat{H} \in P^* \quad \text{such that} \qquad \mu^*(\hat{H}) = \min_{H \in P^*} \mu^*(H)$$

Let us note that the point to set mapping Q^* is isotone in the sense that

$$t' < t'' \Rightarrow Q^*(t') \subset Q^*(t'').$$

In turns out that the problem (I^*) has been constructed in such a way that the following weak duality theorem holds.

Theorem 1. If $p \in P$ and $H \in P^*$, then

$$\alpha < \mu(p) \leq \mu^*(H) < \beta.$$

Proof. For each $p \in P$ and $H \in P^*$ there are t' and t'' such that

$$p \in Q(t') \quad \text{and} \quad \alpha < t' < \beta,$$

$$H \in Q^*(t'') \quad \text{and} \quad \alpha < t'' < \beta.$$

It follows directly from the definition of μ and μ^* that $\alpha < \mu(p)$ and $\mu^*(H) < \beta$. To prove the remaining inequality, let us assume that there are $\bar{p} \in P$ and $\bar{H} \in P^*$ such that $\mu(\bar{p}) > \mu^*(\bar{H})$. It follows from the definition of μ and μ^* that there are t_1 and t_2 belonging to Ω such that

$$t_2 > t_1, \quad \bar{p} \in Q(t_2), \ \bar{H} \in Q^*(t_1).$$

Since $\bar{p} \in P$ and $\bar{H} \in P^*$, we have $\bar{p} \in S$ and $S \subset \bar{H}$. Therefore $\bar{p} \in \bar{H} \cap Q(t_2)$. However, the intersection $\bar{H} \cap Q(t)$ is empty for each $t > t_1$, because $\bar{H} \in Q^*(t_1)$. This is obviously in contradiction with the inequality $t_2 > t_1$.

Let us note that if, for the introductory problem (1'), the set E^* is taken to be the class of those closed halfspaces in R^{m+1} which have common points with the set $\bigcup_{t \in \Omega} Q_t$, then the corresponding dual problem of the form (I*) is equivalent (in the sense mentioned above) to the problem

$$\begin{array}{ll} \text{minimize} & \varphi(y) \\ \text{subject to} & y \geq 0 \ \text{ and } \ \varphi(y) < \infty \end{array}$$

where

$$\varphi(y) := \sup_{x \geq 0} \left(f_o(x) + \sum_{i=1}^{m} y_i f_i(x)\right).$$

It follows from the weak duality theorem that $p \in P$ is optimal whenever there is $H \in P^*$ such that $\mu(p) = \mu^*(H)$. Naturally we are interested whether the converse is also true. Strictly speaking, we are interested in conditions under which the optimality of a point $p \in P$ implies the existence of $H \in P^*$ such that $\mu(p) = \mu^*(H)$. It turns out that such a result can be established under the following two conditions.

Condition 1. The sets S and $\bigcup_{t \in \Omega} Q(t)$, do not intersect.

Condition 2. For each $t \in \Omega$,

$$\left.\begin{array}{l} \bigcap_{\alpha < \tau < t} (S \cap Q(\tau)) \neq \emptyset \\ \\ \bigcup_{\tau < \tau < \beta} (S \cap Q(\tau)) = \emptyset \end{array}\right\} \Rightarrow S^* \cap Q^*(t) \neq \emptyset$$

Theorem 2. If the primal problem (I) satisfies both conditions1 and 2, and if p is an optimal solution of (I), then there is $\hat{H} \in P^*$ such that $\mu(\hat{p}) = \mu^*(\hat{H})$.

Proof. Let $\hat{p}$ be an optimal solution of (I) and let $\hat{t}$ denote the optimal value, i.e.

$\hat{t} = \sup\{\, t \mid p \in Q(t)\,\}$. It follows from the definition of a supremum that for each $t < \hat{t}$ there is $\bar{t}$ such that $t < \bar{t} \le \hat{t}$ and $\hat{p} \in Q(\bar{t})$. By monotonicity of Q, we have $\hat{p} \in Q(t)$. Therefore

$$\bigcap_{\alpha < \tau < \hat{t}} (S \cap Q(\tau) \neq \emptyset$$

and condition 1 guarantees that $\hat{t} < \beta$. Now we show that $\bigcup_{\hat{t} < \tau < \beta} (S \cap Q(\tau)) = \emptyset$

Indeed, if this is not true, then there is $\bar{t}$ such that $\hat{t} < \bar{t} < \beta$ and $(S \cap Q(\bar{t}) \neq \emptyset$. Taking $\bar{p} \in (S \cap Q(\bar{t})$, we obtain $\bar{p} \in P$ such that

$$\mu(\bar{p}) = \sup\{\, t \mid p \in Q(t))\,\} \ge \bar{t} > \hat{t} = \mu(\hat{p}),$$

which is in contradiction with the optimality of $\hat{p}$. To complete the proof, we observe that, by condition 2, the set $S^* \cap Q^*(\hat{t})$ is nonempty and that for $\hat{H} \in S^* \cap Q^*(\hat{t})$, we have

$$\mu^*(\hat{H}) = \inf\{\, t \mid \hat{H} \in Q^*(\hat{t})\,\} \le \hat{t} = \mu(\hat{p}).$$

Since the weak duality theorem guarantees that $\mu^*(\hat{H}) \ge \mu(\hat{p})$, we have $\mu^*(\hat{H}) = \mu(\hat{p})$.
In addition to the previous two conditions let us introduce the following stronger conditions.

Condition 3. For each $t \in \Omega$,

$$\left.\begin{array}{l} P \neq \emptyset \quad \text{and } P^* \neq \emptyset \\ \bigcup_{t < \tau < \beta} (S \cap Q(\tau) = \emptyset \end{array}\right\} \Rightarrow \quad S^* \cap Q^*(t) \neq \emptyset.$$

Condition 4. If $S \cap Q(\tau) \neq \emptyset$ for all $\alpha < t < \tau$, then

$$\bigcap_{\alpha < \tau < t} (S \cap Q(\tau) \neq \emptyset$$

Theorem 3. If the conditions (3) and (4) are fulfilled, and if both P and P^* are nonempty, then both (I) and (I*) have optimal solutions and the corresponding optimal values are equal.

Proof. Since P and P^* are nonempty, we have by the weak duality theorem

$$\alpha < \sup_{p \in P} \mu(p) \le \inf_{H \in P^*} \mu^*(H) < \beta$$

Let us set

$$t_0 := \sup_{p \in P} \mu(p)$$

$$t^*_0 := \inf_{H \in P^*} \mu^*(H)$$

First we observe that

$$\bigcup_{\tau > t_0} (S \cap Q(\tau) = \emptyset$$

Indeed, if this union is nonempty, then there are $\bar{p}$ and $\bar{t}$ such that $t_0 < \bar{t} < \beta$ and $\bar{p} \in (S \cap Q(\bar{t}))$. Therefore $\bar{p} \in P$ and $\mu(p) \geq \bar{t} > t_0$, which is an obvious contradiction. Condition 3 now guarantees the existence of $\bar{H}$ from $S^* \cap Q^*(t_0)$. Consequently $\bar{H} \in P^*$ and

$$\inf_{H \in P^*} \mu^*(H) \leq \mu^*(\bar{H}) \leq t_0.$$

Therefore

$$\sup_{p \in P} \mu(p) = \inf_{H \in P^*} \mu^*(H) = \mu^*(\bar{H}).$$

It remains to prove the existence of $\bar{p} \in P$ such that

$$\mu(\bar{p}) = \sup_{p \in P} \mu(p).$$

It is obvious that for each $\tau < t_0$ there is $p_\tau \in P$ such that $\tau < \mu(p_\tau) \leq t_0$. However

$$\mu(p_\tau) - \sup \{ t | \ p_\tau \in Q(t) \}.$$

Now it is clear that for each $\tau < t_0$ there is $\bar{t}$ such that

$$\tau < \bar{t} \leq t_0 \quad \text{and} \quad p_\tau \in Q(\bar{t}).$$

Therefore $(S \cap Q(\tau) \neq \emptyset$ for all $\tau < t_0$ and, by condition 4, we have

$$\bigcap_{\alpha < \tau < t_0} S \cap Q(\tau) \neq \emptyset$$

Now we can easily show that an optimal solution of (I) exists. Since there is $\bar{p}$ such that $\bar{p} \in S$ and $\bar{p} \in Q(\tau)$ for each $\tau < t_0$, it is obvious that $\sup \{ t | \bar{p} \in Q(t) \}$ cannot be less than

t_0 and therefore

$$\mu(\bar{p}) = \sup \{ t | \bar{p} \in Q(t) \} = t_0 .$$

Various applications of this scheme to optimization problems with real valued objective functions in finite and infinite dimensional vector spaces are presented in the Rubinstein paper [1].

3. Vector optimization

Before extending the Rubinstein scheme to optimization problems with vector valued objective functions we recall some concepts concerning sets in partially ordered linear spaces.

Let Y be a partially ordered linear space, i.e. a real linear space ordered by a reflexive antisymmetric transitive relation $\geq$ which is invariant under traslation and nonnegative scalar multiplication. We denote the corresponding positive cone $\{ y \in Y | y \geq 0 \}$ by Y_+ and define Y_{++} to be the set $Y_+ \setminus 0 \{ 0 \}$. Instead of $t'' - t' \in Y_{++}$ we write $t' < t''$ or $t'' > t'$. If V and W are subsets of Y such that $v > w$ for all $v \in V$ and $w \in W$, then we write $V > W$. The core and the algebraic closure of a set S in Y is denoted by S^0 and $\bar{S}$ respectively. A subset Ω of Y is said to be order-convex if

$$(\Omega - Y_+) \cap (\Omega + Y_+) = \Omega .$$

If S is a subset of Y and Ω is order-convex, then an element y of Y is called a supremal element of S with respect to Ω if the following two conditions are satisfied:

(a) $$y \in \Omega \cap \overline{S - Y_+} ,$$

(b) $$(y + Y_{++}) \cap \Omega \cap \overline{S - Y_+} = \emptyset.$$

The set of all supremal elements of a set S with respect to Ω is denoted by $\mathrm{Sup}_\Omega S$ and the set of all maximal elements of S (i.e. maximal with respect to the partial order $\geq$) is denoted by Max S. $\mathrm{Inf}_\Omega S$ and Min S stand for the set of all infimal elements of S with respect to Ω and the set of all minimal elements of S. The definition of an infimal element is completly parallel, i.e. an element $y \in Y$ is said to be an infimal elements of S with respect to Ω if $y \in \Omega \cap \overline{S + Y_+}$ and $(y - Y_{++}) \cap \Omega \cap \overline{S + Y_+} = \emptyset$. The set of all infimal elements of S with respect to Ω and the set of all minimal elements of S is denoted by $\mathrm{Inf}_\Omega S$ and Min S respectively.

Our extension of the Rubinstein scheme is based on the following replacements. The interval Ω in R^1 we have been considering in the scalar case is replaced by an order-convex set Ω in Y. The set S and the mapping Q are replaced by a point to set mapping $A : \Omega \to 2^E$ such that

$$t' < t'' \Rightarrow A(t') \supset A(t'').$$

Here $t' < t''$ means $t'' - t' \in Y_{++}$, and E is again an arbitrary nonempty set.
Given Ω and A, we define $P \subset E$ and $\mu : P \to \Omega$ as follows:

$$\mu(P) := \bigcup_{t \in \Omega} A(t) ,$$

$$\mu(p) := \sup \{ t \in \Omega \mid p \in A(t) \} .$$

Now we are ready to introduce the vector optimization problems we shall be interested in. We shall consider the following two problems.

(M) Find $\hat{p} \in P$ such that

$$\mu(\hat{p}) \cap \operatorname{Max} \mu(P) \neq \emptyset.$$

(S) Find $\hat{p} \in P$ such that

$$\mu(\hat{p}) \cap \operatorname{Sup}_{\Omega} \mu(P) \neq \emptyset.$$

Here, of course, $\mu(P)$ stands for $\bigcup_{p \in P} \mu(P)$. Taking $Y := R^1$, $Y_+ := R^1_+$, $A(t) := S \cap Q(t)$ we can easily see that problem (I) is a special case of problem (M).
To construct the dual problems to the primal problem (M) and (S), we replace the set E^* by a nonempty set F, and the set S^* and the mapping Q^* by a point to set mapping $B : \Omega \to 2^F$ such that

$$t' < t'' \Rightarrow B(t') \subset B(t'').$$

For such a mapping B, we define $D \subset F$ and $\upsilon : D \to \Omega$ by

$$D := \bigcup_{t \in \Omega} B(t) ,$$

$$\upsilon(d) := \{ t \in \Omega \mid d \in B(t) \}.$$

The dual problems are then defined to be the problems

$$(\mathrm{M}^*)\qquad \begin{array}{l}\text{Find } \hat{d} \in D \text{ such that}\\ \upsilon(\hat{d}) \cap \operatorname{Min} \upsilon(D) \neq \emptyset.\end{array}$$

$$(\mathrm{S}^*)\qquad \begin{array}{l}\text{Find } \hat{d} \in D \text{ such that}\\ \upsilon(\hat{d}) \cap \operatorname{Inf}_{\Omega} \upsilon(D) \neq \emptyset.\end{array}$$

A natural extension of the weak duality theorem of the previous section is the assertion that there are no $p \in P$ and $d \in D$ such that $\mu(p) > \upsilon(d)$. Since there has been no connection between A and B so far (except they have a common domain), we can hardly expect that the weak duality holds without additional conditions. Let us consider the following condition.

Condition 1. If t' and t'' from Ω are such that $t' > t''$ and $B(t'') \neq \emptyset$, then $A(t') \neq \emptyset$.

Theorem 4. Condition 1 is satisfied if and only if there are no $p \in P$ and $d \in D$ such that $\mu(p) > \upsilon(d)$.

Proof. If there are $\bar{p} \in P$ and $\bar{d} \in D$ such that $\mu(\bar{p}) > \upsilon(\bar{d})$, then there are t' and t'' in Ω such that

$$\bar{p} \in A(t'),\ \bar{d} \in B(t'')$$

and

$$\{ t \in \Omega \mid \bar{p} \in A(t) \} = \mu(\bar{p}) > \upsilon(\bar{d}) = \{ t \in \Omega \mid \bar{d} \in B(t) \}.$$

Therefore

$$t' > t'',\ B(t'') \neq \emptyset,\ A(t') \neq \emptyset,$$

which says that Condition 1 is not satisfied. Conversely, if Condition 1 is not satisfied, then there are x and y from Ω such that

$$x > y,\ B(y) \neq \emptyset,\ A(x) \neq \emptyset.$$

Let p^* be from $A(x)$ and d^* from $B(y)$. Obviously $x \in \mu(p^*)$ and $y \in \upsilon(d^*)$. Since $x > y$, it cannot be true that there are no $p \in P$ and $d \in D$ such that $\mu(p) > \upsilon(d)$.

Theorem 5. If Condition 1 is satisfied, then

$$\operatorname{Max} \mu(P) \cap \operatorname{Min} \upsilon(D) = \mu(P) \cap \upsilon(D).$$

Proof. First we observe that the inclusion

$$\mathrm{Max}\ \mu(P) \cap \mathrm{Min}\ \upsilon(D) \subset \mu(P) \cap \upsilon(D).$$

is an immediate consequence of the definitions of maximal and minimal elements. To prove the reverse inclusion, let us consider an arbitrary element x of $\mu(P) \cap \upsilon(D)$. If x is not a maximal element of $\mu(P)$, then there is $y \in \mu(P)$ such that $y > x$. Since $x \in \upsilon(D)$, this contradicts Theorem 4. If x is not a minimal element of $\upsilon(D)$, then there is $z \in \upsilon(D)$ such that $x > z$. Since $x \in \mu(P)$, we again have a contradiction with Theorem 4.

Condition 2. If t^* belongs to Ω^0, then $B(t) \neq \emptyset$ for all $t > t^*$ whenever $A(t) \neq \emptyset$ for all $t > t^*$.

Theorem 6. If Condition 1 and Condition 2 are satisfied and if $Y_{++} - Y^0{}_{+} \neq \emptyset$, then

$$\Omega^0 \cap \mathrm{Sup}_\Omega\ \mu(P) = \Omega^0 \cap \mathrm{Inf}_\Omega\ \upsilon(D).$$

Proof. We shall prove only the inclusion

$$\Omega^0 \cap \mathrm{Sup}_\Omega\ \mu(P) \subset \Omega^0 \cap \mathrm{Inf}_\Omega\ \upsilon(D).$$

The reverse inclusion can be verified analogously.
Let y^* belong to $\Omega^0 \cap \mathrm{Sup}_\Omega\ \mu(P)$. It follows that

(i) $$y^* \in \overline{\mu(P) - Y_+}\ ,$$

(ii) $$\overline{\mu(P) - Y_+} \cap \Omega \cap (y^* + Y_{++}) = \emptyset.$$

Let $y \in \Omega$ be an arbitrary element such that $y > y^*$. Since y belongs to $\Omega \cap (y^* + Y_{++})$, it is obvious from (ii) that $y \notin \overline{\mu(P) - Y_+}$. It follows that $A(y) = \emptyset$, because if there is $p \in A(y)$, then $y \in \mu(p)$ and

$$\mu(p) \subset \mu(P) \subset \mu(P) - Y_+ \subset \overline{\mu(P) - Y_+}\ .$$

Since $A(y) = \emptyset$ for all $y \in \Omega$ such that $y > y^*$, we have, by Condition 2, $B(y) \neq \emptyset$ for all $y \in \Omega$ such that $y > y^*$. We can now easily show that

$$y^* \in \overline{\upsilon(D) + Y_+}\ .$$

Indeed, if $y^* \notin \overline{\upsilon(D) + Y_+}$, then there is $y \in \Omega$ such that $y > y^*$ and $y \notin \upsilon(D) + Y_+$. It follows that $B(y) = \emptyset$, which is a contradiction.
It remains to prove that

$$(y^* - Y_{++}) \cap \Omega \cap \overline{\upsilon(D) + Y_+} = \emptyset.$$

To do so we prove that $y^* - u \notin \overline{\upsilon(D) + Y_+}$ for each $u > 0$ such that $y^* - u \in \Omega$. Let u be such that $u > 0$ and $y^* - u \in \Omega$. It follows from (ii) that $y^* - u$ belongs to $(\mu(P) - Y_+)^0$. Now we observe that

$$(\mu(P) - Y_+)^0 \cap (\upsilon(D) - Y_+)^0 = \emptyset.$$

Indeed, if this is not true, then there is z such that $z = x - u = y + v$ for some $x \in \mu(P)$, $y \in \upsilon(D)$, $u > 0$, $v > 0$. This implies $x > y$, which contradicts Theorem 4. Since $y^* - u \in (\mu(P) - Y_+)^0$ and the core of $\mu(P) - Y_+$ does not intersect the core of $\upsilon(D) + Y_+$, it is obvious that $y^* - u \notin \overline{\upsilon(D) + Y_+}$.

4. Application to vector fractional programming

Let g be a mapping of a nonempty subset S of a real linear space L into a partially ordered linear space $(Z, \geq)$ and let

$$u^j_{k_j} \quad \text{and} \quad v^j_{k_j} \quad (k_j = 1, 2, \ldots, m_j \, ; \, j = 1, 2, \ldots, n)$$

be real valued functions defined on nonempty subsets $U^j_{k_j}$ and $V^j_{k_j}$ of L.

We consider the vector optimization problem of the form

$$\begin{array}{ll} \text{maximize} & f(x) \\ \text{subject to} & x \in X \end{array}$$

with X and f defined as follows:

$$X = S \cap \Big[\bigcap_{j=1}^{n} \bigcap_{k_j=1}^{m_j} (U^j_{k_j} \cap V^j_{k_j}) \Big] \cap \{ x \in S \mid g(x) \geq 0 \},$$

$$f(x) = \left[\frac{\sum u^1_{k_1}(x)}{\sum u^1_{k_1}(x)}, \ldots\ldots, \frac{\sum u^n_{k_n}(x)}{\sum v^n_{k_n}(x)} \right]$$

Here we assume that summation takes place over $k_j = 1, 2, \ldots, m_j$, and that for each $x \in X$ and each j

$$\sum_{k_j=1}^{m_j} v^j_{k_j}(x) \neq 0 .$$

In order to obtain the form (M) or (S) of the previous section we set

$$E := L \times (R \times R^{m_1}) \times (R \times R^{m_2}) \times \cdots\cdots \times (R \times R^{m_n}) \times Z$$

and, for $t = (t_1, t_2, \ldots, t_n)$ from $Y := R^n$, define $Q^j_{k_j}(t)$ as the set of those

$$p = (x\,; r^1_0, \ldots\ldots, r^1_{m_1}; r^n_0, \ldots\ldots, r^n_{m_0}; z)$$

from E for which

$$x \in S \cap U^j_{k_j} \cap V^j_{k_j},$$

$$z \le g(x),$$

$$r^j_0 \le \sum_{i=1}^{m_j} r^j_i,$$

$$r^j_{k_j} \le u^j_{k_j}(x) - t_j v^j_{k_j}(x).$$

Furthermore, if Ω is a nonempty order-convex subset of Y, we define $A : \Omega \to 2^E$ by

$$A(t) = Q \cap \Big[\bigcap_{j=1}^{n} \bigcap_{k_j=1}^{m_j} Q^j_{k_j}(t)\Big]$$

where

$$Q = L \times (R_+ \times R^{m_1}) \times \cdots\cdots \times (R_+ \times R^{m_n}) \times Z_+$$

It turns out that if $v^j_{k_j}(x)$ is nonnegative for all feasible j, k_j and x, then

(a) $t' < t'' \Rightarrow A(t') \supset A(t'')$,

(b) $\mathrm{Sup}_\Omega\, f(x) = \mathrm{Sup}_\Omega\, \mu(P)$,

(c) $\mathrm{Max}\, f(x) = \mathrm{Max}\, \mu(P)$, provided $Y_+ = R^n_+$.

To construct the dual, we define F to be the set of those

$$d = (\bar{x}^1_1, \bar{z}^1_1, \bar{r}^1, \ldots\ldots, \bar{x}^1_{m_1}, \bar{z}^1_{m_1}, \bar{r}^1; \ldots\ldots; \bar{x}^n_1, \bar{z}^n_1, \bar{r}^n, \ldots\ldots, \bar{x}^n_{m_n}, \bar{z}^n_{m_n}, \bar{r}^n)$$

for which

$$\bar{x}^j_{k_j} \in L^*, \quad \bar{z}^j_{k_j} \in Z^*_+,$$

$$(\bar{r}^1, \bar{r}^2, \ldots\ldots, \bar{r}^n) \in R^n_+ \setminus \{0\},$$

$$\sum_{j=1}^{n} \sum_{k_j=1}^{m_j} x^j_{k_j} = 0,$$

and define $B : \Omega \rightarrow 2^F$ by

$$B(t) = \{ d \in F \mid \sum_{j=1}^{n} \sum_{k_j=1}^{m_j} s^j_{k_j}(d, t) \leq 0$$

where

$$s^j_{k_j}(d, t) = \sup [\, \bar{x}^j_{k_j}(x) + \bar{z}^j_{k_j}(g(x)) + \bar{r}^j (u^j_{k_j}(x) - t_j v^j_{k_j}(x)) \,]$$

with the supremum taken over $S \cap U^j_{k_j} \cap V^j_{k_j}$.

In turns out that if $v^j_{k_j}(x)$ is nonnegative for all feasible j, k_j and x, then

$$t' < t'' \Rightarrow B(t') \supset B(t'').$$

If, in addition $Y_+ = \{ y \in R^n \mid y > 0 \}$, then

$$\left.\begin{array}{l} t' > t'' \\ B(t'') \neq \emptyset \end{array}\right\} \Rightarrow A(t') = \emptyset.$$

Further elaboration can be found in [3].

References

[1] **Rubinstein G.S.**, Duality in Mathematical Programming and Some Questions of Convex Analysis (in Russian), Uspechi mat. nauk (1970), pp.171-201.

[2] **Tran Quoc Chien**, Duality Theory in Vector Optimization, Thesis, Charles University, Prague, 1985.

[3] **Tran Quoc Chien**, Fenchel-Lagrange Duality in Vector Fractional Programming via Abstract Duality Scheme, Kybernetika 22 (1986), pp.289-319.

CHAPTER IV

APPLICATIONS OF GENERALIZED CONVEXITY IN MANAGEMENT SCIENCE AND ECONOMICS

GENERALIZED CONVEXITY IN ECONOMICS: SOME EXAMPLES

W. Eichhorn

Abstract

Generalized convexity and generalized concavity play an increasingly important role in economics. This is shown by several examples from production and utility theory, from the theory of measurement in economics and from the theory of aggregation.

1. Introduction

Chapter 4 in Avriel, Diewert, Schaible and Zang [1988] contains a wealth of applications of generalized concavity to economics. It is shown there, among other things, where the main types of generalized concave functions are used in economics. At the same time, three models of constrained choice are studied there that cover the fundamentals of modern competitive microeconomic theory.
Here we add only a few further examples. These examples show, among others,

(i) that the quasiconcave functions are of great interest in production and utility theory, but that there exist quasiconcave increasing functions that do not satisfy the law of diminishing returns or Gossen's first law, respectively, and that there are functions satisfying these laws without being quasiconcave (see Section 2),

(ii) that there are generalized convex functions, namely, the Schur-convex functions which are fundamental in the theory of the measurement of inequality in economics (see Section 3, first part),

(iii) that the solutions of special cases of the so-called equation of measurement are quasiconcave functions (see Section 3, second part),

(iv) that the (disappointingly small) set of solutions of the problem of aggregation in production theory contains important subsets whose elements are quasiconcave functions (see Section 4).

2. Examples from production and utility theory

In what follows, $\mathbb{R}$, $\mathbb{R}_+$ and $\mathbb{R}_{++}$ are the sets of the real numbers, nonnegative real numbers and positive real numbers, respectively.
Well-known classes of functions F which play an important role in production and utility theory are the CD-type functions (Cobb and Douglas [1928]) given by

$$F: IR_+^n \to IR_+$$

(2.1) $$F(x) = cx_1^{a_1}x_2^{a_2}\dots x_n^{a_n},$$

$$c \in IR_{++},\ a_i \in IR_{++}\ (i = 1,\dots, n) \text{ constants}$$

and the ACMS-type functions (Arrow, Chenery, Minhas and Solow [1961]) given by

$$F: IR_{++}^n \to IR_+$$

(2.2) $$F(x) = (b_1x_1^{-r}+\dots+b_nx_n^{-r})^{-s/r},$$

$$0 \neq r \in IR,\ s \in IR_{++},\ b_i \in IR_{++}\ (i = 1,\dots, n) \text{ constants.}$$

They are

- homogeneous of degree $a_i+\dots+a_n$ or s, respectively,
- concave, whenever $a_1+\dots+a_n \le 1$ or $s \le 1, r \ge -1, r \neq 0$, respectively,
- not concave, but quasiconcave, whenever $a_1+\dots+a_n > 1$ or $s > 1$, respectively.

It is well known that a linearly homogeneous production function $F: IR_+^2 \to IR_+$ with constant elasticity of substitution, i.e.,

$$\frac{F_xF_y}{xy} \quad \frac{xF_x+yF_y}{F_x^2F_{yy}-2F_xF_yF_{xy}+F_y^2F_{xx}} =: \sigma = \text{const},$$

$$F_x := \frac{\partial F(x,y)}{\partial x},\quad F_y := \frac{\partial F(x,y)}{\partial y},\quad F_{xy} := \frac{\partial^2 F(x,y)}{\partial x\, \partial y},\ \dots,$$

is a CD production function (i.e., $n = 2$, $a_1 + a_2 = 1$ in (2.1)) in case $\sigma = 1$ or an ACMS production function (i.e., $n = 2$, $s = 1$, $r > -1$, $r \neq 0$ in (2.2)) in case $\sigma = 1/(1 + r)$.
Note that in the case in which $c = 1$, $n = 2$, $a_1 + a_2 = 1$, the function value (2.1) is a mean value of order 1, in the case in which $s = 1$, $n = 2$, $b_1 + b_2 = 1$, the function value (2.2) is a mean value of order $-r = 1 - 1/\sigma$; see Hardy, Littlewood and Polya [1934, p. 13].
The example

$$F: IR_+^2 \to IR_+,\quad f(x,y) = x^2y^2$$

shows that there exist quasiconcave CD-type functions that do not satisfy the law of (eventually) diminishing returns or Gossen's first law, respectively.

A production or utility function $F : IR_+^n \to IR_+$ satisfies these laws if each' northeastern' product of (cardinal) utility curve

$$\left\{ (\xi, \eta) \;\middle|\; \begin{array}{l} \eta = F(x + \xi(\mu_1 e^1 + \ldots + \mu_n e^n)), \\ \xi \in IR : x + \xi(\mu_1 e^1 + \ldots + \mu_n e^n) \in IR_+^n \end{array} \right\}$$

where e_i = i - th unit vector and where
$x \in IR_+^n$ and $\mu_i \in IR_+^n$, $i = 1, \ldots, n$, $\mu_1 + \ldots + \mu_n = 1$
are arbitrarily chosen and fixed,

has the following property: There exists a point ξ_0 such that the curve is restricted quasiconcave for all $\xi \geq \xi_0$, i.e.,

- is quasiconcave for all $\xi \geq \xi_0$, whenever the curve is not monotonically increasing for all $\xi \geq \xi_0$.
- is concave for all $\xi \geq \xi_0$, whenever the curve is not monotonically increasing for all $\xi \geq \xi_0$.

The class C of generalized concave functions so defined should be a topic of further research because of the important role that it plays in production and utility theory.
It should be noted that this class C intersects with the class Q of all quasiconcave functions without being a subclass of it. For instance, the function $F : IR_+^2 \to IR_+$ represented by the isoquants (or difference curves, resp.) in Figure 1 belongs to class C but not to class Q, since the graph of F above the segment from A to B is not quasiconcave.

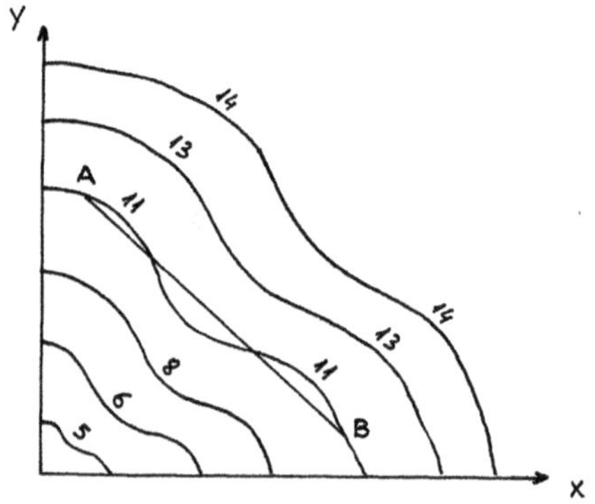

Figure 1: Isoquants of a function of class C.

An example of a class $C^* \subseteq C$ which is contained in class Q is given by

$$F : IR_{++}^n \to IR_{++},$$

$$F(x) = \frac{c x_1^{a_1} x_2^{a_2} \dots x_n^{a_n}}{c_1 x_1^{b} + c_2 x_2^{b} + \dots + c_n x_n^{b}}$$

$a_i \in IR_{++}$, $0 < a_i < 1$, $c_i \in IR_{++}$, $i = 1, \dots, n$.
$b \in IR_{++}$, $b > 1$, $c \in IR_{++}$, $a_1 + a_2 + \dots + a_n \leq 1$.

The quasiconcavity of these functions follows from Proposition 8 in Schaible [1989].
An example of a linearly homogeneous function $F : IR_{++}^2 \rightarrow IR_{++}$ that neither belongs to class C nor to class Q is given by

(2.3) $$F(x_1, x_2) = \frac{x_1^{1.4} x_2^{1.6}}{x_1^2 + x_2^2}$$

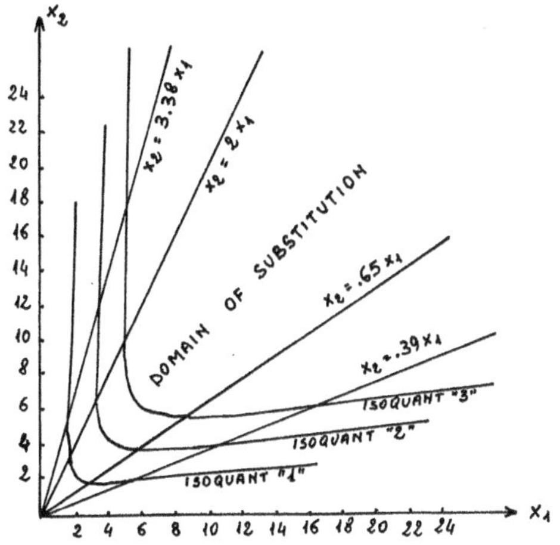

Figure 2: Three isoquants of the function F given by (2.3).

Nevertheless, the mappings $x_1 \rightarrow F(x_1, x_2)$ and $x_2 \rightarrow F(x_1, x_2)$ are quasiconcave and satisfying, in a production theoretical interpretation, the law of eventually diminishing returns as defined above.

In concluding this section it should be emphasized that up to now there is not known any system of sufficiently general axioms for production functions $F: IR_+^n \rightarrow IR_+$ that implies quasiconcavity.

As we have shown in this volume (Eichhorn and Leopold [1989]) the last statement is also true in the case of Shephard's axioms for production correspondences. His axiom A.8, i.e., quasiconcavity of the output correspondences, is not implied by his axioms A.1 - A.7, and this being so, although axiom A.7 requires the convexity of the output sets, i.e., the quasiconcavity of the involved input correspondences.

3. Examples from the Theory of Measurement in Economics

A class of generalized convex functions, which is the corner-stone in the theory of measurement of inequality, is the class of the Schur-convex functions. A function $F: IR_+^n \rightarrow IR_+$ is called Schur-convex if

(2.4) $$F(xB) \leq F(x) \quad \text{for all} \quad x \in IR_+^n \quad \text{and all } B,$$

where B is a doubly stochastic (n, n)-matrix, i.e., the elements of B are nonnegative real numbers and the sum of the elements in each row as well as in each column of B equals one.

The importance of this class of functions for the measurements of inequality follows from the facts that property (2.4)

- implies symmetry of F (the permutation matrices are special B's) and
- is equivalent to Dalton's [1920] "principle of transfers" together with symmetry; see Marshall and Olkin [1979, p. 55].

This principle states in the case in which the components $x_1, \ldots, x_n$ of x are interpreted as incomes of n households: If a part of the income of an income recipient is taken away from him and is given to an income recipient with higher income, the income inequality $F(x)$ has grown.

Note, in this connection, that the extreme values of the components of xB in (2.4), which are convex combinations of the components of x, are closer (or at least as close) together than the extreme values of the components of x.

Seen from this point of view, equation (2.4) says that the inequality of the components of xB is smaller than or equal to the inequality of the components of x.

It is well known (see Dasgupta, Sen and Starrett [1973]) that every symmetric quasiconvex function is Schur-convex. The converse is not true.

Special classes of Schur-convex functions, which play an important role in income

inequalitymeasurement theory, are dealt with in Eichhorn [1988].
As can be seen from the papers Aczél [1988], Aczél, Roberts and Rosenbaum [1986] and Eichhorn and Gleißner [1988] the solutions of many special cases of the so-called equation of measurement are quasiconcave functions.

The equation of measurement is

$$u(Ax + p) = R(A, p)\, u(x) + P(A, p)$$

where $u: X \to V$ is an unknown function from a nonempty subset X of a real vector space U into a real vector space V, where the nonempty subsets $Y \ni p$ and $S \ni A$ of $L(U)$ (the algebra of linear operators of U into itself) are such that

$$(Ax+p) \in X \quad \text{for all } x \in X,\ p \in Y,\ A \in S \subsetneq L(U),$$

and where $R : S \times Y \to \mathrm{IR}$ and $P : S \times Y \to V$ are 'given functions'.
For instance, all solutions of

$$u(r_1 x_1, \dots, r_n x_n) = R(r_1, \dots, r_n)\, u(x_1, \dots, x_n) + P(r_1, \dots, r_n),$$

$$u : \mathrm{IR}_{++}{}^n \to \mathrm{IR},\ R : \mathrm{IR}_{++}{}^n \to \mathrm{IR}_{++},\ P : \mathrm{IR}_{++}{}^n \to \mathrm{IR},$$

which are nondecreasing in an arbitrarily small ε-neighbourhood of a point $x^* \in \mathrm{IR}_{++}{}^n$, are of the form

(3.1) $$u(x_1, \dots, x_n) = c x_1{}^{a_1} x_2{}^{a_2} \dots x_n{}^{a_n} + b,$$

$$c \in \mathrm{IR}_+,\quad a_i \in \mathrm{IR}_+,\quad b \in \mathrm{IR},\ \text{arbitrary constants},$$

and of the form

(3.2) $$u(x_1, \dots, x_n) = c_1 \log x_1 + \dots + c_n \log x_n + b,$$

$$c_i \in \mathrm{IR}_+,\ b \in \mathrm{IR},\ \text{arbitrary constants};$$

see Aczél [1988] or Aczél, Roberts and Rosenbaum [1986].

That the functions given by (3.1) and (3.2) are quasiconcave follows from Propositions 6 and 8 in Schaible [1989]. The functions given by (3.2) even are concave.

4. Examples from the Theory of Aggregation

The problem of aggregation in production theory, say, is: Determine all systems of functions

$$F : \mathbb{R}_+^M \rightarrow \mathbb{R}_+ , \; f_1 : \mathbb{R}_+^M \rightarrow \mathbb{R}_+ , \dots , f_J : \mathbb{R}_+^M \rightarrow \mathbb{R}_+ ,$$

$$G : \mathbb{R}_+^J \rightarrow \mathbb{R}_+ , \; g_1 : \mathbb{R}_+^J \rightarrow \mathbb{R}_+ , \dots , g_M : \mathbb{R}_+^J \rightarrow \mathbb{R}_+$$

for which the equation

$$(4.1) \qquad F(g_1(\,|x_1), \dots , g_M(\,|x_M)) = G(f_1(\underline{x}_1), \dots , f_J(\underline{x}_J))$$

is identically satisfied in all variables. Equation (4.1) in which

$$|x_m \text{ denotes the vector } \begin{pmatrix} x_{1m} \\ \cdot \\ \cdot \\ \cdot \\ x_{Jm} \end{pmatrix} \qquad (m = 1, \dots , M)$$

and

$$\underline{x}_j \text{ denotes the vector } (x_{j1}, \dots , x_{jM}) \; (j = 1, \dots , J)$$

can be illustrated by means of the following scheme (see Pokropp [1972, p. 29 ff]):

$$\begin{array}{ccccc} x_{11}, \dots , x_{1m}, \dots , x_{1M} & & \overset{f_1}{\rightarrow} & & y_1 \\ \vdots \quad \vdots \quad \vdots & & & & \vdots \\ x_{j1}, \dots , x_{jm}, \dots , x_{jM} & & \overset{f_j}{\rightarrow} & & y_j \\ \vdots \quad \vdots \quad \vdots & & & & \vdots \\ x_{J1}, \dots , x_{Jm}, \dots , x_{JM} & & \overset{f_J}{\rightarrow} & & y_J \\ \vdots \quad \vdots \quad \vdots & & & & \vdots \\ \downarrow g_1 \quad \downarrow g_m \quad \downarrow g_M & & & & \downarrow G \\ x_1, \dots , x_m, \dots , x_M & & \overset{F}{\rightarrow} & & Y \; . \end{array}$$

There are connections between this "equation of aggregation" (Pokropp [1978]) and the equation of measurement considered in section 3. Let, for instance, the inputs x_{jm} be that much disaggregated that the m-th input is qualitatively identical in all J production sectors, then x_m is the sum of the elements of the vector $| x_m$ $(m = 1, \dots, M)$.
If the output of each of the J production sectors is the monetary value y_j of the goods produced in the respective sector j, then

$$G(f_1(\underline{x}_1), \dots, f_J(\underline{x}_J)) = f_1(\underline{x}_1) + \dots + f_J(\underline{x}_J).$$

Hence, the following special case od equation (4.1) has to be solved:

$$(4.2) \qquad F(\underline{x}_1 + \dots + \underline{x}_J) = f_1(\underline{x}_1) + \dots + f_J(\underline{x}_J).$$

Writing $\underline{x}_1 =: \underline{x}$, $\underline{x}_2 =: p$, $F =: u$ and putting $\underline{x}_3 = \dots = \underline{x}_J = 0$ we get

$$(4.3) \qquad u(\underline{x} + \underline{p}) = f_1(\underline{x}) + f_2(\underline{p}) + c \qquad (c := f_3(\underline{0}) + \dots + f_J(\underline{0})).$$

This becomes

$$(4.4) \qquad u(\underline{x} + \underline{p}) = u(\underline{x}) + f_2(\underline{p}) - d \qquad (d := f_2(\underline{0})).$$

since from (4.4) with $\underline{p} = \underline{0}$ we get $f_1(\underline{x}) = u(\underline{x}) - d - c$.

We have shown that the problem of aggregation leads, at least in an important special case, to the problem of solving an equation which is the equation of measurement in the case, where

$$A = 1 = \text{unit matrix}, R(A, p) \equiv 1, \quad P(A, p) = f_2(p) - d.$$

It is well known that a system of regular (e.g., locally monotonic) functions $F, f_1, \dots, f_J$ satisfy (4.2) if and only if

$$F(\underline{x}) = \underline{a}\underline{x} + b = a_1 x_1 + \dots + a_M x_M + b,$$

$$f_1(\underline{x}) = \underline{a}\underline{x} + b_1, \dots, f_J(\underline{x}) = \underline{a}\underline{x} + b_J$$

$$(a_1, \dots, a_M, b_1, \dots, b_J, b = b_1 + \dots + b_J \text{ real constants});$$

see Aczél]1966] or Eichhorn [1978]. Obviously, the solution of equation (4.2) are (affine) linear, i.e., convex, concave, quasiconcave and quasiconvex.

It is also well known (see, for instance, Pokropp [1972] or [1978] that every function h which belongs to a set

$$F, g_1, \dots, g_M, G, f_1, \dots, f_J$$

of continuous, increasing solutions of (4.1) has an additively separable structure, i.e., can be written as follows:

$$h(y_1, \dots, y_n) = h_0(h_1(y_1) + \dots + h_n(y_n)), \tag{4.5}$$

where h_i: $IR_+ \rightarrow IR_+$, $i = 0, \dots, n$.

According to Proposition 6 in Schaible [1989] functions of this type are quasiconcave, if h_0 is increasing and quasiconcave and if the functions $h_1, \dots, h_n$ are concave. Here it should be pointed out that concavity with respect to variation of a single input is often assumed in production theory.

References

/1/ **Aczél J.**, Lectures on Functional Equations and Their Applications. Academic Press, New York 1966.

/2/ **Aczél J.**, 'Cheaper by the Dozen': Twelve Functional Equations and Their Applications to the 'Laws of Science' and to Measurement in Economics, in: Measurement in Economics, edited by W.Eichhorn. Physica-Verlag, Heidelberg 1988, pp. 3-17.

/3/ **Aczél J., Roberts F.S., and Rosenbaum Z.**, On Scientific Laws without Dimensional Constants, Journal of Mathematical Analysis and Applications 119 (1986), pp.389-416.

/4/ **Arrow K.J., Chenery H.B., Minhas B.S., and Solow R.M.**, Capital-labour Substitution and Economic Efficiency, Review of Economics and Statistics 43 (1961), pp. 225-250.

/5/ **Avriel M., Diewert W.E., Schaible S., and Zang J.**, Generalized Concavity. Plenum Press, New York and London 1988.

/6/ **Cobb C.W., and Douglas P.H.**, A Theory of Production, American Economic Review 18 (1928), Suppl., 139-165.

/7/ **Dalton H.**, The Measurement of the Inequality of Incomes, Economic Journal 30 (1920), pp.348-361.

/8/ **Dasgupta P., Sen A., and Starrett D.**, Notes on the Measurement of Inequality, Journal of Economic Theory 6 (1973), pp. 180-187.

/9/ **Eichhorn W.**, Functional Equations in Economics. Addison-Verlag, Reading 1978.

/10/ **Eichhorn W.**, On a Class of Inequality Measures, Social Choice and Welfare 5 (1988), pp. 171-177.

/11/ **Eichhorn W., and Gleißner W.**, The Solutions of Important Special Cases of the Equation of Measurement, in: Measurement in Economics, edited by W. Eichhorn. Physica-

Verlag, Heidelberg 1988, pp. 29-37.

/12/ **Eichhorn W., and Leopold U.**, Logical Remarks on Shephard's Axioms of Production Theory, this volume.

/13/ **Hardy G.H., Littlewood J.E., and Polya G.**, Inequalities. Cambridge University Press, Cambridge 1934 (reprinted 1967).

/14/ **Marshall A.W., Olkin I.**, Inequalities: Theory of Majorization and Its Application. Academic Press, New York 1979.

/15/ **Pokropp F.**, Aggregation von Produktionsfunktionen. Springer-Verlag, Berlin-Heidelberg-New York 1972.

/16/ **Pokropp F.**, The Functional Equation of Aggregation, in: Eichhorn W., Functional Equations in Economics. Addison-Wesley, Readings 1978, pp. 122-139.

/17/ **Schaible S.**, Introduction to Generalized Convexity, this volume.

LOG-CONVEXITY AND GLOBAL PORTFOLIO IMMUNIZATION

L.Montrucchio- L.Peccati

Abstract

We discuss the role of log-convexity in immunizing cash-flows. We deal both with the standard case, where the cash-flow does not depend on the market interest rate and with the indexed case. Further the shock in the market force of interest can vay over time.

Introduction

The immunization problem for an asset portfolio concerns with the possibility to choose its composition so that its value at a given epoch is protected against shocks on the market interest rate. A key role in immunization is played by the so-called "Duration" of the portfolio.

A lot of results has been obtained, through various techniques with various degrees of generality (see, for instance: Bierwag-Kaufman-Toevs (1984)). Our aim here is to show how greater generality and simplicity can be achieved by using convexity tools in a functional space approach. Furthermore, in this frame a new and deeper insight into the intimate nature of the problem is also attained.

We have kept our exposition on a very synthetic level. For details and pratical implications see Montrucchio-Peccati (1988 a,b), Buffa-Peccati (1989).

In section 1 we set down basic notations and concepts. In section 2 the case of nonindexed assets is considered. In section 3 we present a general theoretical solution to the problem for nonindexed assets. In sections 4 and 5 we turn our attention to indexed assets and in section 6 we collect some conclusions.

1. Portfolios and Their Values

A fairly general way to define a portfolio starts describing the output distribution on the relevant time set. This distribution on IR_+ can be any positive finite Lebesgue-Stieltjes measure μ on IR_+ (Castagnoli-Peccati (1973)). If A is any μ-measurable set: $A \subseteq IR_+$, then $\mu(A)$ is the money output during A. A Portfolio will be described through a pair (μ, δ), where $\delta: IR_+ \to IR_+$ is in $L^1(0, +\infty)$ (L^1 for short) and it can be thought as a variable force of interest. We shall also use the following shorthand notation: $F(t, T) = \exp \int_t^T \delta(u)\, du$. We

shall also suppose that μ has a compact support.
The value at epoch T of the portfolio (μ, δ) is defined by:

$$V(T; \delta) = \int F(t, T)\, \mu(dt)$$

where $\int$ without domain specification means here and in the following "integral on IR_+ ".
The present value of the portfolio (μ, δ) is $V(0; \delta)$ and will be denoted by $P(\delta)$.
The immunization problem consists in the search for a T such that $V(T; \delta)$ favourably reacts to any shift $h \in L^1$ of the force of interest. Usually it is suggested to choose:

(1) $$T = \int tF(t, 0)\, \mu(dt) / P(\delta)$$

the so-called duration of the portfolio.
In the next section we try to investigate the "rationality" behind this choice having in mind Wald's criterion of minimax (see also Bierwag-Khang (1979) and Prisman (1986)).

2. Duration and Minimax

Consider V as a function of δ defined over L^1. In principle the immunization problem should be that of finding $T \in IR_+$ such that $V(\delta + h; T) \geq V(\delta; T)$ $\forall h \in L^1$ (case of complete immunization).
Note that it is not restrictive to conceive shock functions in $L^1(0, +\infty)$, even if any $h \in L^1_{loc}(0, +\infty)$ can be thought as a shock. This is due to the fact that μ has a compact support. And so the effect of a $h \in L^1_{loc}$ agrees with the one of $h\chi_{[0, \tau]} \in L^1(0, +\infty)$ with τ large enough, where χ_A denotes the indicator function of the set A. An analogous remark could be made about the meaning of $\delta \in L^1$.
Obviously it is impossible to get $V(\delta + h; T) \geq V(\delta; T)$ $\forall h \in L^1$ but the case when μ is concentrated at T. It follows that the desidered inequality will generally be satisfied only for shocks in a subset of L^1. Significant properties of this subset will be identified through the implication:

(2) $$V(\delta + h; T) \geq V(\delta; T) \Leftrightarrow \ln P(\delta + h) \geq \ln P(\delta) - \int_0^T h(u)du$$

which is trivially verified by using the relation $V(\delta; T) = P(\delta)\, F(0, T)$. Furthermore $V(\cdot\, ; T)$ turns out to be convex on L^1 and Fréchet differentiable. It follows that $\ln P(\delta)$ is differrentiable too and its gradient $\nabla \ln P(\delta)(u) = k(u)$, belonging to $L^\infty(0, +\infty)$, is given by:

$$(3) \qquad k(u) = -(1-P(\delta)) \int_{t \geq u} F(t,0)\,\mu(dt)$$

Actually also ln P(δ) is convex but this is not essential now (see later).
In the following we shall say that the portfolio (μ , δ) is immunized at time T for a shock function $h \in L^1$ if $V(\delta + h; T) \geq V(\delta; T)$. Having in mind the definition of k given in (3), we can prove:

Proposition 1 - Let μ be not concentrated at a point and let T be fixed in IR_+ . There exists a closed half-space of L^1:

$$\langle k + \chi_{[0,T]}, h \rangle \geq 0$$

of shock functions, such that (μ , δ) is immunized for them. Further the one-codimensional hyperplane H_T of equation $\langle k + \chi_{[0,T]}, h \rangle = 0$ is charcaterized as the maximal vector subspace of L^1 of shock functions for which (μ , δ) is immunized.
The half space $\langle k + \chi_{[0,T]}, h \rangle < 0$ is characterized by the property: $V(\delta + \lambda h; T) < V(\delta ; T)$ for $\lambda < 0$ small enough. The identically one function $h = \chi_{[0,+\infty]}$ belongs to H_T iff T is the duration defined by (1).

Proof. Fix $T \in IR_+$ and take a vector subspace $S \subset L^1$ such that $V(\delta + h; T) \geq V(\delta ; T)$ $\forall h \in S$. By (2) it follows that lnP(δ) is subdifferentiable at δ over the linear variety $S + \{ \delta \}$ with subdifferential given by $-\chi_{[0,T]}$. As lnP(δ) is differentiable, its differential agrees with the subdifferential. Thus $k + \chi_{[0,T]}$ must be "orthogonal" to S, i.e.; $\langle k + \chi_{[0,T]}, h \rangle = 0$ $\forall h \in S$. So we have $S \subset H_T$ where H_T is the hyperplane of equation $\langle k + \chi_{[0,T]}, h \rangle = 0$. Note that it is an hyperplane because $k = -\chi_{[0,T]}$ iff μ is concentrate at T. The gradient of V(. ; T) is easily computed as:

$$(4) \qquad \nabla V(\delta ; T) = V(\delta ; T)\,(\chi_{[0,T]} + k)$$

and, thus, μ is immunized for the shocks belonging to $\langle k + \chi_{[0,T]}, h \rangle \geq 0$ because $V(\delta ; T)$ is convex.
Note further that the functions $h \in H_T$ are characterized by:

$$(5) \qquad h \in H_T \Leftrightarrow -\int_0^T h(u)\,du = \frac{d}{d\lambda} \ln P(\delta + \lambda h)\,|_{\lambda=0}$$

i.e.;

$$(6) \qquad \int_0^T h(u)\,du = \int F(t,0)\left[\int_0^T h(u)\,du\right] \mu(dt)/ P(\delta)$$

Therefore if $h \equiv 1$ belongs to H_T then, from (6), we get: $T = \int tF(t, 0)\,\mu(dt)/P(\delta)$ which is the duration given in (1) and this concludes the proof.

□

The next proposition justifies the practical interest of truncating the investment in the portfolio at a point T within the support of the distribution μ. Let us indicate by $[0, T_0]$ the convex hull of the support of the portfolio measure μ.

Proposition 2 - (i) For any $h \in L^1$ there exists a $T \in [0, T_0]$ such that $h \in H_T$.
(ii) If H_T contains an essentially positive shock h, then $T \in [0, T_0]$.

Proof. The statements follow from (6) which can be written as:

$$(7) \qquad [\int_0^T h(u)\,du\,] \int F(t, 0)\,\mu(dt) = \int F(t, 0)(\int_0^t h(u)\,du)\,\mu(dt)$$

The equality (7) is nothing but the integral intermediate value theorem. Thus, given an h, a number $T \in [0, T_0]$ will exist such that (7) is fulfilled. Also (ii) is easily derived.

□

Remark - From Propositions 1 and 2 it follows that any choice of $T \in [0, T_0]$ has good requirements to be chosen. The privileged choice of the duration given by (1) is determined by its well-known property to immunize a portfolio for flat shocks $h = \lambda \chi_{\mathbb{R}_+}$

Let us analyse the best choice of T in $[0, T_0]$ following the minimax criterion.
As $V(\,.\,; T)$ is convex, we have:

$$V(\delta + h; T) \geq V(\delta\,; T) + \langle \nabla V(\delta\,; T), h \rangle$$

By (4) we get:

$$(8) \qquad \Delta_h V/V \geq \langle k + \chi_{[0,T]}, h \rangle \qquad \forall h \in L^1$$

where $\Delta_h V = V(\delta + h; T) - V(\delta\,; T)$. We get then estimate:

$$(9) \qquad \Delta_h V/V \geq -\,\| k + \chi_{[0;T]} \|_{L^\infty} \,.\, \| h \|_{L^1}$$

A good criterion to minimize the loss will be that to choose a time $T \in [0, T_0]$ such that $\| k + \chi_{[0;T]} \|_{L^\infty}$ is minimized.
It is not difficult to see that this criterion leads to pick a T different from the usual duration (1), because it is related to the median value of the distribution. Suppose for simplicity's sake that the portfolio outflow measure μ is continuous. Then it is easy to check

that minimizing $\| k + \chi_{[0;T]} \|_{L^\infty}$ is equivalent to find a number $T \in [0, T_0]$ such that $\max(|k(t)|, |k(t)+1|)$ has the largest value and this, in turn, is equivalent to take T as the median value of the distribution defined by the elements of probability $(1/P(\delta))\, F(t, 0)\, \mu(dt)$. For that T the value of $\| k + \chi_{[0;T]} \|_{L^\infty}$ will equal 1/2 and we obtain the estimate:

$$\Delta_h V/V \geq -1/2 \, \| h \|_{L^1}$$

Note that the duration (1) is the mean value of the distribution above defined.
Different results on lower bounds for $\Delta_h V/V$ are known in literature. See Fong-Vasicek (1983) and Montrucchio-Peccati (1988 a).

3. A Theoretical Solution to the Immunization problem

It is perhaps useful to show how to solve completely the immunization problem by enlarging the space of strategies (i.e. ; by using mixed strategies). Instead of choosing a time $T \in [0, T_0]$ we now choose a probability measure σ over $[0, T_0]$. The function $V(\delta\,;T)$ will become $\int V(\delta\,;T)\,\sigma(dt)$, which has several interpretations (even nonprobabilistic).

Proposition 3 - For any δ fixed in $L^1(0, T_0)$ there exists one and only one probability measure σ on $[0, T_0]$ which immunizes completely the portfolio (μ, δ), i.e. ; such that $\int V(\delta + h; T)\,\sigma(dt) \geq \int V(\delta\,; T)\,\sigma(dt)\ \forall h \in L^1(0, T_0)$.
Moreover σ is given by $\sigma(dt) = cF^2(t, 0)\,\mu(dt)$, where c is the normalizing coefficient.

Proof - Compute: note that $\int V(\delta\,;T)\,\sigma(dt)$ is convex for any σ and it can be written as $P(\delta)\int F(0, T)\,\sigma(dT)$. Its logarithm is $\ln \int V(\delta\,;T)\,\sigma(dT) = \ln P(\delta) + \ln \int F(0, T)\,\sigma(dT)$. Differentiating and imposing the condition $\nabla \ln \int V(\delta\,;T)\,\sigma(dT) = 0$ one gets easily the result.

□

4. Immunizing Market-Rate Dependent Cash-Flows: Log-Convexity

In this section we investigate a generalization of the theory dealt with in the previous sections. We suppose that the portfolio, identified to the pair (μ, δ) is characterized by the dependence of μ on δ. We shall write in this case $\mu = \mu_\delta$. This case arises for instance in presence of financial indexation of bonds yelds (see Morgan (1986), Landskroner-Ruthenberg (1988), Buffa-Peccati (1989), Montrucchio-Peccati (1988 b)). Of course the convexity and also differentiability properties of $V(\delta\,;T)$ may fail to be true in this new context. Let us give just two examples:
Ex. A) $\mu_\delta = \mu_0 + f(\delta)\,\mu_1$, where μ_0, μ_1 are measures independent of δ and $f(\delta)$ is a scalar function.

Ex. B) Put:

$$\mu_\delta = \sum_i f_i(\delta)\, \varepsilon_{T_i}$$

where ε_{T_i} is the Dirac measure at T_i and the functions $f_i(\delta)$ are scalar functions ($i = 1, 2, \dots, N$). To be precise, $f_i(\delta)$ has usually a simpler structure, such as:

$$f_i(\delta) = \varnothing_i\Big(\int_0^{T_i} \delta(u)du\Big) \; ; \; f_i(\delta) = \varnothing_i\Big[\int_{T_i-\varepsilon}^{T_i} \delta(u)du\Big]$$

or else. In other words $f_i(\delta)$ can be expressed as $f_i(\delta) = \varnothing_i(l_i(\delta))$ where l_i are linear forms and $\varnothing_i$: $\mathbb{R} \to \mathbb{R}$.

As one can see also here the functionals that are involved are of the type:

$$V(\delta\,; T) = P(\delta)\, F(0, T)$$

which can be written as:

$$V(\delta\,; T) = P(\delta) \exp \langle \chi_{[0,T]}, \delta \rangle$$

with $P(\delta) = \int F(t, 0)\, \mu_\delta(dt)$.

The next propositions will play a central role to understand the relationship among several forms of convexity of $P(\delta)$ and $V(\delta\,; T)$.

Proposition 4 - Let X be a normed space and X^* be its dual. Let $f: X \to \mathbb{R} \cup \{+\infty\}$ a positive function. The function f is log-convex iff $f(x)\exp\langle x^*, x\rangle$ are convex for any $x^* \in X^*$. Furthermore, if f is continuous at x_0, then one has $\partial \ln f(x_0) = \partial f(x_0) / f(x_0)$.

Proof - There is no loss of generality to prove this statement for real functions $f(t)$ defined on a real interval. Recall that a function $f(t)$ is convex iff the second difference $\Delta^2_h f(t) = f(t+2h) - 2f(t+h) + f(t) \geq 0$ everywhere. Apply this characterization to the family $f(t)\exp(\lambda t)$. It becomes:

$$f(t+2h)\exp(2\lambda h) - 2\, f(t+h)\exp(\lambda h) + f(t) \geq 0$$

This is equivalent to $f^2(t+h) - f(t+2h)\, f(t) \leq 0$, which is, in turn, equivalent to $\Delta^2_h \ln f(t) \geq 0$. The converse is trivial.

The relation $\partial \ln f(x) = \partial f(x) / f(x)$ follows easily by the chain rule (see 2.3.9., (ii) in Clarke (1983)).

□

Whenever f is differentiable the last proposition can be slightly generalized.

Proposition 5 - Suppose f differentiable and positive. If $f(x) \exp \langle x^*, x \rangle$ are all pseudo-convex, then $\ln f(x)$ is convex.

Proof - Fix an element x_0 in X. We prove that $\ln f(x)$ is subdifferentiable at x_0.
The gradient of $f(x) \exp \langle x^*, x \rangle$ at x_0 is: $\exp \langle x^*, x \rangle [\nabla f(x_0) + x^* f(x_0)]$. If $x^* = -\nabla f(x_0)/f(x_0)$, the gradient of $f(x) \exp \langle x^*, x \rangle$ is zero. Because $f(x) \exp \langle x^*, x \rangle$ is pseudo-convex it follows that $f(x) \exp \langle x^*, x \rangle \geq f(x_0) \exp \langle x^*, x_0 \rangle$. By taking the logarithms, we get:

$$\ln f(x) \geq \ln f(x_0) + \langle x - x_0, x^* \rangle$$

and this completes the proof.

□

Some interesting corollaries can be stated:

Corollary 6 - If $f(x) \exp \langle x^*, x \rangle$ are all pseudo-convex and $f(x)$ is differentiable, then $f(x) \exp \langle x^*, x \rangle$ are all convex.
As a direct consequence of Proposition 4, we get a well-known Artin's theorem, which is equivalent to the Holder inequality (see Marshall-Olkin (1979)).

Corollary 7 - The sum of log-convex functions is log-convex.
It is difficult to prove directly that the functions $P(\delta)$ and $V(\delta; T)$, we have encountered before are log-convex. The above propositions lead to:

Corollary 8 - $P(\delta)$ and $V(\delta; T)$ for a portfolio (μ, δ) are log-convex whenever μ is independent of δ.

Proof - $P(\delta) \exp \langle k, \delta \rangle$ are trivially convex for any $k \in L^\infty$. Thus the statement follows from Proposition 5.

Now we turn back to the portfolio μ_δ. The next proposition is a general statement on the possibility of defining a duration like (1).

Proposition 9 - The portfolio (μ_δ, δ) is immunized for flat shocks $h = \lambda \chi_{IR_+}$ iff the function $Q(\lambda) = \ln [P(\delta + \lambda \chi_{IR_+}]$, $\lambda \in IR$, is subdifferentiable at $\lambda = 0$. As duration any element $T \in -\partial Q(0)$ may be picked.

Proof - Trivial because the relation (2) holds still for our portfolios.

We see an application to Example A. In this case we have $P(\delta) = P_0(\delta) + f(\delta) P_1(\delta)$ with

$$P_i(\delta + \lambda \chi_{IR_+}) = \int \exp \int_t^c (\delta(u) + \lambda \chi_{IR_+})\, du)\, \mu_i(dt) \qquad i = 0, 1$$

If we suppose $f(\delta + \lambda\chi)$ to be log-convex on IR then $\ln P(\delta + \lambda\chi)$ is surely subdifferentiable at $\lambda = 0$.
In fact, $P_0(\delta + \lambda\chi)$ is log-convex and differentiable. The function $f(\delta + \lambda\chi)\, P_1(\delta + \lambda\chi)$ is log-convex. In this case, it is not difficult to compute the duration which turns out to be any element: $T \in P_0(\delta)/ P(\delta)\, \tau_0 + f(\delta)\, P_1(\delta)/ P(\delta)\, \tau_1 - P_1(\delta)/ P(\delta)\, \partial f(0)$

where $\partial f(0)$ is the subgradient of $f(\delta + \lambda\chi)$ at $\lambda = 0$ and τ_0 and τ_1 are the durations of μ_0 and μ_1 respectively.

□

5. Global Results on Immunization: The Differentiable Case

We end this paper by giving a result which extends Proposition 1 to portfolios (μ_δ, δ). Two things are in order: differentiability and convexity.
The former has been investigated in Section 4, we 'face now the differentiability of the portfolio (μ_δ, δ).

Definition - Given a measure μ_δ, we shall say that μ_δ is differentiable at δ (in a weak sense) if there exists a measure $\mu^*(dt \times du)$ on $[0, T_0] \times IR_+$ which is finite but non necessarily positive such that the following conditions hold:

(i) the projection measure $\int_{t=0}^{t=T_0} \mu^*(dt \times du)$ is absolutely continuous with respect to the Lebesgue measure du and with density belonging to L^∞.

(ii) $\forall f \in \mathcal{C}^0(0, T_0)$ and $\forall h \in L^1$:

$$\int f(t)\, \mu_{\delta+h}(dt) = \int f(t)\, \mu_\delta(dt) + \iint f(t)\, h(u)\, \mu^*(dt \times du) + o(\| h \|)$$

as $\| h \| \to 0$, where the norm is the one of L^1.
Note that, under (i), the projection measure is a measure over IR_+ which is absolutely continuous with respect to the Lebesgue measure and with density in L^∞ for any f. This follows from Fubini theorem. In this way the integral $\iint f(t)\, h(u)\, \mu^*(dt \times du)$ makes sense.

Example - If μ_δ is the one considered in Ex. B, μ_δ is differentiable whenever $f_i(\delta)$ are Frécher differentiable and one has:

$$\mu^*(dt \times du) = \sum_{i=1}^{N} \nabla f_i(\delta)\; du \otimes \varepsilon_{T_i}(dt)$$

Similarly the derivative of $\mu_0 + f(\delta)\, \mu_1$ is:

$$\mu^*(dt \times du) = \nabla f(\delta)\, du \otimes \mu_1(dt)$$

supposing $f : L^1 \to IR$ differentiable at δ.

We can state and prove the following:

Proposition 10 - If μ_δ is differentiable, then $P(\delta)$ and $V(\delta; T)$ are Fréchet differentiable over $\delta \in L^1$. The derivative evaluated at δ along the direction $h \in L^1$ is given by:

$$D_h P(\delta) = \int F(t, 0) \int_t^c h(s)\, ds\, \mu_\delta(dt) + \iint h(u) F(t, 0)\, \mu^*(dt \times du)$$

The above Proposition is trivially verified by using the definition of differentiability. Note also how it is possible to write the gradient of $\ln P(\delta)$ as:

$$\nabla \ln P(\delta) = \tilde{k}(u) = 1/P(\delta) . \left[-\int_{t \geq u} F(t, 0)\, \mu_\delta(dt) + \frac{d}{du} \int_{t=0}^{T_0} F(t, 0)\, \mu^*(dt \times du) \right]$$

where $\frac{d}{du}$ means the derivative of the measure:

$$\int_0^{T_0} F(t, 0)\, \mu^*(dt \times du)$$

with respect to the Lebesgue measure du.
The previous formula is directly comparable to (3) of Section 2.
We are in position to state the main theorem of this section:

Theorem 11 - Let (μ_δ, δ) be a portfolio which is differentiable. If $P(\delta)$ is log-convex then the same statement of Proposition 1 of section 2 holds where k(u) is replaced by k(u) given above. This unless $k + \chi_{[0, T]} \equiv 0$ for some T. In which case the portfolio is completely immunized.

Proof - It is the same of Proposition 1. Note that the log-convexity of $P(\delta)$ implies that all the $V(\delta; T). \exp \langle \chi_{[0, T]}, \delta \rangle$ are all convex and differentiable.

□

Let us apply our results to the two Ex. A-B.
Example A - In this case $\mu_\delta = \mu_0 + f(\delta)\, \mu_1$.
Suppose $f(\delta)$ log-convex and differentiable. We saw that $P(\delta)$ is log-convex and μ_δ is differentiable. Proposition 11 applies. The function k is given by:

$$\tilde{k}(u) = \frac{1}{P(\delta)} \left[-\int_{t \geq u} F(t, 0)\ \mu_\delta(dt) + \nabla f(\delta) \int F(t, 0)\ \mu(dt) \right]$$

Example B - If $f_i(\delta)$ are differentiable and log-convex then Proposition 11 applies. In fact μ_δ is differentiable and:

$$P(\delta) = \sum_i f_i(\delta)\, P_i(\delta)$$

where:

$$P_i(\delta) = F_i(T, 0)$$

The functions $f_i(\delta)$ are log-convex and $P_i(\delta)$ are log-convex too (see Corollary 7). Thus $P(\delta)$ is log-convex by Theorem 11. The function k is given by:

$$\tilde{k}(u) = \frac{1}{P(\delta)} \left[-\sum_{T_i \leq u} f_i(\delta)\, F(T_i, 0) + \sum_i F(T_i, 0)\, \nabla f_i(\delta) \right]$$

6. Concluding Remarks

The results put in evidence in this paper are:

(1) The classical duration is not a minimax choice criterion, contrary to what is usually recognized in literature. In Section 2 it is shown that the minimax criterion leads to the choice of a median for the distribution.

(2) In Sections 4 and 5 we have investigated the immunization whenever the portfolio cash-flow depends on the rate of interest. We have shown that log-convexity assumption for the present value $P(\delta)$ arises in a very natural way. Note that the proof of Proposition 1 rests basically on pseudo-convexity of $V(\delta; T)$. From Corollary 6 it follows that whether $P(\delta)$ is not log-convex, some of $V(\delta; T)$ possibly fail to be pseudo-convex and the global theory, developed in Section 5 does not hold any more. Note that also in this case the usual duration looses its importance. The best choice turns out to minimize the norm of $k + \chi_{[0, T]}$ which may diverge a lot from duration criterion.

(3) Our results give great generality to the treatment of problems in the so called semi-deterministic approach to immunization theory (see M. De Felice - F. Moriconi (1985-1987)). Many interesting propositions can be derived by specializing some of our assumption. In particular, by studying special cases of μ_δ, conclusions of some relevance for the practice of indexation can be obtained.

References

/1/ **Bierwag G.O., Kaufman G.G., Toevs A. (eds.)** Innovations in Bond Portfolio Management: Duration Analysis and Immunization, JAI Press, (1984) London.

/2/ **Bierwag G.O., Khang C.**, An Immunization strategy is a Minmax Strategy, Journal of Finance, 34, (1979), pp.389-414.

/3/ **Buffa P., Peccati L.**, Immunizzazione di un investimento in CCT, Banche e Banchieri (1989).

/4/ **Castagnoli E., Peccati L.**, Alcune Osservazioni sulla Classificazione degli Investimenti, Giornale degli Economisti e Annali di Economia, marzo-aprile (1973).

/5/ **Chance D.**, Floating Rate Notes and Immunization, Journal of Financial and Quantitative Analysis, 18, (1983), pp. 365-380.

/6/ **Clarke F.H.**, Optimization and Nonsmooth Analysis, J.Wiley & Sons, (1983) New York.

/7/ **De Felice M.**, Modelli per la Gestione del Rischio di Tasso nell'Attività di Intermediazione finanziaria, Atti del Convegno su "I Metodi Quantitativi per le Applicazioni Finanziarie", Monte dei Paschi, Siena (1987), pp. 174-205.

/8/ **De Felice M., Moriconi F.**, Modelli Stocastici della Struttura a Termine dei Tassi di Interesse, Liguori, Napoli (1985).

/9/ **Fong H.G., Vasicek O.**, The Tradeoff Between Return and Risk in Immunized Portfolios', Financial Analysis Journal, Sept.-Oct., (1983), pp. 73-78.

/10/ **Grobel R.A.**, Understanding the Duration of Floating-Rate Notes, Salomon Brothers Inc., New York (1987).

/11/ **Landskroner Y., Ruthenberg D.**, Bank Duration and Immunization Under Variable Interest Rate, Presented at the 15th Annual Meeting of European Finance Association, Istanbul, (1988)

/12/ **Marshall A.W., Olkin I.**, Inequalities: Theory of Majorization and Its Applications, Academic Press, New York, (1979).

/13/ **Montrucchio L., Peccati L.**, A Note on Shiu-Fisher-Weil Immunization Theorem, Atti del XII Convegno AMASES, Palermo, (1988 a).

/14/ **Montrucchio L., Peccati L.**, Immunization of Market-rate Dependent Cash-flows, Presented at the Fourth Meeting of EURO Group on Financial Modelling, London (1988 b).

/15/ **Morgan G.E.**, Floating Rate Securities and Immunization: Some Further Result's, Journal of Financial and Quantitative Analysis, 21 (1986), pp. 87-94.

/16/ **Prisman E.Z.**, Immunization as a maxmin strategy, Journal of Baking and Finance, 10, (1986), pp. 491-509.

/17/ **Shiu E.S.**, On the Fisher-Weil Immunization Theorem, Insurance Mathematics and Economics, 6 (1987) pp. 259-266.

IMPROVED ANALYSIS OF THE GENERALIZED CONVEXITY OF A FUNCTION IN PORTFOLIO THEORY

J.P. Crouzeix - J.A. Ferland - S. Schaible

Abstract

Portfolio selection problems characterized by lognormal markets and power utility functions give rise to a convex program. This can be approximated by a numerically tractable, but often nonconvex program.
Sufficient conditions for the generalized convexity of the objective function were recently derived by Schaible and Ziemba using criteria for functions on the entire space R^n. The analysis is improved upon in this paper by taking advantage of criteria for pseudoconvex functions on affine subsets of R^n. Such criteria became known only very recently.

1. The Model

In [4] Ohlson and Ziemba consider the problem of determining a portfolio $x \in R^n$ that maximizes the expected utility:

$$(1) \qquad U(y) = \int_0^\infty \dots \int_0^\infty u(\rho^T y)\, dF(\rho) \to \text{Max}$$

$$\text{subject to } e^T y = 1,\ y \geq 0$$

where $e = (1, \dots, 1)^T \in R^n$ and T denotes transposition. The return is assumed to follow a lognormal distribution, i.e., $\log \rho \sim N(\mu, \Sigma)$ with mean μ and variance-covariance matrix Σ. Furthermore, suppose that the investors utility function is a power function, i.e.

$$u(w) = \begin{cases} w^\gamma & \gamma \in (0, 1) \\ -w^\gamma & \gamma \in (-\infty, 0). \end{cases}$$

The expected utility $U(y)$ is a concave function. Hence the model in (1) can be solved by a convex programming algorithm. But this means that at each iteration of such an algorithm a

multiple integral of order n has to be evaluated. Hence even for small sizes of n model (1) is numerically not tractable.
Ohlson and Ziemba [4] developed the following approximating model

$$(2) \qquad \widetilde{U}(y) = \log d^T y - \phi \log (y^T C y) \to \text{Max}$$

$$\text{subject to } e^T y = 1, \ y \geq 0$$

where $d = (\exp(\mu_i + 1/2\, \sigma_{ii}))$, $D = \text{diag}(d_i)$, $C = DLD$, $L = (\exp \sigma_{ij})$

$$\phi = \frac{1-\gamma}{2(2-\gamma)} \in (0, 0.5).$$

Then C is a symmetric, positive definite matrix.
The approximating model (2) is numerically tractable, but the objective function U is no longer concave in general. As long as $\widetilde{U}$ is (quasi) pseudoconcave the problem can be solved essentially like a concave program.
Thus, the question arises for which values of μ, Σ and γ, that is for which values of d, C and ϕ the new objective function $\widetilde{U}$ is at least generalized concave in some sense. Several answers to that question were given in Schaible and Ziemba [5]. To derive these conditions criteria for the (quasi) pseudoconcavity of C^2 - functions (see [1]) were used in the analysis.
In the present survey we report on improved results. They stem from the observation that generalized concavity of $\widetilde{U}$ is not needed on all of the nonnegative orthant but only on its restriction to then affine subspace $\{ y \in R^n \mid e^T y = 1 \}$.
In order to conduct such an analysis we need criteria for the pseudoconvexity of functions on affine subspaces. Once presented we will apply these criteria to a standardized version of (2):

$$(3) \qquad f(x) = \phi \log x^T x - \log b^T x \to \text{Min}$$

$$\text{subject to } x \in S$$

where $S = \Omega \cap K \cap E$, $\Omega = \{ x \in R^n \mid b^T x > 0 \}$, K is a closed convex cone and $E = \{ x \in R^n \mid a^T x = 1 \}$. Note that problem (3) results from applying the transformation $x = C^{1/2} y$ to (2). Hence $a, b \neq 0$ and a, b are not colinear. Throughout the remainder of this paper we summarize some major results in [3]. For proofs and additional results the reader is referred to [3].

2. Pseudoconvexity of functions on affine subspaces

Le f be a twice continuously differentiable function on a convex subset Ω of R^n and E be an affine subspace of R^n. In this section we give necessary and sufficient conditions for the convexity and pseudoconvexity of f on $\Omega \cap E$. These results are extensions of classical ones where $E = R^n$. We assume that the interior of Ω, int (Ω), is not empty.
An affine subspace E of R^n having dimension p can be specified as follows: there exists a point $x^o \in E$ and $(n - p)$ linearly independent vectors $a^1, a^2, \ldots, a^{n-p}$ in R^n such that

$$E = \{ x \in R^n \mid A^T(x - x^o) = 0 \}$$

where $A = (a^1, a^2, \ldots, a^{n-p})$ is an $n \times (n-p)$ matrix.
For convex functions we have

Proposition 1: The restriction of f to E is convex on $S = \Omega \cap E$ if and only if for all $x \in S$ $h^T\nabla^2 f(x) h \geq 0$ whenever $A^Th = 0$.
Furthermore, if for all $x \in S$

$$h^T\nabla^2 f(x) h > 0 \quad \text{whenever} \quad A^Th = 0 \text{ and } h \neq 0, \text{ then } f \text{ is strictly convex on } S.$$

We now present criteria for pseudoconvex functions on E.

Proposition 2: The restriction of f to E is pseudoconvex on $S = \Omega \cap E$ if and only if for all $x \in S$

i) $h^T\nabla^2 f(x) h \geq 0$ whenever $\nabla f(x)^Th = 0$ and $A^Th = 0$, and
ii) x is a local minimum of f on S whenever $\nabla f(x) = Az$ for some $z \in R^{n-p}$.

Furthermore we have

Proposition 3: Assume that Ω is a compact convex set of R^n. If for all $x \in \Omega$, $h^T\nabla^2 f(x) h > 0$ whenever $\nabla f(x)^Th = 0$, $A^Th = 0$, $h \neq 0$, then there exists $\alpha > 0$ such that the function $\exp(\alpha f(x))$ is strongly convex on Ω, i.e. f is strongly convex transformable.

The conditions in the preceding propositions are specified in terms of the positive (semi) definiteness of quadratic forms restricted to subspaces. Useful characterizations of such conditions can be given in terms of bordered matrices. Let M and N be $n \times m$ matrices, respectively. Assume that M is symetric and rank $(N) = m'$ where $0 \leq m' \leq m < n$. The bordered matrix P associated with M and N is the following $(n + m) \times (n + m)$ symmetric matrix

$$P = \begin{bmatrix} M & N \\ N^T & 0 \end{bmatrix}$$

The following result was recently obtained by Chabrillac and Crouzeix [2]. It involves the inertia of P. In (P), which denotes the number of positive, negative and null eigenvalues of P, respectively.

Proposition 4:
a) P has at least m' negative eigenvalues.
b) The following are equivalent (positive definiteness):
  i) $h^TM\ h > 0$ whenever $N^Th = 0, h \neq 0$,
  ii) $\text{In}\ (P) = (n, m', m - m')$.
c) The following are also equivalent (positive semidefiniteness):
  i) $h^TM\ h \geq 0$ whenever $N^Th = 0$,
  ii) P has only m' negative eingenvalues.

The following formula related to the Schur complement will be useful to compute the inertia of a partioned symmetric matrix

$$Q = \begin{bmatrix} Q_{11} & Q_{12} \\ Q^T_{12} & Q_{22} \end{bmatrix}$$

where Q_{11}, Q_{22} are symmetric and Q_{11} is nonsingular:

(4) $$\text{In}\ Q = \text{In}\ (Q_{11}) + \text{In}\ (Q_{22} - Q^T_{12}\ Q_{11}^{-1}\ Q_{12})$$

We refer to Q_{11} as the pivot of the complement.

3. Pseudoconvexity in the Portfolio Selection Problem

We now return to the standardized portfolio selection problem (3). In view of Proposition 2 and 3 we are led to determine under what conditions

(5) $$h^T\nabla^2 f(x)\ h \geq 0 \quad \text{whenever} \quad \nabla f(x)^T h = 0 \quad \text{and} \quad a^Th = 0\ (h \neq 0).$$

For the function $f(x) = \emptyset \log x^Tx - \log b^Tx$ the conditions in (5) become

(6) $$h^TM\ h \geq 0\ (> 0) \quad \text{whenever} \quad N^Th = 0\ (h \neq 0)$$

where $$M = I - 2(1 - \varnothing)\,\frac{xx^T}{\| x \|^2}$$

and $$N = \left[\ \frac{2\varnothing}{\| x \|^2}\, x - \frac{1}{(b^T h)}\, b,\ a\ \right]$$

M has $n - 1$ eigenvalues equal to 1 and the other one equals to $2\varnothing - 1$.
Since $\varnothing \in (0, 0.5)$, it follows that $\mathrm{In}\,(M) = (n-1, 1, 0)$. We take M as the pivot of the Schur complement of

$$P = \begin{bmatrix} M & N \\ N^T & 0 \end{bmatrix}$$

In view of (4)

(7) $$\mathrm{In}\,(P) = \mathrm{In}\,(M) + \mathrm{In}\,(R)$$

where $R = -N^T M^{-1} N$ with

(8) $$r_{11}(x) = \frac{2(1 - \varnothing)}{\| x \|^2} - \frac{\| b \|^2}{(b^T x)^2}$$

(9) $$r_{12}(x) = r_{21}(x) = \frac{a^T b}{b^T x} - \frac{2a^T x}{\| x \|^2}$$

(10) $$r_{22}(x) = \frac{2(1 - \varnothing)}{1 - 2\varnothing}\ \frac{(a^T x)^2}{\| x \|^2} - \| a \|^2$$

To analyze the psuedoconvexity of the restriction of f to E on $S = \Omega \cap E$ two different cases have to be considered in view of Proposition 2.

Case 1: $x \in S$ is such that $\nabla f(x)$ and a are not colinear.
In this case rank $(N) = 2$, and then (ii) of Proposition 2 does not occur at x. Hence, the necessary and sufficient condition reduces to $R(x)$ having at most one negative eigenvalue.

Moreover, since P has at least m'= 2 negative eigenvalues, R(x) must have at least one negative eigenvalue.
Hence, the condition reduces to R(x) having exactly one negative eigenvalue.
This implies

$$(11) \qquad r_{11}(x)\ r_{22}(x) - r_{12}(x)^2 \le 0$$

On the other hand, sufficient (but not necessary) conditions are given by

$$(12) \qquad r_{11}(x) \ge 0$$

or

$$(13) \qquad r_{22}(x) \ge 0$$

Case 2: $x \in S$ is such that $\nabla f(x)$ and a are colinear. In this case rank(N) = 1, and in view of Proposition 2 pseudoconvenxity of f is related to the existence of a local minimum of f at x. Using the classical second order optimality conditions written in terms of the properties of the bordered matrix P this reduces to R(x) having no negative eingenvalue. The matrix P is singular since the columns of N are colinear. Hence R(x) has at least one null eigenvalue, i.e.

$$(14) \qquad \det R(x) = r_{11}(x)\ r_{22}(x) - r_{12}(x)^2 = 0$$

Thus, $r_{11}(x)$ and $r_{22}(x)$ are of the same sign, and the other eigenvalue of R(x) is $r_{11}(x) + r_{22}(x)$. Since R(x) has no negative eigenvalue, necessarily

$$(15) \qquad r_{11}(x) + r_{22}(x) \ge 0,$$

implying $r_{11}(x) \ge 0$ and $r_{22} \ge 0$, because both have the same sign.
On the other hand, sufficient conditions for x to be a local minimum of f on S are

$$(16) \qquad r_{11}(x) > 0$$

or

$$(17) \qquad r_{22}(x) > 0.$$

Referring to the preceding results and to the elements of R(x) given in (8), (9) and (10) we now characterize some sets S where either necessary or sufficient conditions for pseudoconvexity of the restriction of f to E hold.
From (11), (14) and (15) we derive necessary conditions.
In particular

(18) $$S \subset T = \{ x \in R^n \mid r_{11}(x)\, r_{22}(x) - r_{12}(x)^2 \le 0 \}$$

is necessary for f to be pseudoconvex on S.
A sufficient condition is

(19) $$\mathrm{int}\,(S) \subset \widetilde{T} = \{ x \in R^n \mid r_{11}(x)\, r_{22}(x) - r_{12}(x)^2 < 0 \}$$

Combining (12) and (16), another sufficient condition is given by

(20) $$r_{11}(x) > 0 \quad \text{for all} \quad x \in S.$$

Set $T_1 = \{ x \in R^n \mid r_{11}(x) > 0 \}$. Referring to (8)

(21) $$T_1 = \{ x \in R^n \mid \frac{b^T x}{\|b\|\,\|x\|} > \frac{1}{\sqrt{2(1-\phi)}} \} = \{ x \in R^n \mid \cos(b, x) > \frac{1}{\sqrt{2(1-\phi)}} \}$$

where $\cos(b, x)$ denotes the cosinus of the angle between b and x. T_1 is an open convex cone. Then

(22) $$S \subset T_1$$

is a sufficient condition for f to be pseudoconvex on S.
Similarly, combining (13) and (17) another sufficient condition is given by

(23) $$r_{22}(x) > 0 \quad \text{for all} \quad x \in S.$$

Set $T_2 = \{ x \in R^n \mid r_{22}(x) > 0 \}$. Referring to (10)

(24) $$T_2 = \left\{ x \in R^n \mid \frac{a^T x}{\|a\|\,\|x\|} > \sqrt{\frac{1-2\phi}{2(1-\phi)}} \right\} = \left\{ x \in R^n \mid \cos(a, x) > \sqrt{\frac{1-2\phi}{2(1-\phi)}} \right\}$$

T_2 is an open convex cone. Then

(25) $$S \subset T_2$$

is a sufficient condition for f to be pseudoconvex on S.

Remark: For the special case $E = R^n$ condition (19) corresponds to the main condition in [5].

References

[1] **Avriel M., Diewert W.E., Schaible S. and Zang I.**, Generalized Concavity, Plenum Press, New York and London 1988.

[2] **Chabrillac Y. and Crouzeix J.P.**, Definiteness and Semidefiniteness of quadratic forms revisited, Linear Algebra Appl. 63, 1984, pp.283-292.

[3] **Crouzeix J.P., Ferland J.A. and Schaible S.**, Generalized Convexity of Functions on Affine Subspaces and Applications, Publication n.654, Départment d'Informatique et de Recherche Opérationelle, Université de Montréal, 1988.

[4] **Ohlson J.A. and Ziemba W.T.**, Portfolio Selection in a Lognormal Market when the Investor has a Power Utility Function. Journal of Financial and Quantitative Analysis 11, 1976, pp. 57-71.

[5] **Schaible S. and Ziemba W.T.**, Generalized Concavity of a Function in Portfolio Theory, Z. Oper. Res. 29 (5) 1985, pp.161-186.

ON SOME FRACTIONAL PROGRAMMING MODELS OCCURRING IN MINIMUM-RISK PROBLEMS

I. M. Stancu-Minasian - S. Tigan

This paper deals with an extension of the minimum-risk criterion considered by B. Bereanu [5, 6, 7] and A. Charnes and W. W. Cooper [11] in the linear case to the nonlinear mathematical programming case. In what follows the minimum-risk criterion is applied to some special classes of nonlinear problems as are, for instance, linear Tchebysheff problems, bottlenech transportation problems, max-min (linear or linear fractional) problems with linked constraints, max-min bilinear programming problems. It is shown that, under certain hypotheses, these stochastic problems are equivalent to deterministic fractional problems. The last section examines the vectorial minimum-risk problem. The ideas discussed in this paper string together the developments given by the authors in [33-38, 43-45].

1. Introduction

Consider the linear stochastic programming problem

(1.1) $$\min c'x$$

subject to

(1.2) $$Ax = b;\ x \geq 0$$

where $A = (a_{ij})$, $(i = 1, \dots, m;\ j = 1, \dots, n)$ is an $m \times n$ $(m < n)$ matrix of rank m; 0 and x are n-vectors; b is an m-vector and c is an n-dimensional random vector.
In connection with problem (1.1)-(1.2), the notion of Minimum-Risk Solution was indenpendently introduced in stochastic programming by B. Bereanu [5, 6, 7] and A. Charnes and W. W. Cooper [11]. Thus, one considers a level k of resources and expenditures and then one is satisfied if the objective function takes at most the value k.

Definition 1.1. $\overline{x}(k)$ is said to be a minimum-risk solution of level k for problem (1.1)-(1.2) if $\overline{x}(k) \in S = \{x \mid Ax = b, x \geq 0\}$ and

(1.3) $$P\{\omega \mid c'(\omega)\,\overline{x}(k) \leq k\} = \max_{x \in S} P\{\omega \mid c'(\omega)\,x \leq k\}$$

If the set S is regular (i.e. non-empty and bounded) and does not contain the origin, then there exists at least one solution $\bar{x}(k)$ to (1.3). One assumes that $0 \notin S$ only to rule out some trivial cases. Thus, if $0 \in S$, then, for $x = 0$, $P\{\omega | c'(\omega) x \le k\} = 1$ if $k \ge 0$ (i.e. $x = 0$ is trivially optimal), while for $k \le 0$ $P\{\omega | c'(\omega) x \le k\} = 0$ so that $x = 0$ cannot be optimal. There are also cases when the minimum-risk problem appears as (in the case when (1.1) is a maximum problem)

$$(1.3') \qquad \max_{x \in S} \; P\{\omega | c'(\omega) x \le k\}$$

These situations will be tackled in a similar manner.
For the particular case presented below (the simple randomization case) the solution of the minimum-risk problem does not depend on the distribution function of c and is obtained by solving a fractional programming problem. It is obvious that the probability in (1.3') will depend on the distribution of c.

Theorem 1.1. [6] If $c(\omega) = c_1 + t(\omega) c_2$, $c_2 > 0$ and the distribution function $T_t(z)$ of $t(\omega)$ is continuous, then the minimum-risk solution of (1.3') does not depend on $T_t(z)$ and is obtained by solving the following fractional programming problem:

$$\max_{x \in S} \frac{-k + c_1' x}{c_2' x}$$

Proof. Indeed, $\{\omega | (c_1 + c_2 t(\omega))' x \ge k\} = \{\omega | t(\omega) \ge (k - c_1' x)/(c_2' x)\}$
Hence

$$P\{\omega | (c_1 + c_2 t(\omega))' x \ge k\} = P\{\omega | t(\omega) \ge (k - c_1' x)/(c_2' x)\} = 1 - T_t((k - c_1' x)/(c_2' x))$$

and

$$\max_{x \in S} \; P\{\omega | (c_1 + c_2 t(\omega))' x \ge k\} = 1 - T_t(\max (c_1' x - k)/(c_2' x))$$

The proof is now obvious considering that T_t is continuous and non-decreasing.

Since deterministic equivalents are known only for stable probability laws, among which the normal distribution is widely used, the costs c_i are ordinarily treated as normal independent random variables. If c has the mean value vector $\bar{c}$ and the covariance matrix V, then $c' x$ is of type $N(\bar{c}' x, x' V x)$. Then one can state

Theorem 1.2 [6] The minimum-risk solution of problem (1.3') is given by the solution of the following fractional programmin problem

$$(1.4) \qquad \max_{x \in S} \ (\bar{c}'x - k) / (\sqrt{x'Vx})$$

Proof. Indeed,

$$P\{\omega \mid c'(\omega)x \geq k\} = P\{\omega \mid (c'(\omega)x - \bar{c}'x)/(\sqrt{x'Vx}) \geq (k - \bar{c}'x)/(\sqrt{x'Vx})\} = 1 - \phi((-\bar{c}'x + k)/(\sqrt{x'Vx}))$$

and

$$\max_{x \in S} P\{\omega \mid c'(\omega)x \geq k\} = 1 - \phi(\max_{x \in S} (\bar{c}'x - k)/(\sqrt{x'Vx}))$$

(where $\phi(\cdot)$ is the probability distribution function of the standard normal variable $N(0,1)$). Note that if x^* is the solution of problem (1.4) and $z(x^*) = (\bar{c}'x^* - k)/(\sqrt{x^{*\prime}Vx^*})$ is the value of the objective function, then we obtain $P\{\omega \mid c'(\omega)x^* \geq k\}$ since $z(x) \sim N(0,1)$.

B. Bereanu [5] uses the Zoutendijk's [49] method of feasible directions to solve (1.4), whereas C. Bergthaller [9] adopts the transformation method to change (1.4) into a quadratic programming problem.
An extension of the minimum-risk problem to the case when the objective function is a product, or a ratio, of linear functions was given by I. M. Stancu-Minasian [28], [29].

Theorem 1.3. [28], [29]. The minimum-risk solution of the stochastic programming problem

$$\max_{x \in S} P\{\omega \mid c'(\omega)x \,.\, d'x \leq k\} \quad (d = (d_1, \dots, d_n)')$$

is given by the solution of the following nonlinear fractional programming problem

$$\max_{x \in S} (k - \bar{c}'x \,.\, d'x)/(d'x \sqrt{x'Vx})$$

If d is also a random variable (say normal with mean $\bar{d}$ and covariance matrix W), then the problem becomes considerably more involved.

Theorem 1.4. [28], [29]. If c and d are independent random variables, then the solution of the stochastic programming problem $\max_{x \in S} P\{\omega \mid c'(\omega)x.\, d'x \leq k\}$ is identical with the solution of the fractional programming problem

$$\max_{x \in S} \ (k - \bar{c}' x. \bar{d}' x)/(\sqrt{(\bar{c}' x)^2 (x' Wx) + (x' Vx)(\bar{d}' x)^2 + (x' Vx)(x' Wx)})$$

Theorem 1.5. [28], [29]. The minimum-risk solution of the stochastic programming problem

$$\max_{x \in S} \ P\{ \omega | (c'(\omega)x)/(d' x) \le k \}$$

is given by the solution of the nonlinear fractional programming problem

$$\max_{x \in S} \ (k. d' x - \bar{c}' x)/(\sqrt{x' Vx})$$

This paper deals with an extension of the minimum-risk criterion considered by B. Bereanu [5], [6], [7] and A. Charnes and W.W. Cooper [11] in the linear case to the nonlinear mathematical programming case.
In what follows the minimum-risk criterion is applied to some special classes of nonlinear problems as are, for instance, linear Tchebysheff problems, bottleneck transportation problems, max-min (linear or linear fractional) problems with linked constraints, max-min bilinear programming problems. It is shown that, under certain hypotheses, these stochastic problems are equivalent to deterministic fractional problems. The last section examines the vectorial minimum-risk problem. The ideas discussed in this paper string together the developments given by the authors in [33] - [38], [43] - [45].
First, we shall discuss a general model of nonlinear minimum-risk problem and then give a treatment of the stated problems.

2. General Formulation of Nonlinear Minimum-Risk Problem

The general problem described here can be mathematically expressed as
Problem P:

(2.1) Minimize $f(a(t), x)$ subject to $x \in S$, where the set $S \subseteq R^n$, the real vector-valued function $a : A \to B$, the real-valued function $f : B \times S \to R$, the sets $A \subseteq R$ and $B \subseteq R^m$ are assumed to be deterministic and known (in particular, B may be the whole space R^n, i.e. $B \equiv R^n$).

We shall adopt the following assumption concerning this problem.
(A1) There exists a function $g : S \times C \to C$, $C \subseteq R$ such that for all $x \in S$ and for all $t \in A$,

$$f(a(t), x) \le w \Leftrightarrow t \le g(x, w)$$

where w is an arbitrary real number in C.
In fact, the above condition ensures the separation of t from the inequality $f(a(t), x) \le w$.
Now, we consider that $t(\omega)$ is a random variable on a given probability space (Ω, K, P) with a continuous and strictly increasing distribution function $T(z)$.
Throughout this paper, we consider stochastic programs with simple randomization (see [4], [8]), that is programs, the random coefficients of which are affine functions of a single random variable.
The minimum-risk approach to problem (P) is to find the solution of the
Problem PRM:

$$\max_{x \in S} P\{\omega \mid f(a(t(\omega)), x) \le w\}. \tag{2.2}$$

Then, as has already been done by I. M. Stancu-Minasian [31], we can establish a relationship between the minimum-risk problem associated with Problem P corresponding to the level w(Problem PRM) and a deterministic problem which does not depend on the distribution function of the random variable $t(\omega)$.
From assumption (A1) above, it follows that

$$P\{\omega \mid f(a(t(\omega)), x) \le w\} = P\{\omega \mid t(\omega) \le g(x, w)\} = T(g(x, w)).$$

Further, according to the fact that T is an increasing function, we have:

$$\max_{x \in S} P\{\omega / f(a(t(\omega)), x) \le w\} = \max_{x \in S} T(g(x, w)) = T(\max_{x \in S} g(x, w)).$$

We have

Theorem 2.1. If the distribution function $T(z)$ of $t(\omega)$ is continuous, strictly increasing, then the minimum-risk solution corresponding to the level w of Problem PRM does not depend on $T(z)$ and can be determined by solving the deterministic
Problem PA:

$$\max_{x \in S} g(x, w). \tag{2.3}$$

The next sections of this paper will illustrate this technique for some special classes of mathematical programming problems, as are linear Tchebysheff problems, bottleneck transportation problem and max-min linear (bilinear and linear fractional) problems.

3. The Linear Tchebysheff Problem

Consider the Tchebysheff problem

(3.1) $$\min_{x \in S} \max_{1 \le i \le r} \{ Z_i(x) \},$$

where $Z_i(x) = (c^i)' x + \alpha^i$ $(i = 1, \dots, r)$, c^i $(i = 1, \dots, r)$ are constant vectors and α^i $(i = 1, \dots, r)$ are scalar constants.

Let now c^i and α^i assume random values: $c^i(\omega) = c_1{}^i + t(\omega)\, c_2{}^i$; $\alpha^i(\omega) = \alpha_1{}^i + t(\omega)\, \alpha_2{}^i$, where $c_1{}^i$, $c_2{}^i$ $(i = 1, \dots, r)$ are constant vectors, $\alpha_1{}^i$, $\alpha_2{}^i$ $(i = 1, \dots, r)$ are scalar constants and $t(\omega)$ is a randoml variabile.

We assume that

(3.2) $$c_2{}^{i'} x + \alpha_2{}^i \neq 0 \text{ for all } x \in S, \text{ and } i \in M = \{1, \dots, r\}$$

The minimum-risk approach to the Tchebysheff problem (3.1) consists in finding the optimal solution of the following programming problem:

(3.3) $$v(w) = \max_{x \in S} P\{ \omega | \max_{i \in M} [c^i(\omega)' x + \alpha^i(\omega)] \le w \}$$

Let $z_i''(x) = c_2{}^{i'} x + \alpha_2{}^i$, $g_i(x, w) = (w - c_1{}^{i'} x - \alpha_1{}^i)/(c_2{}^{i'} x + \alpha_2{}^i)$ for all $x \in S$ and $i \in M$. Then, we have

$$F(x, w) = P\{ \omega | \max_{i \in M} [(c_1{}^i + t(\omega)\, c_2{}^i)' x + \alpha_1{}^i + t(\omega)\, \alpha_2{}^i] \le w \} =$$

$$= P\{ \omega | (c_1{}^i + t(\omega)\, c_2{}^i)' x + \alpha_1{}^i + t(\omega)\, \alpha_2{}^i \le w, \forall i \in M \} =$$

$$= \begin{cases} P\{ \omega | t(\omega) \le g_i(x, w), \forall i \in M \} & \text{if } z_i''(x) > 0 \\ P\{ \omega | t(\omega) \ge g_i(x, w), \forall i \in M \} & \text{if } z_i''(x) < 0 \end{cases} =$$

$$= \begin{cases} P\{ \omega | t(\omega) \le \min_{i \in M} g_i(x, w), \forall i \in M \} & \text{if } z_i''(x) > 0 \\ P\{ \omega | t(\omega) \ge \max_{i \in M} g_i(x, w), \forall i \in M \} & \text{if } z_i''(x) < 0 \end{cases}$$

$$= \begin{cases} T(\min_{i \in M} g_i(x, w)) & \text{if } z_i''(x) > 0 \\ 1 - T(\max_{i \in M} g_i(x, w)) & \text{if } z_i''(x) < 0 \end{cases}$$

Hence, the problem becomes

$$\max_x F(x, w) = \begin{cases} \max_x T\left(\min_{i \in M} g_i(x, w) \right) & \text{if } z_i''(x) > 0 \\ 1 - \min_x T\left(\max_{i \in M} g_i(x, w) \right) & \text{if } z_i''(x) < 0 \end{cases}$$

If the distribution function T(z) of the random variable $t(\omega)$ is continuos and strictly increasing, then

$$v(w) = \max_x F(x, w) = \begin{cases} T\left(\max_x \min_{i \in M} g_i(x, w) \right), & \text{if } z_i''(x) > 0 \\ 1 - T\left(\min_x \max_{i \in M} g_i(x, w) \right)^{*}, & \text{if } z_i''(x) < 0 \end{cases}$$

The following theorem is immediate.

Theorem 3.1. If assumption (3.2) holds and the distribution T(z) of $t(\omega)$ is continuous and strictly increasing, then the minimum-risk solution of problem (3.3) does not depend on T(z) and can be obtained by solving the deterministic piecewise linear fractional programming problems.

$$\text{(3.4)} \qquad \max_x \min_{i \in M} \; (w - c_1^{i'} x - \alpha_1^i)/(c_2^{i'} x + \alpha_2^i) \qquad \text{if } z_i''(x) > 0$$

or

$$\text{(3.5)} \qquad \min_x \max_{i \in M} \; (w - c_1^{i'} x - \alpha_1^i)/(c_2^{i'} x + \alpha_2^i) \qquad \text{if } z_i''(x) < 0$$

Remark 3.1 The piecewise linear fractional programming problems (3.4) or (3.5) can be solved by use of a parametric algorithm similar to Dinkelbach's algorithm [15] for fractional programming;

Remark 3.2 If $c^i(\omega) = c_1^i + t^i(\omega) c_2^i$, $\alpha^i = 0$, $i \in M$ and if $t^i(\omega)$ $(i \in M)$ are independent random variables whose distribution function T_i is continuos and strictly increasing, then the minimum-risk solution of problem (3.3) depends on T_i as follows:

$$\max_x F(x, w) = \max_x \prod_{i=1} T_i\left((w^i - c_1^{i'} x)/(c_2^{i'} x) \right);$$

Remark 3.3 When the functions Z_i $(i \in M)$ defining the objective fcuntion Z are nonlinear (linear fractional, linear plus fractional or linear plus indefinite functions), the minimum-risk approach to the Tchebysheff problem can be restated and solved under appropriate hypotheses in a manner similar to the linear case.

4. The Bottleneck Transportation Problem

We consider here the bottleneck transportation problem

$$\text{minimize } t^0(X) = \max \{ t_{ij} | (i, j) \in L(X) \}$$

subject to

(4.1) $$\sum_{j=1}^{n} x_{ij} = a_i, \; i \in M = \{1, 2, \ldots, m\}$$

(4.2) $$\sum_{i=1}^{m} x_{ij} = b_j, \; j \in Q = \{1, 2, \ldots, n\}$$

(4.3) $$x_{ij} \geq 0, \; (i, j) \in M \times Q,$$

(4.4) $$\sum_{i=1}^{m} a_i = \sum_{j=1}^{n} b_j$$

where

a_i is the available amount at the ith supply point,

b_j is the requirement at the jth demand point,

t_{ij} is the transportation time from supply point i to demand point j,

x_{ij} is the amount to be transportated from the ith supply point to the jth demand point,

$X = (x_{ij})$ is an element (an $m \times n$ matrix) of the set S (defined by (4.1)-(4.3) of all feasible solutions,

$L(X) = \{ (i, j) \in M \times Q | x_{ij} > 0 \}$ is the positive graph of X, and all of data are integers (or, equivalently, rationals). The transportation time is independent of the amount of commodity shipped from suppliers to consumers.

Let now t_{ij} assume random values:

(4.5) $$t_{ij} = t_{ij}' + t(\omega) t_{ij}'', \; \forall (i, j) \in M \times Q,$$

where t_{ij}', t_{ij}'' $(i \in M, j \in Q)$ are constants and $t(\omega)$ is a random variable on a probability space (Ω, K, P), with the continuous strictly increasing distribution function $T(z)$. The minimum-risk problem corresponding to level z, associated to the bottleneck transportation problem (4.1)-(4.4) consists in finding the optimal solution of the following problem: Find

$$(4.6) \qquad v(z) = \max P\{\,\omega \mid \max_{(i,j)\in L(X)} (t_{ij}' + t(\omega)\, t_{ij}'') \le z\,\},$$

subject to constraints (4.1)-(4.4).
We assume that

$$(4.7) \qquad t_{ij}'' \ge 0 \text{ for all } (i,j) \in M \times Q \text{ and there exists at least } (i,j) \in M \times Q \text{ such that } t_{ij}'' > 0.$$

We shall show that, under assumption (4.7), the minimum-risk problem (4.6) can be solved by a deterministic bottleneck transportation problem which does not depend on the distribution function of the random variable $t(\omega)$.

Theorem 4.1. If assumption (4.7) holds and if the distribution function T is continuous and strictly increasing, then the minimum-risk solution of problem (4.6) does not depend on T and can be obtained by solving the following bottleneck transportation problem:

$$\max_{X\in S} \min_{(i,j)\in L(X)} g_{ij},$$

where

$$g_{ij} = \begin{cases} (z-t_{ij}')\,/\,t_{ij}'', & \text{if } \quad t_{ij}'' \ne 0, \\ +\infty, \ \text{if } z-t_{ij}' \ge 0 & \text{and } \quad t_{ij}'' = 0, \\ -\infty, \ \text{if } z-t_{ij}' < 0 & \text{and } \quad t_{ij}'' = 0. \end{cases}$$

Proof. From (4.5) we get:

$$F(X,z) = P\{\,\omega \mid \max_{(i,j)\in L(X)} (t_{ij}' + t(\omega)t_{ij}'') \le z\,\} = P\{\,\omega \mid (t_{ij}' + t(\omega)t_{ij}'') \le z,\ \forall (i,j) \in L(X)\,\}.$$

Hence, according to (4.7), it results that

$$F(X,z) = P\{\,\omega \mid t(\omega) \le g_{ij},\ \forall (i,j) \in L(X)\,\} = P\{\,\omega \mid t(\omega) \le \min_{(i,j)\in L(X)} g_{ij}\,\} = T(\min_{(i,j)\in L(X)} g_{ij}).$$

Then problem (4.6) becomes

$$\max_{x\in S} F(X, z) = \max_{x\in S} T(\min_{(i,j)\in L(X)} g_{ij}).$$

hence, by virtue of the assumption that T is continuous and strictly increasing, we get:

$$v(z) = \max_{x \in S} F(X, z) = T(\max_{x \in S} \min_{(i,j) \in L(X)} g_{ij}).$$

Thus, the theorem is proven.

Remark 4.1. If assumption (4.7) is replaced by:

(4.7') $t_{ij}'' \leq 0$, for all $(i,j) \in M \times Q$ and there exists at least $(i,j) \in M \times Q$ such that $t_{ij}'' < 0$,

then theorem 4.1 can be restated and proved in a similar manner:

Theorem 4.1'. If assumption (4.7') holds and if the distribution function T is continuous and strictly increasing, then the minimum-risk solution of problem (4.6) does not depend on T and can be obtained by solving the following bottleneck transportation problem:

$$\min_{x \in S} \max_{(i,j) \in L(X)} g_{ij}',$$

where

$$g_{ij}' = \begin{cases} (z-t_{ij}') / t_{ij}'' , & \text{if} \quad t_{ij}'' \neq 0, \\ -\infty, \quad \text{if } t_{ij}' = 0 & \text{and} \quad z-t_{ij}' \leq 0, \\ +\infty, \quad \text{if } t_{ij}'' = 0 & \text{and} \quad z-t_{ij}' > 0. \end{cases}$$

We now assume that

(4.8) $$t_{ij} = t_{ij}' + t_{ij}^*(\omega)t_{ij}'', \qquad \forall (i, j) \in M \times Q,$$

where $t_{ij}^*((i, j) \in M \times Q)$ are independent random variables with continuous and strictly increasing distribution functions T_{ij}.
Also, we assume that:

(4.9) $$t_{ij}'' > 0, \qquad \forall (i, j) \in M \times Q.$$

In this case, the minimum-risk solution of problem (4.6) depends on T_{ij}. Indeed, as in the previous case, we have:

$$F(X, z) = P\{\omega / \max_{(i,j) \in L(X)} (t_{ij}' + t_{ij}^*(\omega)t_{ij}'') \leq z\} =$$
$$= P\{\omega / ((t_{ij}' + t_{ij}^*(\omega)t_{ij}'') \leq z, \ \forall (i, j) \in L(X)\} =$$
$$= P\{\omega / t_{ij}^*(\omega) \leq g_{ij}, \ \forall (i, j) \in L(X)\} = \prod_{(i,j) \in L(X)} T_{ij}(g_{ij}).$$

where $g_{ij} = (z-t_{ij}')/((t_{ij}''))$

Hence,

(4.10)
$$\max_{X \in S} F(X, z) = \max_{X \in S} \prod_{(i,j) \in L(X)} T_{ij}(g_{ij})$$

This problem is equivalent with the following optimization problem:

(4.11)
$$\max_{X \in S} \ln \Big(\prod_{(i,j) \in L(X)} T_{ij}(g_{ij})$$

or

(4.12)
$$\max_{X \in S} \sum_{(i,j) \in L(X)} \ln (T_{ij}(g_{ij})).$$

Hence, we have

Theorem 4.2. If assumption (4.9) holds an if the distribution functions T_{ij} are continuous and strictly increasing, then the minimum-risk solution of the following stochastic bottleneck transportation problem

$$\max P \{ \omega | \max_{(i,j) \in L(X)} (t_{ij}' + t_{ij}^*(\omega)\, t_{ij}'') \le z \}$$

subject to constraints (4.1)-(4.4), can be obtained by solving the nonlinear transportation problem (4.10) or, equivalently, the transportation problem (4.12).

Problem (4.12) is a fixed charge transportation problem which, however, depends on the distribution functions T_{ij} of the random variables t_{ij}^*.

Remark 4.2. A similar approach can be applied to the classical bottleneck assignment problem or to the bottleneck assignment problem with an additional constraint (see [39]), or to some optimization problems in graphs as the maximum capacity path problem (see [17]).

Mention should be made that another type of stochastic transportation and bottleneck assigment problems was studied by U. Yechiali [47], [48].

5. The Max-Min Problem with Linked Constraints

Consider the following linear max-min problem:

(5.1)
$$\max_{x} \min_{y} F(x, y) = cx + dy$$

subject to:

(5.2) $$Ax + By \le b, \quad x \ge 0, \quad y \ge 0,$$

where $x \in R^p$, $y \in R^q$ are variable vectors, whereas $c \in R^p$, $d \in R^q$, $b \in R^m$ and $A \in R^{m\times p}$, $B \in R^{m\times q}$ are given vectors and matrices, respectively.
The max-min problems (5.1)-(5.2) are encountered in several situations. Thus, for instance, they may arise in production and allocation problems in which profits or revenues can be modeled as linear functions and an element of competition is also present. Finding approximate globale maximizing points of convex functions over linear polyhedra leads to max-min problems. Other applications of linear max-min problems can be found in [14], [21].
J. E. Falk [16] showed that (5.1)-(5.2) is equivalent to a non-convex programming problem involving the maximization of a piecewise linear convex function over a linear polyedron and that the optimal solution occurs at a vertex of its feasible region. Cook a.o. [13], I. M. Stancu-Minasian and V. N. Patkar [32] and St. Tigan [41], [42] generalized these results to include the case in which $F(x, y)$ is a fractional function.
The following definitions are as in [16]. We recall here some of them. Denote by

$$S = \{ (x, y) \in R^p \times R^q \mid Ax + By \le b, x \ge 0, y \ge 0 \}$$

the feasible region of problem (5.1)-(5.2), and let S_V stand for the set of all vertices of the set S.
Let $Pr(X)$ and $Pr(Y)$ be the orthogonal projections of S on R^p and R^q respectively, i.e.

$$Pr(X) = \{ x \ge 0 \mid \exists y \ge 0 \text{ for which } Ax + By \le b \}$$

and

$$Pr(Y) = \{ y \ge 0 \mid \exists x \ge 0 \text{ for which } Ax + By \le b \}$$

It is clear that $Pr(X)$ and $Pr(Y)$ are linear polyhedra in R^p and R^q respectively.

Definition 5.1. A point y^0 is optimal with respect to x^0 if (x^0, y^0) is feasible for (5.1)-(5.2) (i.e. $(x^0, y^0) \in S$) and if y^0 is an optimal solution of the problem

$$\min_{y \in Q(x^0)} (cx^0 + dy) \qquad \text{where } Q(x^0) = \{ y \ge 0 \mid By \le b - Ax^0 \}$$

Definition 5.2. A point $((x^0, y^0)$ is an optimal solution of (5.1)-(5.2) if:

(i) y^0 is optimal with respect to x^0,

and

(ii) $cx^0 + dy^0 \geq \min_{y \in Q(x)} (cx+dy), \quad \forall\, x \in Pr(X).$

The minimum-risk approach to problem (5.1)-(5.2) corresponding to the level w is to find the optimal solution of the following programming problem:

$$(5.3) \qquad v(w) = \max_{x \in Pr(X)} P\{\omega \mid \min_{y \in Q(x)} (c(\omega)\, x + d(\omega)\, y \geq w\}$$

Let $c(\omega) = c_1 + t(\omega)c_2$ and $d(\omega) = d_1 + t(\omega)d_2$, where c_1, c_2, d_1, d_2 are constant vectors, whereas $t(\omega)$ is a random variable on a probability space (Ω, K, P) with continuous and strictly increasing distribution function $T(z)$.
Let us assume that:

(I) S is a nonempty and compact set,
(II) $g(x, y) = c_2x + d_2y \neq 0$, for all $(x, y) \in S$

From (5.2) and II, by the continuity of g, it is clear that only one of the cases
(II.1) $g(x, y) > 0$, for all $(x, y) \in S$
or
(II.2) $g(x, y) < 0$, for all $(x, y) \in S$
holds.
One can establish a relationship between the minimum-risk problem (5.3) associated with problem (5.1)-(5.2) corresponding to the level w and a deterministic problem which does not depend on the distribution function of the random variable $t(\omega)$.

Theorem 5.1. If assumption (II) holds and the distribution function T is continuous and strictly increasing, then the minimum-risk solution of problem (5.3) does not depend on T and can be determined by solving one of the following deterministic linear-fractional max-min problems:

P.1. $\max_{x \in Pr(X)} \min_{y \in Q(x)} (c_1x + d_1y - w)/(c_2x + d_2y)$, subject to $(x, y) \in S$ (when (II.1) holds)

or

P.2. $\max_{x \in Pr(X)} \min_{y \in Q(x)} (w - c_1x - d_1y)/(c_2x + d_2y)$, subject to $(x, y) \in S$ (when (II.2) holds)

Proof: First, suppose that (II.1) holds. Denote $f(x, y, w) = (w - c_1x - d_1y)/(c_2x + d_2y)$ and $F(x, w) = P\{\omega \mid \min_{y \in Q(x)} (c(\omega)\, x + d(\omega)\, y \geq w\}$
Obviuosly, it is true that

$$F(x, w) = P\{\omega \mid (c(\omega)x + d(\omega)y \geq w, \ \forall y \in Q(x)\} =$$
$$= P\{\omega \mid (c_1 + t(\omega)c_2)x + (d_1 + t(\omega)d_2)y \geq w, \ \forall y \in Q(x)\} =$$
$$= P\{\omega \mid t(\omega) \geq f(x, y, w), \ \forall y \in Q(x)\} =$$
$$= P\{\omega \mid t(\omega) \geq \max_{y \in Q(x)} f(x, y, w)\} = 1 - T(\max_{y \in Q(x)} f(x, y, w)).$$

Hence, the problem will be

$$\max_{x \in Pr(X)} F(x, w) = 1 - \min_{x \in Pr(X)} T(\max_{y \in Q(x)} f(x, y, w)).$$

Further, according to the fact that T is a continuous and strictly increasing function, we have

$$v(w) = \max_{x \in Pr(X)} F(x, w) = 1 - T(\min_{x \in Pr(X)} \max_{y \in Q(x)} f(x, y, w)).$$

Hence, if assumption (II.1) holds, it results that problem (5.3) is equivalent with the linear-fractional max-min problem P1.
In case when assumption (II.2) is verified, we get similarly

$$v(w) = \max_{x \in Pr(X)} F(x, w) = T(\max_{x \in Pr(X)} \min_{y \in Q(x)} f(x, y, w)).$$

Then every minimum-risk solution of problem 5.3) can be obtained by solving the max-min problem P2.

6. The Max-Min Bilinear Programming

The following max-min problem was studied in [1], [2], [25].

Problem PM. Find

(6.1) $$\max_{x \in X} \min_{y \in Y} (xAy + dx + ay)$$

where

(6.2) $$X = \{x \in R^n \mid Bx \leq b, x \geq 0\}$$
$$Y = \{y \in R^m \mid Ey \leq e, y \geq 0\}.$$

x and y are variable vectors, whereas $A \in R^{n \times m}$, $B \in R^{p \times n}$, $E \in R^{q \times m}$, $d \in R^n$, $a \in R^m$, $b \in R^p$ and $e \in R^q$ are given matrices and vectors, respectively. As can be seen, the constraints are separate in problem (6.1)-(6.2).

Definition 6.1. Given a function $f : X \times Y \to R$, $X \times Y \subseteq R^{n \times m}$, we say that $(x', y') \in X \times Y$ is an optimal solution for the max-min problem $\max_{x \in X} \min_{y \in Y} f(x, y)$,

if (x', y') verifies the following conditions:

(i) $f(x', y') = \min_{y \in Y} f(x', y)$

(ii) $f(x', y') \geq \min_{y \in Y} f(x, y),\ x \in X.$

In what follows, we assume that the matrix A and the vectors d and a are random.

$$(6.3)\qquad A(\omega) = A^1 + t(\omega)A^2,\quad d(\omega) = d^1 + t(\omega)d^2,\quad a(\omega) = a^1 + t(\omega)a^2,$$

where A^1, $A^1 \in R^{n \times m}$ are the given matrices, $d^i \in R^n$, $a^i \in R^m$ $(i = 1, 2)$ are the given constant vectors, and $t(\omega)$ is a random variable on a probability space (Ω, K, P), with the continuous and strictly increasing distribution function $T(z)$.

The minimum-risk problem corresponding to level z, associated to PM problem, consists in finding the optimal solution of Problem PRM. Find

$$(6.4)\qquad v(z) = \max_{x \in X} P\{\omega | Z(x, \omega) > z\},$$

where

$$(6.5)\qquad Z(x, \omega) = \min_{y \in Y} (xA(\omega)y + d(\omega)x + a(\omega)y)$$

Definition 6.2. A pair $(x', y') \in X \times Y$ is called minimum-risk solution for the PRM problem, if it verifies the following conditions:

$$(6.6)\qquad P\{\omega | x'A(\omega)y' + d(\omega)x' + a(\omega)y' > z\} = P\{\omega | \min_{y \in Y} x'A(\omega)y + d(\omega)x' + a(\omega)y > z\},$$

$$(6.7)\qquad P\{\omega | x'A(\omega)y'+d(\omega)x'+a(\omega)y' > z\} \geq P\{\omega | \min_{y \in Y} (xA(\omega)y+d(\omega)x+a(\omega)y) > z\},\ \forall\, x \in X$$

Let us denote

$$Z_i(x, y) = xA^iy + d^ix + a^iy,\ (i = 1, 2),\ \forall\ (x, y)\ X \times Y.$$

Further, we shall assume that

(6.8) $$Z_2(x, y) \neq 0, \quad \forall\ (x, y)\ X \times Y,$$

and

(6.9) X and Y are nonvoid and bounded sets.

By the continuity of the function Z_2 and by assumptions (6.8), (6.9), it results that only one of the cases

(i) $Z_2(x, y) > 0,\ \forall\ (x, y)\ X \times Y$
or

(ii) $Z_2(x, y) < 0,\ \forall\ (x, y)\ X \times Y$
holds.

Thus, we shall show that, under assumptions (6.8), (6.9), the minimum-risk PRM problem can be solved by a deterministic max-min problem, which does not depend on the distribution function of the random variable $t(\omega)$.

Theorem 6.1. If conditions (6.8), (6.9) are met and if the distribution function T is continuous and strictly increasing, then the minimum-risk solution of the PRM problem does not depend on T and can be obtained by solving one of the max-min bilinear fractional problems:

<u>Problem PA1</u> $$\max_{x \in X} \min_{y \in Y} (Z_1(x, y)-z)/(Z_2(x, y)), \text{ if } Z_2(x, y) > 0$$

or

<u>Problem PA2</u> $$\max_{x \in X} \min_{y \in Y} (z-Z_1(x, y))/(Z_2(x, y)), \text{ if } Z_2(x, y) < 0$$

Proof: From (6.3) and (6.5), we get

$$\begin{aligned} F(x, z) &= P\{\omega \mid \min_{y \in Y} (xA(\omega)y + d(\omega)x + a(\omega)y) > z\} = \\ &= P\{\omega \mid \min_{y \in Y} (Z_1(x, y) + t(\omega)Z_2(x, y)) > z\} = \\ &= P\{\omega \mid Z_1(x, y) + t(\omega)Z_2(x, y) > z,\ \forall y \in Y\}, \end{aligned}$$

hence, according to (6.8), it results that

$$F(x, z) \begin{cases} P\{\omega \mid t(\omega) > (z-Z_1(x, y))/(Z_2(x, y)),\ \forall y \in Y\}, & \text{if } Z_2(x, y) > 0 \\ P\{\omega \mid t(\omega) < (z-Z_1(x, y))/(Z_2(x, y)),\ \forall y \in Y\}, & \text{if } Z_2(x, y) < 0. \end{cases}$$

Let $g(x, y) = (z-Z_1(x, y))/(Z_2(x, y))$ for any $(x, y) \in X \times Y$.
Then, we have

$$F(x, z) = \begin{cases} P\{\omega \mid t(\omega) > \max_{y \in Y} g(x, y)\}, & \text{if } Z_2(x, y) > 0 \\ P\{\omega \mid t(\omega) < \min_{y \in Y} g(x, y)\}, & \text{if } Z_2(x, y) < 0 \end{cases}$$

$$= \begin{cases} 1 - T(\max_{y \in Y} g(x, y)), & \text{if } Z_2(x, y) > 0 \\ T(\min_{y \in Y} g(x, y)), & \text{if } Z_2(x, y) < 0 \end{cases}$$

Then, the PRM Problem is equivalent to

$$\max_{x \in X} F(x, z) = \begin{cases} \max_{x \in X} (1-T(\max_{y \in Y} g(x, y)), & \text{if } Z_2(x, y) > 0 \\ \max_{x \in X} T(\min_{y \in Y} g(x, y)), & \text{if } Z_2(x, y) < 0 \end{cases}$$

and, hence, by the assumption that T is continuous and strictly increasing, we get

$$v(z) = \max_{x \in X} F(x, z) = \begin{cases} 1 - T(\min_{x \in X} \max_{y \in Y} g(x, y)), & \text{if } Z_2(x, y) > 0 \\ T(\max_{x \in X} \min_{y \in Y} g(x, y)), & \text{if } Z_2(x, y) < 0 \end{cases}$$

Therefore, a minimum-risk solution for the PRM problem can be obtained by solving the PA2 problem or

Problem PA1' $\quad \min_{x \in X} \max_{y \in Y} g(x, y)$

However the PA1' Problem is equivalent to the PA1 problem and this concludes the proof.

Remark 6.1. A similar approach can be applied to the minimum-risk problem associated to a min-max bilinear problem having random coefficients of the form (6.3), that is

<u>Problem PRM'</u> $$v'(z) = \max_{x \in X} P\{\omega \mid \max_{y \in Y} (xA(\omega)y + d(\omega)x + a(\omega)y) \leq z\}$$

Remark 6.2. The minimum-risk Tchebysheff problems studied in Sec. 3 can be considered as minimum-risk problems associated to discrete min-max problems (that is, min-max problems, whose feasible set Y is a finite set).

Remark 6.3. If T is continuous and strictly increasing, the PRM problem can be formulated as a max-min problem:

$$v(z) = \max_{x \in X} \min_{y \in Y} P\{\omega \mid (xA(\omega)y + d(\omega)x + a(\omega)y) > z\}.$$

We shall give a parametrical algorithm for the max-min bilinear-fractional Problem PF. Find

(6.10) $$v = \max_{x \in X} \min_{y \in Y} (xA^1y + d^1x + a^1y + w^1)/(xA^2y + d^2x + a^2y + w^2)$$

where $A^i \in R^{n \times m} (i = 1, 2)$, $d^i \in R^n$, $a^i \in R^m$ and $w^i \in R (i = 1, 2)$ are the given matrices, vectors and real numbers, respectively.
The feasible sets X and Y are polyhedral sets defined by (6.2).
Obviously, the max-min PA1 and PA2 problems are particular cases of the PF problem.
For every $x \in X$ and $y \in Y$, we denote

(6.11) $$z_i(x, y) = xA^iy + d^ix + a^iy + w^i, \quad i = 1, 2$$

(6.12) $$h(x, y) = (z_1(x, y)) / (z_2(x, y))$$

We adopt the following hypotheses:
(H1) X and Y are nonvoid and bounded sets;
(H2) $z_2(x, y) > 0$, for all $(x, y) \in X \times Y$.
For solving the max-min bilinear fractional PF problem, we adopt the parametrical method given in [42], which can be applied to a rather general class of max-min non-linear fractional problems.
The following theorem gives a necessary and sufficient condition for a pair $(x^*; y^*) \in X \times Y$

to be an optimal solution of the PF problem.

Theorem 6.2. Let $x^* \in X$, $y^* \in Y$ and let $t^* \in R$ such that $t^* = h(x^*, y^*) = \min_{y \in Y} h(x^*, y)$.

Then (x^*, y^*) is an optimal solution for the PF problem if and only if $F(t^*) = 0$, where

$$F(t) = \max_{x \in X} \min_{y \in Y} (z_1(x, y) - tz_2(x, y)) \text{ for any } t \in R$$

The proof is similar to that of Theorem 1 from [42].
The algorithm consists in the following steps:

Step 1. Choose $x_0 \in X$ and take $k := 0$

Step 2. Find $y_k \in Y$, which is an optimal solution of the linear fractional problem:

$$\min_{y \in Y} h(x_k, y) \tag{6.13}$$

and take

$$t_k = h(x_k, y_k). \tag{6.14}$$

Step 3. Solve the max-min bilinear problem

$$F(t_k) = \max_{x \in X} \min_{y \in Y} (z_1(x, y) - t_k z_2(x, y)), \tag{6.15}$$

and let (x_{k+1}', y_{k+1}') be an optimal solution of this problem.

Step 4. i) If $F(t_k) = 0$, then, by Theorem 6.2, (x_k, y_k) is an optimal solution for the PF problem, and the algorithm stops. In this case, $v = t_k$.
ii) If $F(t_k) > 0$, then go to Step 5.

Step 5. Take $x_{k+1}' := x_{k+1}'$, $k := k+1$ and go to Step 2.

Ordinarily, this algorithm requires an infinite number of iterations. It can be stopped after a finite number of iterations, by using an approximative stop criterion in Step 4.

Step 4': i) If $F(t_k) \leq \varepsilon$, then stop.
ii) If If $F(t_k) > \varepsilon$, then go to Step 5.
We mention that in Step 4', ε is a very small pre-assigned positive quantity. Its value

depends on the degree of approximation that is required.
The algorithm convergence and the finiteness of the approximate variant is based on

Theorem 6.3. Let $\varepsilon \geq 0$ be a given real number. Under hypotheses H1 and H2, if $F(t_k) \leq \varepsilon$ and if $z_2(x, y) \geq \beta > 0$, for every $(x, y) \in X \times Y$, then, for every optimal solution (x', y') of the PF problem, we have

$$0 \leq h(x', y') - t_k \leq (F(t_k)) / (\beta) \leq (\varepsilon) / (\beta)$$

The proof of this theorem is similar to that of theorem 3 in [42].

Remark 6.4. Step 2 requires to solve a linear-fractional programming problem at each iteration. As objective function alone changes with each iteration, at the $k+1$ iteration the optimal solution y_k obtained at the previous iteration can be taken as the initial solution for problem (6.4).

Remark 6.5. The max-min bilinear problem (6.6) in Step 3 can be solved by two linear programming problems (see, for instance [2], [25]).

Remark 6.6. When A^1 and A^2 are zero matrices, that is, when the PF problem is basically a max-min linear fractional problem, then the parametrical algorithm stops after a finite mumber of iterations [42].

7. The Linear-Fractional Max-Min Problem

We consider the following linear-fractional max-min problem

(7.1) $$\max_x \min_y \ (c x + d y + f) / (e x + g y + h)$$

subject to

(7.2) $$A x + B y \leq b, \ x \geq 0, \ y \geq 0,$$

where $x \in R^p$, $y \in R^q$ are variable vectors, and c, $e \in R^p$, d, $g \in R^q$, $b \in R^m$ and $A \in R^{mxq}$, $B \in R^{mxq}$, f, $h \in R$ are given vectors and matrices, respectively.
Also, we denote by

(7.3) $$S = \{ (x, y) \in R^p \times R^q : A x + B y \leq b, x \geq 0, y \geq 0 \},$$

the feasible region of problem (7.1)-(7.2) and

(7.4) $$Q(x) = \{ y \geq 0 : By \leq b - Ax \},$$

(7.5) $$X = \{ x \geq 0 : Q(x) \neq \emptyset \}.$$

Let us consider now the case when c and d are random vectors on a given probability space (Ω, K, P).

The minimum-risk approach to problem (7.1)-(7.2) corresponding to level w consists in the programming Problem SFP. Find

(7.6) $$V(w) = \max_{x \in X} P\{ \omega : \min_{y \in Q(x)} (c(\omega)x + d(\omega)y + f) / (ex + gy + h) \geq w \},$$

where $c(\omega)$ and $d(\omega)$ are random vectors on a probability space (Ω, K, P).

We denote

$$H(x, y, \omega) = (c(\omega)x + d(\omega)y + f) / (ex + gy + h), \ \forall (x, y) \in S, \omega \in \Omega.$$

Definitions 5.1 and 5.2 hold also for the SFP Problem and so we have

Definition 7.1. A point (x', y') is an optimal solution for the minimum-risk problem SFP if:

(i) $$P\{ \omega : H(x', y', \omega) \geq w \} = P\{ \omega : \min_{y \in Q(x')} H(x', y, \omega) \geq w \}$$

and

(ii) $$P\{ \omega : H(x', y', \omega) \geq w \} \geq P\{ \omega : \min_{y \in Q(x)} H(x, y, \omega) \geq w \}, \ \forall x \in X.$$

We make the following assumptions on the SFP problem:

(H1) S is a nonempty and compact set;

(H2) $ex + gy + h > 0$, for all $(x, y) \in S$.

Next, we show that by an appropriate change of variable, the linear-fractional stochastic max-min SFP problem can be reduced to a linear stochastic max-min problem.

If we take

(7.7) $$t = (1) / (ex + gy + h), \ u = tx, \ v = ty.$$

then to the SFP problem we can associate the linear stochastic max-min SLP Problem. Find

$$V_1(w) = \max_{u} P\{ \omega : \min_{v,t} (c(\omega)u + d(\omega)v + ft) \geq w \}$$

subject to

(7.8) $$Au + Bv - bt \leq 0,$$

(7.9) $$eu + gv + ft = 1,$$

(7.10) $$u \geq 0,\ v \geq 0,\ t \geq 0.$$

Let U be the feasible set of the SLP problem.
Using arguments similar to those employed in [41], we can prove:

Theorem 7.1. If assumption H1 holds, then (u, v, 0) is not a feasible solution for SLP.

Proof. It must be proved that every (u, v, t) verifying (7.8)-(7.10) has $t > 0$. In the contrary case, there exists a pair (u, v) such that $Au + Bv \leq 0$, $u \geq 0$, $v \geq 0$.
Then by (7.9), it follows that (u, v) is not the null vector. But if the pair (x, y) verifies (7.2), then the pair $(x + zu,\ y + zv)$ verifies (7.2) for every $z \geq 0$. But this fact contradicts the boundedness of the feasible set S assumed at H1.

Theorem 7.2. If assumptions H1 and H2 hold an if (u", v", t") is a minimum-risk solution for the SLF problem, then the point (u"/t" , v"/t") is a minimum-risk solution for the SFP problem.

Proof. By Theorem 7.1, it follows that $t'' > 0$. Then from (7.8), we get $A(u'')/(t'') + B(v'')/(t'') \leq b$, which means that $(u''/t'', v''/t'') \in S$.
Since $\{\omega : \min_{(v,t)\in Q'(u)} (c(\omega)u+d(\omega)v+ft) \geq w\} = \{\omega : \min_{y\in Q(u/t)} (c(\omega)u/t+d(\omega)y+f)/(eu/t+gy+h) \geq w\}$,
$(Q'(u) = \{(v, t) : (u, v, t) \in U\})$ for all $t > 0$ and $(u, v, t) \in U$, it results that the point (x", y"), where $x'' = u''/t''$, $y'' = v''/t''$, verifies the conditions of Definition 7.1. Hence, the point (x", y") is an optimal solution for SFP.
Let us now consider the minimum-risk SFP problem with simple randomization, i.e. the random vectors $c(\omega)$ and $d(\omega)$ are

(7.11) $$c(\omega) = c_1 + t_1(\omega)c_2,\quad d(\omega) = d_1 + t_1(\omega)d_2,$$

where $c_i \in R^p$, $d_i \in R^q$ $(i = 1, 2)$ are constant given vectors and $t_1(\omega)$ is random variable on a probability space (Ω, K, P) with a continuous and strictly increasing distribution function T.
Let us further assume that:
(H3) $c_2 x + d_2 y > 0$, for all $(x, y) \in S$.

Theorem 7.3. If assumptions H1, H2, H3 hold and the distribution function T is continuous and strictly increasing, then the optimal solutions of the SFP problem does not depend on T and can be determined by solving the following max-min problem:

(7.12) $\max_{x} \min_{y} \; ((c_1 - ew)x + (d_1 - gw)y - wh + f)) / (c_2 x + d_2 y)$, subject to $(x,y) \in S$.

Proof. Using assimption H2, it follows that

$$P\{\omega | \min_{y \in Q(x)} H(x, y, \omega) \geq w\} = P\{\omega | \min_{y \in Q(x)} (c(\omega)-ew)x+(d(\omega)-wg)y \geq wh-f\}.$$

Hence, the SFP problem is equivalent with a linear stochastic max-min problem having the objective functions $F(x, y) = (c(\omega)-ew)x + (d(\omega)-gw)y$. Thenn using Theorem 5.1, we get the conclusion of the theorem.

8. The Vectorial Minimum-Risk Problem

Several papers [26], [27], [28], [30], [31] by I. M. Stancu-Minasian considered a more general case of the minimum-risk problem, i.e. the case when the probabilities that the values or r objective functions exceed some levels of performance u_k are maximized.

(8.1) $$\underset{x \in S}{V \max} \; \{ f_k(x) = P[\omega \in \Omega : c_k'(\omega)x \geq u_k]; k \in I = \{1, \dots, r\}\}$$

where S is defined as in Sec. 1.
"V max" means here that efficient points are regarded as optimal solutions to (8.1) defined as follows:

Definition 8.1. A point $x^0 \in S$ is said to be efficient for (8.1) if and only if there exists no other point $x' \in S$ such that $f_k(x') \geq f_k(x^0)$ for all $k \in I$, and such that $f_{k_0}(x') > f_{k_0}(x^0)$, for at least one k_0.
We can introduce other models in a manner similar to that suggested in [19], [20]. Thus, a possible objective is to maximize the joint probability.

$$P\{ c_1'(\omega)x \geq u_1, \dots, c_r'(\omega)x \geq u_r \}$$

or to maximize

$$\sum_{k=1}^{r} w_j P(c_k'(\omega) \geq u_k),$$

where w_j are weights associated with each objective function.
For solving problem (8.1), I. M. Stancu-Minasian suggested a method which consists in the sequential solving of some minimum-risk problem (containing also quadratic constraints). Related to this method, the multiple minimum-risk solution concept is introduced and a relaxation-type method for obtaining this solution is given.

Worth mentioning is that J. P. Leclercq [22], [23], [24] obtained a model somewhat similar to problem (8.1), starting from a stochastic problem with random (multinormal or Cauchy) coefficients in the objective function and constraints.
We assume that

$$c_k(\omega) = c_k^1 + c_k^2\, t_k(\omega), \tag{8.2}$$

where $c_k^i \in R^n$ $(i = 1, 2)$ are given constant vectors and $t_k(\omega)$ is a random variable on the probability space (Ω_k, K_k, P_k).
It whould be mentioned that generality is unimpaired if the random variables t_k is assumedly defined on the same probability space (Ω, K, P).
In what follows, we take the concept of efficient points [10], [18], [31] as an optimal solution for problem (8.1).
According to this definition, we can formulate

Definition 8.2. A point $x^0 \in S$ is said to be a vectorial minimum-risk solution for problem (8.1) if it is an efficient solution for the vector-valued problem

$$V \max_{x \in S} \{ f_k(x) : k \in I \}$$

that is (cf.e.g. H. P. Benson and T. L. Morin [3]), if $f_i(x) > f_i(x^0)$ for some $x \in S$ and $i \in I$ implies that there exists at least one $j \in I - \{i\}$ such that $f_j(x) < f_j(x^0)$.
We shall show that in the case when the objective functions are simply randomized, problem (8.1) is equivalent with a deterministic linear fractional programming problem with multiple objective functions. Further, these results will be extended to the fractional case.
Concerning to problem (8.1), we adopt the following asumptions:
(H1) S is a nonempty and compact set,
(H2) $c_k^2 x > 0$ for all $x \in S$ and for all $k \in I$,
(H3) The distribution functions T_k of the random variables $t_k(\omega)$ are continuous and strictly increasing.
We first give a lemma which will be needed in the sequel:

Lemma 8.1. If assumptions (H1)-(H3) are satisfied, then for all $k \in I$, we have

$$f_k(x) = 1 - T_k((u_k - c_k^1 x) / (c_k^2 x)) \quad \text{for all } x \in S \tag{8.3}$$

Proof. The proof is similar to that of Theorem 4.1 [31] (sse also [6]). Thus, from (8.2) and (H2), we have

$$f_k(x) = P\{\omega \in \Omega : (c_k^1 + t_k(\omega) c_k^2) x \geq u_k\} = P\{\omega \in \Omega : t_k(\omega) \geq (u_k - c_k^1 x) / (c_k^2 x)\}$$

Further, according to the fact that T_k is a continuous and strictly increasing function, we have

$$f_k(x) = 1 - T_k((u_k - c_k^1 x) / (c_k^2 x)),$$

i.e. (8.3). The proof is complete now.

Remark 8.1. Since $f_k = (1-T_k)\, o g_k$ for all $k \in I$, and $1-T_k$ is strictly decreasing and g_k, where

$$(8.4) \qquad g_k(x) = (u_k - c_k^1 x) / (c_k^2 x), \text{ for all } x \in S,$$

is quasimonotone, it results that when S is a convex set, f_k is a quasimonotone function. Worth mentioning is that R. Weber [46] considered pseudomonotonic multiobjective programming, which does nor ordinarly include probleḿ (8.1) as a particular case.
We can now state and prove a theorem which relates the multiobjective minimum-risk problem (8.1) to a linear fractional programming problem with several objective functions (studied, for instance, in [13], [40], [46]).

Theorem 8.1. Let assumptions (H1)-(H3) be satisfied. Then a point $x' \in S$ is a vectorial minimum-risk solution for problem (8.1) if and only if x' is an efficient solution for the following linear fractional programming problem with several objective functions.
Problem PF

$$\operatorname*{V\,max}_{x \in S} \ \{ c_k^1 x - u_k) / (c_k^2 x) : k \in I \}.$$

Proof. Indeed, by definition 8.2, $x' \in S$ is a vectorial minimum-risk solution for problem (8.1) if for any $x'' \in S$ and $i \in I$ for which $f_i(x'') > f_i(x')$, there exists a $j \in I - \{i\}$ such that

$$f_j(x'') < f_j(x').$$

But then in virtue of Lemma 8.1 and by (8.4), it results that for all $x'' \in S$ and $i \in I$ for which $T_i(g_i(x'')) < T_i(g_i(x'))$ there exists a $j \in I - \{i\}$ such that $T_j(g_j(x'')) > g_j(x'))$. Since T_i and T_j are strictly increasing, these inequalities are equivalent with $g_i(x'') < g_i(x')$ and $g_j(x'') > g_j(x')$, respectively.
Consequently, this yelds that for all $x'' \in S$ and $i \in I$ for which

$$(c_i^1 x'' - u_i) / (c_i^2 x'') > (c_i^1 x' - u_i) / (c_i^2 x'),$$

there exists a $j \in I - \{i\}$ such that

$$(c_j^1 x'' - u_j) / (c_j^2 x'') < (c_j^1 x' - u_j) / (c_j^2 x')$$

This means that x' is an efficient solution for problem (PF). The conversion part of the theorem can be shown using similar arguments.

Remark 8.2. By Theorem 8.1, we can conclude that the vectorial minimum-risk problem can be solved with a method used for solving the multiobjective linear fractional programming problem (PF). For instance, an interactive procedure to solve (PF) was given by R. Weber [46].

Similarly, we can approach the case of the vectorial minimum-risk problem with linear fractional objective functions, which includes the previously discussed linear case as a particular case.

Thus, we consider the vectorial linear fractional problem

$$\text{(P1)} \quad \underset{x \in S}{V \max} \; \{ Z_k(x) = ((c_k x + \alpha k) / (d_k x + \beta k)) : k \in I = \{1, 2, \ldots, r\} \},$$

where $c_k, d_k (k \in I)$ are given constant vectors in R^n, $\alpha_k, \beta_k (k \in I)$ are scalar constants. We also assume that for all $k \in I$, $d_k x + \beta_k$ does not change is sign (say positive) on the set S.

Let now c_k and d_k assume random values for all $k \in I$:

$$c_k(\omega) = c_k^1 + c_k^2 t_k(\omega), \quad d_k(\omega) = d_k^1 + d_k^2 t_k(\omega)$$

where $c_k^i, d_k^i \in R^n (i = 1, 2)$ are given constant vectors and $t_k(\omega)$ is a random variable on the probability space $\{\Omega, K, P\}$.

As in problem (8.1), we can state the vectorial minimum-risk problem corresponding to levels $u_1, u_2, \ldots, u_r$:

$$\text{(8.5)} \qquad \underset{x \in S}{V \max} \; \{ f_k(x) = P\{\omega \in \Omega : (c_k(\omega)x + \alpha_k)/(d_k(\omega)x + \beta_k) \geq u_k \} : k \in I \}.$$

Regarding problem (8.5), we make the following assumptions:

(HH1) S is a nonempty and compact set;

(HH2) $P\{\omega \in \Omega : (d_k^1 + d_k^2 t_k(\omega)) x + \beta_k > 0\} = 1$, for all $x \in S$ and for all $k \in I$;

(HH3) $(c_k^2 - d_k^2 u_k)x > 0$ for all $x \in S$ and for all $k \in I$;

(HH4) The distribution functions T_k of the random variables t_k are continuous and strictly increasing.

The corresponding results like those of Lemma 8.1 and Theorem 8.1 hold for problem (8.5) as well:

Lemma 8.2. If assumptions (HH1)-(HH2) are satisfied, then for all $k \in I$, we have

$$f_k(x) = 1 - T_k((d_k^1 u_k - c_k^1)x + u_k \beta_k - \alpha_k)/(c_k^2 - d_k^2 u_k)x)) \quad \text{for all } x \in S.$$

Theorem 8.2. Let assumptions (HH1)-(HH4) be satisfied. Then a point $x' \in S$ is a vectorial minimum-risk solution for problem (8.5) if and only if x' is efficient for the following deterministic linear fractional programming problem with several objective functions:

$$\text{(PF1)} \qquad \max_{x \in S} \{ ((c_k{}^1 - d_k{}^1 u_k)x + \alpha_k - u_k\beta_k) / ((c_k{}^2 - d_k{}^2 u_k)x)) : k \in I \} .$$

The proofs are similar to those of Lemma 8.1 and Theorem 8.1 and are omitted.

Remark 8.3. The feasible set S in the formulation of problem (P1) is not necessarily a convex set. For instance, we can consider a discrete feasible set S, as in the pseudoboolean-programming case (see, for instance [10]), and the results of Theorems 8.1 and 8.2 remain valide. In this case, the minimum-risk solutions for problem (8.1) and (8.5) can be obtained by solving some appropriate vectorial fractional programming problems with zero-one variables, using, for instance, the algorithm described by Bina Gupta [18].

References

[1] **Belenkii A.S.**, Finding Sabble Points on Convex Polyhedra in Mini-Max Optimization Problems with Linear Constraints, Avtomat.i Telemech., n°7, (1980) pp.161-167 (Russian); translated as Automat. Remote Control 41 (1980), n°7, part.2, (1981) pp.1017-1022.

[2] **Belenkii A.S.**, Minimax Scheduling Problems with Linear Constraint and Methods of their Solutions. Avtomat.i Telemeh. n°11, (1981) pp.157-170 (Russian); translated as Automat. Remote Control 42, n° 10, (1982) pp.1409-1419.

[3] **Benson H.P., Morin T.L.**, The Vector Maximization Problem: Proper Efficiency and Stability, SIAM J.Appl.Math. 32 (1977) 1, pp. 64-72.

[4] **Bereanu B.**, On Stochastic Linear Programming. I: Distribution Problems. A Single Random Variable. Rev.Roumaine Math. Pures Appl. 8 (4), (1963), pp.683-697.

[5] **Bereanu B.**, Solutii cu Risc Minim in Programarea Liniara. Studii de Statistica. Lucrarile celei de a 3-a Consfatuiri Stiintifice de Statistica 26-30 Dec. (1963). Also in: An. Univ. Bucuresti, Mat. Mec., 13, (1964), pp. 121-140.

[6] **Bereanu B.**, Programme de Risque Minimal en Programmation Linéaire Stochastique. C.R. Acad. Sci. Paris, 259, 5 (1964), pp. 981-983.

[7] **Bereanu B.**, Distribution Problems and Minimum-Risk Solutions in Stochastic Programming. In: Colloquium on Applications of Mathematics to Economics, Budapest (1963), pp.37-42, A. Prékopa (ed.), Publishing House of the Hungarian Academy of Sciences, Budapest 1965.

[8] **Bereanu B.**, Some Numerical Methods in Stochastic Linear Programming Under Risk and Uncertainty. In : Stochastic Programming, M. A. H. Dempster (ed.) Academic Press, (1980), pp.169-205.

[9] **Bergthaller C.**, A Quadratic Equivalent for the Minimum-Risk Problem. Rev. Roumaine Math. Pures Appl. 15, 1, (1970) pp.17-23.

[10] **Bitran G.R.**, Linear Multiple Objective Programs with Zero-One Variables, Math. Programming 13, (1977), pp.121-139.

[11] **Charnes A., Cooper W.W.**, Deterministic Equivalents for Optimizing and Satisfying under Chance Constraints. Operations Research, 11, (1963), pp. 18-39.

[12] **Choo E.U., Atkins D.R.**, Bicriteria Linear Fractional Programming. J.O.T.A. 36 (2), (1982), pp. 203-220.

[13] **Cook W.D., Kirby M.J.L., Mehndiratta S.L.**, A Linear Fractional Max-Min Problem. Operations Res. 23(3), (1975), pp. 511-521.

[14] **Danskin J.M.**, The Theory of Max-Min, Springer, Berlin 1967.

[15] **Dinkelbach W.**, On Nonlinear Fractional Programming. Management Sci.13 (7), (1967), pp.492-498.

[16] **Falk J.E.**, A Linear Max-Min Problem. Math. Programming, 5, (1973), pp.169-188.

[17] **Gondran M., Minoux M.**, Graphes et Algorithmes, Ed. Eyrolles, Paris, 1979.

[18] **Gupta B.**, Finding the Set of all Efficient Solutions for the Linear Fractional Multiobjective Program with Zero-One Variables. Opsearch 18(4), (1982), pp.204-214.

[19] **Hendrix G.C., Stedry A.**, The Elementary Redundancy-Optimization Problem: A Case Study in Probabilistic Multiple-Goal Programming. Operations Res. 22, (1974), pp.639-653.

[20] **Kaplan R., Soden J.**, On the Objective Function for the Sequential P-Model of Chance-Constrained Programming. Operations Res. 19(1), (1971), pp.105-114.

[21] **Konno H.**, Bilinear Programming. Part. II. Applications of Bilinear Programming. Technical Report n°71-10, Operations Research House, Stanford, CAlifornia, 1971.

[22] **Leclercq J.P.**, Résolution de Programmes Linéaires Stochastiques par des Techniques Multicritères. Thèse de doctorat Faculté des Sciences de l'Université de Namur, 1979.

[23] **Leclercq J.P.**, La Programmation Linéaire Stochastique: une Approache Multicritère. Partie I-ère. Formulation. Partie II-ème: Un Algorithme Interactif Adapté aux Distributions Multinormales. Cahiers Centre Etudes Rech. Opér. 23, (1), (1981) pp.31-41; ibidem 23(2), (1981), pp.121-132.

[24] **Leclercq J.P.**, Stochastic Programming: an Interactive Multicriteria Approach. European J. Oper. Res. 10(1), (1982), pp.33-41.

[25] **Owen G.**, Teoria Jocurilor (traducere din lb.engleza) Ed. Tehnica, Bucuresti, 1972.

[26] **Stancu-Minasian I.M.**, Nota Asupra Programdrii Stochastice cu mai Multe Functii-Obiectiv. Avind Vectorul c Aleator. Stud.Cerc.Mat.27(4), (1975), pp.453-459.

[27] **Stancu-Minasian I.M.**, Asupra Problemei de Risc Minim Multiplu. I: Cazul a Doua Functii Obietiv. II: Cazul a r(r>2) Functii Obiectiv. Stud.Cerc.Mat.28 (1976), 5, pp. 617-623; ibidem 28 (1976), 6, pp.723-734.

[28] **Stancu-Minasian I.M.**, Criterii Multiple in Programarea Stohastica. Doctoral thesis. Centrul de Statistica Matematica, Bucuresti, 1976.

[29] **Stancu-Minasian I.M.**,Asupra Problemei lui Kataoka. Stud. Cerc. Mat. 28,1, (1976),

pp.95-111.

[30] **Stancu-Minasian I.M.**, On the Multiple Minimum Risk Problem. Bull. Math. Soc.Sci. Math. R.S. Roumanie (N.S.) 23 (71), 4, (1979), pp.427-437.

[31] **Stancu-Minasian I.M.**, Stochastic Programming with Multiple Objective Functions. Ed. Academiei, Bucaresti and D. Reidel Publishing Company, Dordrecht, Boston, Lancester, Tokio, 1984.

[32] **Stancu-Minasian I.M., Patkar V.N.**, A Note on Nonlinear Fractional Max-Min Problem. Nat. Acad. Sci. Letters (India) 8(2), (1985), pp.39-41.

[33] **Stancu-Minasian I.M., Tigan St.**, The Minimum Risk Approach to Special Problems of Mathematical Programming. The Distribution Function of the Optimal Value. Rev. Anal. Numer. Théor. Approx. 13(2), (1984), pp.175-187.

[34] **Stancu-Minasian I.M., Tigan St.**, The Vectorial Minimum-Risk Problem. Proceedings of the Colloquium on Approximation and Optimization, Cluj-Napoca, October 25-27, (1984), pp.321-328.

[35] **Stancu-Minasian I.M., Tigan St.**, The Minimum-Risk Approach to the Bottleneck Transportation Problem. Itinerant Seminar on Functional Equations, Approximation and Convexity, Cluj-Napoca, (1985), pp. 203-208.

[36] **Stancu-Minasian I.M., Tigan St.**, The Minimum Risk Approach to Max-Min Bilinear Programming. An. Stiint. Univ. "Al.I.Cuza" Iasi Sect. I-a Mat.(N.S.) Tome XXXI, s.II.b, (1985), pp.205-209.

[37] **Stancu Minasian I.M., Tigan St.**, The Stochastic Linear-Fractional Max-Min Problem. Itinerant Seminar on Functional Equations, Approximation and Convexity, Cluj-Napoca, (1987), pp.275-280.

[38] **Stancu-Minasian I.M., Tigan St.**, Criteriul Riscului Minim in Programarea Stohastica. Lucrarile Sesiunii Stiintifice a Centrului de Calcul al Universitatii Bucuresti, 20-21 februarie 1987, pp.392-397.

[39] **Tigan St.**, Sur un Problème d'Affectation. Mathematica 11(34), 1, (1969), pp.163-166.

[40] **Tigan St.**, Sur le Problème de Programmation Vectorielle Fractionnaire. Math.-Rev. Analyse Numér. Théor. Approx. 4, (1975), pp. 99-103.

[41] **Tigan St.**, On the Max-Min Nonlinear Fractional Problem. Rev. Anal. Numér. Théor. Approx. 9(2), (1980), pp. 283-288.

[42] **Tigan St.**, A Parametrical Method for Max-Min Nonlinear Fractional Problems. Itinerant Seminar on Functional Equations, Approximation and Convexity 19-21 mai 1983, Cluj-Napoca, pp.175-184.

[43] **Tigan St., Stancu-Minasian I.M.**, Criteriul Riscului Minim Pentru Problema Cebisev. Lucrarile celui de al IV-lea simpozion "Modelarea cibernetica a proceselor de productie" 26-28 mai 1983, ASE-Bucuresti, vol. I, pp.338-342.

[44] **Tigan St., Stancu-Minasian I.M.**, The Stochastic Bottleneck Transportation Problem. Rev. Anal. Numér. Théor. Approx. 12(2), (1985), pp.153-158.

[45] **Tigan St., Stancu Minasian I.M.**, The Stochastic Max-Min Problem. Cahiers Centre Etudes Rech. Opér. 27(3-4), (1985), pp.247-254.

/46/ **Weber R.**, Pseudomonotonic Multiobjective Programming. Cahiers Centre ›Etudes Rech. Opér. 25, (1983), pp. 115-128.

/47/ **Yechiali U.**, A Stochastic Bottleneck Assignment Problem, Management Sci., 14(11), (1968), pp. 732-734.

/48/ **Yechiali U.**, A Note on a Stochastic Production Maximizing Transportation Problem. Naval Res. Logist. Quart., 18(3), (1971), pp. 429-431.

/49/ **Zoutendijk G.**, Method of Feasible Directions. Elsevier, Amsterdam, 1960.

QUASI CONVEX LOWER LEVEL PROBLEM AND APPLICATIONS IN TWO LEVEL OPTIMIZATION

P. Loridan - J. Morgan

Abstract

In this work, some results given in [14] on approximate solutions for two level optimization problems are improved by using strictly quasi convexity assumptions on the lower level problem and only "sequential" compacity and continuity hypothesis. As applications the case of increasing sequences and an approximation method using external penalization technics on the lower and on the upper level are considered.

1. Introduction

The role of game theory in modern economic theory and as basic modeling tool in political science, sociology, operations research and other fields is becoming each day more relevant. From the point of view of optimization, game theory is concerned with different agents eventually in conflict which want to optimize their own cost funtions. In this paper we are interested by Stackelberg solutions in non zero sum non cooperative games, first introduced by H. Von Stackelberg in 1934 ([20]) in the context of economic competition. Stackelberg differential games were first treated in [7], [18] and [19] and have been extensively studied in recent years. For the dynamic version, see for example [6] and [3] for application to economic models.

In this paper we are interested, in the case in which the reaction set is not a singleton and the considered topological spaces do not necessarily verify the first axiom of countability, to results which are in the same line that convergence and stability results already given in mathematical programming with respect to data perturbations by variational convergence and Γ^-- sequential convergence ([8], [9], [16], [21], [22]) and to their application to some approximation methods using classical optimization techniques such that external penalization. For motivating such a study we refer to some previous paper ([12], [13],...). In section 2 we recall some convergence results which can be obtained without strictly quasi convexity assumptions ([13]). But, in order to avoid some restrictive hypothesis, sufficient conditions will be given in section 3 by introducing an ε-regularized problem when strictly quasi convexity assumption on the lower level problem is satisfied.

Moreover an existence result for approximate solution is given whereas the initial two level problem may fail have a solution. Such results give some improvement to the ones obtained in [14].

In section 4 the case of increasing sequences is considered and in section 5, as an

application, an approximation method for constrained problems using external penalization simultaneously in the lower and in the upper level is studied.
An application to approximation using interior penalty functions, will be considered in a separate paper.

2. The Two-Level Optimization Problem: Formulation and convergence results

First, we recall the formulation considered in [13].

2.1 Formulation, Notations and Definitions

Let X and Y be two topological spaces and f_i, $i = 1, 2$, two functionals defined on $X \times Y$ and valued in $R \cup \{+\infty\}$. We consider the following two-level optimization problem (S):

$$
S \left\{
\begin{array}{l}
\text{Find } \bar{x} \in X \text{ such that} \\[2ex]
\sup\limits_{y \in M_2(\bar{x})} f_1(\bar{x}, y) = \inf\limits_{x \in X} \sup\limits_{y \in M_2(x)} f_1(x, y) \\[2ex]
\text{where } M_2(x) \text{ is the set of optimal solutions to the lower level problem} \\[2ex]
P(x) \quad \left\{
\begin{array}{l}
\text{Minimize } f_2(x, y) \\[1ex]
\text{subject to } y \in Y
\end{array}
\right.
\end{array}
\right.
$$

Let us introduce the following notations:

$$
w_1(x) = \sup_{y \in M_2(x)} f_1(x, y) \ , \quad v_1 = \inf_{x \in X} w_1(x) \ , \quad v_2(x) = \inf_{y \in Y} f_2(x, y)
$$

In the sequel, for the sake of simplicity, we shall assume that v_1 and $v_2(x)$, for any $x \in X$, are finite numbers.

Definition 2.1.

(i) Any $\bar{x} \in X$ verifying $v_1 = w_1(\bar{x})$ is called a Stackelberg solution to (S).

(ii) Any pair $(\bar{x}, \bar{y}) \in X \times Y$ verifying $v_1 = w_1(\bar{x})$ and $\bar{y} \in M_2(\bar{x})$ is called a Stackelberg equilibrium pair.

Remark 2.1. Let us bear in mind that (S) may fail to have a solution even if the decision variables range over a compact set ([6], [13]). So, we shall consider approximate solutions to (S).

Let $\eta > 0$ be a given number. The set of η - Stackelberg solutions will be denoted by $M_1(\eta)$:

$$M_1(\eta) = \{ x \in X \;/\; w_1(x) \le v_1 + \eta \}$$

For $\varepsilon \ge 0$ and for any $x \in X$, we shall also consider: $M_2(x, \varepsilon) = \{ y \in Y \;/\; f_2(x, y) \le v_2 + \varepsilon \}$.
For $\varepsilon = 0$ we shall $M_2(x, 0) = M_2(x)$.

2.2 A Theoretical Approximation Scheme: Description and Main Assumptions.

For the convenience of the reader, we recall the general approach introduced in [13] together with the main assumptions for the leader and the follower. We consider a sequence of two-level optimization problems (S_n), $n \in N$.

$$(S_n) \begin{cases} \text{Find } \bar{x}_n \in X \text{ such that} \\ \sup\limits_{y \in M_{2,n}(\bar{x}_n)} f_{1,n}(\bar{x}_n, y) = \inf\limits_{x \in X} \sup\limits_{y \in M_{2,n}(x)} f_{1,n}(x, y) \\ \text{where } M_{2,n}(x) \text{ is the set of optimal solutions to the lower level problem} \\ P_n(x): \\ \qquad P_n(x) \begin{cases} \text{Minimize } f_{2,n}(x, y) \\ \text{subject } y \in Y \end{cases} \end{cases}$$

with $f_{1,n}$ and $f_{2,n}$ defined on $X \times Y$ and valued in $R \cup \{+\infty\}$.

Main Assumptions

(L1) For any $(x, y) \in X \times Y$, for any sequence (x_n, y_n) converging to (x, y) in $X \times Y$, we have:

$$\liminf_{n \to +\infty} f_{1,n}(x_n, y_n) \ge f_1(x, y)$$

(L2) For any $x \in X$, there exists a sequence x_n converging to x in X such that, for any $y \in Y$ and for any sequence y_n converging to y in Y:

$$\limsup_{n \to +\infty} f_{1,n}(x_n, y_n) \le f_1(x, y)$$

(F1) For any $(x, y) \in X \times Y$, for any sequence (x_n, y_n) converging to (x, y) in $X \times Y$, we have:

$$\liminf_{n \to +\infty} f_{2,n}(x_n, y_n) \geq f_2(x, y)$$

(F2) For any $(x, y) \in X \times Y$, for any sequence x_n converging to x in X, there exists a sequence y_n such that

$$\limsup_{n \to +\infty} f_{2,n}(x_n, y_n) \leq f_2(x, y)$$

Remark 2.2. Such an approach takes into account various aspects, especially:

(i) approximation and stability results, by analogy with those already obtained in mathematical programming (for example, see [1], [21], [22])

(ii) approximation methods such as penalty techniques ([11], [17]): see Section 5.

Comments about the main assumptions can be found in [12], [13].

2.3 Convergence Results

By analogy with the notattions previously introduced, for any $x \in X$, $n \in N$ and for any $\varepsilon \geq 0$, we let

$$v_{2,n}(x) = \inf_{y \in Y} f_{2,n}(x, y)$$

$$M_{2,n}(x, \varepsilon) = \{ y \in Y \,/\, f_{2,n}(x, y) \leq v_{2,n}(x) + \varepsilon \}$$

$$M_{2,n}(x, 0) = M_{2,n}(x)$$

For the upper level problems, we let:

$$w_{1,n}(x) = \sup_{y \in M_{2,n}(x)} f_{1,n}(x, y) \qquad v_{1,n}(x) = \inf_{x \in X} w_{1,n}(x)$$

and, for $\eta > 0$, $M_{1,n}(\eta) = \{ x \in X \,/\, w_{1,n}(x) \leq v_{1,n} + \eta \}$. Then, with the previous assumptions concerning the lower level problems, we get the following proposition:

Proposition 2.1. For any $x \in X$ and any sequence x_n converging yo x in X, for any $\varepsilon \geq 0$, we have:

(i) $\limsup M_{2,n}(x_n, \varepsilon) \subset M_2(x_n, \varepsilon)$ that is: any accumulation point of a sequence $y_n \in M_{2,n}(x_n, \varepsilon)$ belongs to $M_2(x_n, \varepsilon)$

(ii) $\limsup_{n \to +\infty} v_{2,n}(x_n) \leq v_2(x)$

(iii) if, moreover, Y is sequentially compact, the:
$$\lim_{n \to +\infty} v_{2,n}(x_n) = v_2(x)$$

Remarks 2.3.

(i) The result i) is given in a sequential sense. In the case where Y is a first countable space, we recall that the topological definition of $\limsup M_{2,n}(x_n, \varepsilon)$ is equivalent to the sequential one.

(ii) The previous results have been given in [13]. Another proof of the result iii) can be found in [14].

From now, we shall assume that Y is sequentially compact. Furthermore, in the sequel, assumption (F2) will be replaced by the following:

($\overline{\text{F}}$2) For any $(x, y) \in X \times Y$, for any sequence x_n converging to x in X, there exists a sequence y_n converging to y in Y such that
$$\limsup_{n \to +\infty} f_{2,n}(x_n, y_n) \leq f_2(x, y)$$

Then, we get a new result:

Proposition 2.2. For any $x \in X$ and any sequence x_n converging to x in X, for any $\varepsilon > 0$, we have:
$$M_2(x) \subset \liminf M_{2,n}(x_n, \varepsilon)$$

that is for any $y \in M_2(x)$ there exists a sequence $y_n \in M_{2,n}(x_n, \varepsilon)$ which converges to y in Y.

Proof. Let x_n be a sequence converging to x in X, $g(y) = f_2(x, y)$, $g_n(y) = f_{2,n}(x_n, y)$. From (F1) and ($\overline{\text{F}}$2), we observe that g_n sequentially Γ^- converges to g ([8], [21], [22]). This notion is analogous with the sequential definition of epiconvergence ([2]). Then, from the compacteness of Y, the proof can be drived by using arguments similar to those used in [1], p. 136.
A direct proof of the previous (sequential) result can be found in [14]

□

Remark 2.4. In general the previous proposition does not hold for $\varepsilon = 0$. Sufficient conditions for obtaining such a result have been given in [13]. In such a case, we have proved that $w_{1,n}$ sequentially Γ^- converges to w_1 and, with the compacteness of Y, we have got the following result: $\lim\sup M_{1,n}(\eta) \subset M_1(\eta)$ for any $\eta > 0$. But these conditions are somewhat restrictive or difficult to verify (see [13]).

Here, we shall give new approximation results for a large class of two-level problems without assuming the condition: $M_2(\eta) \subset \lim\inf M_{2,n}(x_n)$.
Instead of (S), we shall consider an ε-regularized problem. Under quasi convexity assumptions, the main result for the lower level problems will be:

$$M_2(x, \varepsilon) \subset \lim\text{in } M_{2,n}(x_n, \varepsilon)$$

(for $\varepsilon > 0$ and any sequence x_n converging to x in X). Such a property is really the key for obtaining new approximation results. This is the purpose of the next section.

3. The ε-Regularized Problem: Definitions and Convergence Results

3.1 Approximate Solutions and ε-Regularized Problem

Let $\varepsilon \geqslant 0$. In the upper level problem, we shall consider:

$$w_1(x, \varepsilon) = \sup_{y \in M_2(x,\varepsilon)} f_1(x, y) \quad , \quad v_1(\varepsilon) = \inf_{x \in X} w_1(x, \varepsilon)$$

For $\varepsilon = 0$, we shall write $w_1(x)$ instead of $w_1(x, 0)$ and $v_1(0) = v_1$.
First we extend the definition of approximate solutions to (S).

Definition 3.1. Let $\eta \geq 0$, $\varepsilon \geq 0$. $x^* \in X$ is an (η, ε)-approximate solution to (S) if and only if $w_1(x^*, \varepsilon) \leq v_1(\varepsilon) + \eta$. The set of (η, ε)-approximate solutions to (S) will be denoted by $M_1(\eta, \varepsilon)$ (with $\varepsilon > 0$ and $\eta > 0$).

Remarks 3.1:

(i) for $\varepsilon = 0$ and $\eta > 0$, we get the notion of η-Stackelberg solution introduced in [13] and we let $M_1(0, \eta) = M_1(\eta)$ (see the previous section).

(ii) for $\varepsilon = \eta = 0$, we let $M_1(0, 0) = M_1$. From definition 2.1, M_1 is the set of Stackelberg solutions to (S).

Fom the definition of (η, ε)-approximate solutions to (S) we are led to the following notion of ε-regularized two-level problem (S_ε):

$$(S_\varepsilon)\quad \begin{cases} \text{Let } \varepsilon > 0 \text{ be a given number. Find } x_\varepsilon \in X \text{ such that:} \\ \sup\limits_{y \in M_{2,n}(x_\varepsilon, \varepsilon)} f_1(x_\varepsilon, y) = \inf\limits_{x \in X} \sup\limits_{y \in M_2(x,\varepsilon)} f_1(x, y) \\ \text{where } M_2(x, \varepsilon) \text{ is the set of } \varepsilon\text{-approximate solutions to } P(x) \text{, previously defined} \end{cases}$$

This formulation is similar to the one introduced in [15].

Remark 3.2. Under additional quasi convexity assumptions in the lower level problems $P(x)$, $x \in X$, we shall prove, in the next section, that (S_ε) has at least one solution (contrary to (S)). From this point of view, for $\varepsilon > 0$, the set $M_1(\varepsilon, 0)$ can be considered as a set of regularized Stackelberg solutions and, for $\eta > 0$, the set $M_1(\varepsilon, \eta)$ as a set of η-approximate regularized Stackelberg solutions, generalizing the notions considered in [13]. Let us also notice that an (η, ε)-approximation to (S) is an approximate solution to (S_ε).

3.2. Convergence Results

In the next sections the following notations will be used for the approximate upper level problems corresponding to $f_{1,n}$ and $M_{2,n}(x, \varepsilon)$, $n \in N$, $x \in X$, $\varepsilon \geq 0$,

$$w_{1,n}(x, \varepsilon) = \sup_{y \in M_{2,n}(x,\varepsilon)} f_{1,n}(x, y)$$

$$w_{1,n}(x, 0) = w_{1,n}(x)$$

$$v_{1,n}(\varepsilon) = \inf_{x \in X} w_{1,n}(x, \varepsilon)$$

$$v_{1,n}(0) = v_{1,n}$$

and, for $\eta \geq 0$, we shall consider the sets:

$$M_{1,n}(\varepsilon, \eta) = \{ x \in X / w_{1,n}(x, \varepsilon) \leq v_{1,n}(\varepsilon) + \eta \}$$

Proposition 3.1. ([14]). If (L_1), (L_2), (F_1), $(\overline{F}_2)$ are satisfied and Y is sequentially compact then

i) For any $x \in X$, there exists a sequence x_n converging to x in X, such that, for any $\varepsilon \geq 0$

$$\limsup_{n \to \infty} w_{1,n}(x_n, \varepsilon) \leq w_1(x, \varepsilon)$$

and

$$\limsup_{n \to \infty} v_{1,n}(\varepsilon) \leq v_1(\varepsilon)$$

ii) For any $\varepsilon > 0$, for any $x \in X$ and any sequence x_n converging to x in X we have:

$$\liminf_{n\to\infty} w_{1,n}(x_n, \varepsilon) \geq w_1(x)$$

iii) Moreover, if there exists a sequence of real positive numbers α_n converging to zero when $n\to\infty$ and such that:

$$\limsup_{n\to\infty} M_{1,n}(\varepsilon, \alpha_n) \neq \emptyset \quad \text{then}$$

$$v_1 \leq \limsup_{n\to\infty} v_{1,n}(\varepsilon), \text{ for any } \varepsilon > 0$$

Remark 3.3. Comments and other results, with sequences of real positive numbers ε_n converging to zero, can be found in [14].

Remark 3.4. Unfortunately we have neither the sequential Γ^- convergence of $w_{1,n}(\cdot\,,\varepsilon)$ to $w_1(\cdot\,)$, nor the sequential Γ^- convergence of $w_{1,n}(\cdot\,,\varepsilon)$ to $w_1(\cdot\,,\varepsilon)$, which is a serious drawback for obtaining approximation results concerning approximate solutions. In [13] we have got the sequential Γ^- convergence of $w_{1,n}(\cdot\,)$ to $w_1(\cdot\,)$ by introducing the following assumption:

(H) For any $x \in X$ and any sequence x_n converging to x in X, we have

$$M_2(x) \subset \limsup_{n\to\infty} M_{2,n}(x_n)$$

With such condition, the following can be proved

$$\limsup_{n\to\infty} M_{1,n}(\eta) \subset M_1(\eta) \text{ for } \eta > 0$$

with

$$M_{1,n}(\eta) = \{ x \in X / w_{1,n}(x) \leq v_{1,n} + \eta \}$$

In the following we shall give some results for approximating $(\eta\,,\varepsilon)$-approximate solutions to (S) without assuming (H). The key for obtaining these results is to consider (S_ε) instead of (S) together with quasi-convexity assumptions in the lower level problem.

3.3. Strictly Quasi Convex Lower Level Problem

Let $\varepsilon > 0$, $x \in X$ and let us denote

$$N_2(x, \varepsilon) = \{ y \in Y / f_2(x, y) < v_2(x) + \varepsilon \}$$

then:

Proposition 3.3. ([14]) If (F1), ($\overline{F}_2$) are satisfied and Y is sequentially compact then for any $\varepsilon > 0$, $x \in X$ and for any sequence x_n converging to x in X we have

$$N_2(x, \varepsilon) \subset \liminf_{n\to\infty} M_{2,n}(x_n,\varepsilon)$$

From now on we suppose that Y is a non empty convex subset in a vector topological space V and, endowed with the induced topology, is sequentially compact then:

Proposition 3.4. If (F1), ($\overline{F}_2$) are satisfied and the function $y \to f_2(x, y)$ is strictly quasi convex ([2], [4]) and sequentially lower semi continuous on Y sequentially compact for any $x \in X$ then, for any $x \in X$ and $\varepsilon > 0$ we have

$$M_2(x, \varepsilon) = \overline{N_2(x, \varepsilon)}^S \text{ (sequential clousere of } N_2(x, \varepsilon))$$

and for any $x \in X$ and any sequence x_n converging to x in X we have

$$M_2(x, \varepsilon) \subset \liminf_{n\to\infty} M_{2,n}(x_n,\varepsilon)$$

Proof. Let $x \in X$ and $\varepsilon > 0$. Since the function $y \to f_2(x, y)$ is sequentially lower semi conituous on Y, $M_2(x, \varepsilon)$ is a sequentially closed set and $\overline{N_2(x, \varepsilon)}^S \subset M_2(x, \varepsilon)$. Now we want to show that:

$$M_2(x, \varepsilon) \subset \overline{N_2(x, \varepsilon)}^S$$

that is any $u \in M_2(x, \varepsilon)$ is the limit of a sequence $u_n \in N_2(x, \varepsilon)$. It suffices to consider the case where u satisfies

$$f_2(x, u) = v_2(x) + \varepsilon.$$

Let us consider $v \in N_2(x, \varepsilon)$ and $u_n = \lambda_n v + (1 - \lambda_n)u$ with $\lambda_n \to 0$ as $n \to \infty$.
As $n \to \infty$ $u_n \to u$ and from strctly quasi convexity of $y \to f_2(x, y)$ ([2]) we have

$$f_2(x, u_n) < \max (f_2(x, u), f_2(x, v)) \qquad (f_2(x, u) \neq f_2(x, v))$$

So $f_2(x, u_n) < v_2(x) + \varepsilon$ and $u_n \in N_2(x, \varepsilon)$.

The last part of proposition 3.4 follows from the sequential closedness of the set $\liminf_{n\to\infty} M_{2,n}(x, \varepsilon)$ (X_n, ε), by using proposition 3.3.

Remark 3.4. The same result is obtained in [14] but here only strictly quasi convexity and sequential assumptions are required.
By similar arguments to the one used in [14] it can be also easily proved:

Proposition 3.5. If assumptions (F_1), $(\overline{F}_2)$, (L_1), (L_2) are satisfied and the function $y \to f_2(x, y)$ is strictly quasi convex and sequentially lower semicontinuous for any $x \in X$ on Y sequentially conpact then

(i) Let $\varepsilon > 0$. For any $x \in X$ and any sequence x_n converging to x in X, we have

$$\liminf_{n \to \infty} w_{1,n}(x_n, \varepsilon) \geq w_1(x_n, \varepsilon)$$

(ii) Let $\varepsilon > 0$ and $\eta \geq 0$ then $\liminf_{n \to \infty} M_{1,n}(\varepsilon, \eta) \subset M_1(\varepsilon, \eta)$

Remark 3.5. If moreover X is sequentially compact then we can prove that $\lim_{n \to \infty} v_{1,n}(\varepsilon) = v_1(\varepsilon)$

Proposition 3.6. With the same assumptions of Proposition 3.5 we have: for any $\varepsilon > 0$ and $\eta \geq 0$:

$$\limsup_{n \to \infty} E_n(\varepsilon, \eta) \subset E(\varepsilon, \eta)$$

where $E(\varepsilon, \eta)$ is the set on (η, ε)-Stackelberg equilibrium that is:

$$E(\varepsilon, \eta) = \{ (x, y) \in X \times Y / x \in M_1(\varepsilon, \eta), y \in M_2(x, \varepsilon) \}$$

and
$$E_n(\varepsilon, \eta) = \{ (x, y) \in X \times Y / x \in M_{1,n}(\varepsilon, \eta), y \in M_{2,n}(x, \varepsilon) \}$$

3.4 Existence Results

Proposition 3.7. If all the assumptions of proposition 3.5 are satisfied:

(i) The ε-regularized problem (S_ε) has at least one solution $\overline{x}_\varepsilon$, that is $M_1(\varepsilon, 0)$ is non empty

(ii) If moreover the function $y \to f_1(x, y)$ is sequentially upper semicontinuous on Y, for any $x \in X$, then for any $\varepsilon > 0$: there exists at least one equilibrium pair $(\overline{x}_\varepsilon, \overline{y}_\varepsilon) \in X \times Y$ with $\overline{y}_\varepsilon \in M_2(\overline{x}_\varepsilon, \varepsilon)$ such that

$$f_1(\overline{x}_\varepsilon, \overline{y}_\varepsilon) = \inf_{x \in X} \sup_{y \in M_2(x, \varepsilon)} f_1(x, y)$$

Proof. From propositions 3.1 and 3.5 we know that $w_{1,n}(\cdot, \varepsilon)$ Γ^- sequentially converges to $w_1(\cdot, \varepsilon)$. So from the compactness of X, we can deduce

$$M_1(\varepsilon, 0) \neq \emptyset.$$

Let us consider $\bar{x}_\varepsilon \in M_1(\varepsilon, 0)$. We know that $M_2(x, \varepsilon)$ is sequentially closed in the sequentially compact space Y, for any $x \in X$, $\varepsilon > 0$ and the function $y \to f_2(x, y)$ is sequentially upper semicontinuous of Y for any $x \in X$. So, there exists $\bar{y}_\varepsilon \in M_2(\bar{x}_\varepsilon, \varepsilon)$ such that $w_1(\bar{x}_\varepsilon) = f_1(\bar{x}_\varepsilon, \bar{y}_\varepsilon)$.

The following is an improvement of proposition 3.1.

Proposition 3.8. For $\varepsilon > 0$ we have $\lim\limits_{n \to \infty} v_{1,n}(\varepsilon) = v_1(\varepsilon)$

Remark 3.6. Other improvements about lower semicontinuity by inclusion can be obtained by using similar arguments to the one used in [14].

4. Increasing Sequences

Now, we give sufficient conditions in order that our main assumptions (F_1), $(\bar{F}_2)$, (L_1), (L_2) of section 2 are satisfied.

Proposition 4.1. Let $f_{i,n}$ and f_i, $i = 1, 2$, such that

(4.1) $(f_{i,n})_{n \in N}$ is an increasing sequence and $\sup f_{i,n} = f_i$, $i = 1, 2$

(4.2) the functional $f_{i,n}$ is sequentially lower semicontinuous for $i = 1, 2$ and $n \in N$

(4.3) the functional f_2 is such that for any $(x, y) \in X \times Y$ and any sequence x_n converging to x in X there exists a sequence y_n converging to y in Y such that

$$\limsup_{n \to \infty} f_2(x_n, y_n) \leq f_2(x, y)$$

(4.4) the functional f_1 is such that for any $x \in X$ there exists a sequence x_n converging to x in X such that for any $y \in Y$ and any sequence y_n converging to y in Y we have:

$$\limsup_{n \to \infty} f_1(x_n, y_n) \leq f_1(x, y)$$

then the assumptions (F_1), $(\bar{F}_2)$, (L_1) and (L_2) are satisfied.

Remark 4.1. (4.3) (respectively (4.4)) is satisfied if the functional $x \to f_2(x, y)$ (respect. the functional $y \to f_1(x, y)$) is sequentially upper semicontinuous for any $y \in Y$ (respect. for any $x \in X$).

Definition 4.1. ([16]) $\overline{f}^S$ is called the sequentially lower semicontinuous regularized of f if:

$$\overline{f}^S = \sup \{ g : g \leq f,\ x_n \to x \text{ implies lin inf } g(x_n) \geq g(x) \}$$

Remark 4.2. It is obvious that $f = \overline{f}^S$ if and only if f is sequentially lower semicontinuous.

Proposition 4.1 can be obviously deduced from the following lemmas.

Lemma 4.1. Let f_n be an increasing sequence of functionals defined on $X \times Y$ and valued in $R \cup \{+\infty\}$, then the sequence $(f_n)_{n \in N}$ and $f = \sup_n \overline{f}^S$ satisfy assumption (F1).

Proof. Let $(x, y) \in X \times Y$ and (x_n, y_n) converging to (x, y) from

$$\overline{f}_m^S(x_n, y_n) \leq f_m(x_n, y_n) \leq f_n(x_n, y_n) \text{ for } n \geq m$$

wre obtain

$$\overline{f}_m^S(x_n, y_n) \leq \liminf_{n \to \infty} \overline{f}_m^S(x_n, y_n) \leq \liminf_{n \to \infty} f_m(x_n, y_n) \leq \liminf_{n \to \infty} f_n(x_n, y_n)$$

Hence

$$f(x, y) = \sup_n \overline{f}_m^S(x, y) \leq \liminf_{n \to \infty} f_n(x_n, y_n)$$

Remark 4.3. A topological version of such a result can be found in [1].

Lemma 4.2. Let f_n be an increasing sequence of functionals defined on $X \times Y$ and valued in $R \cup \{+\infty\}$ such that $f = \sup_n f_n$ satisfy (4.3) then the sequence $(f_n)_{n \in N}$ and f satisfy $(\overline{F}_2)$.

Proof. Let $(x, y) \in X \times Y$ and a sequence (x_n) converging to x in X. From (4.3) there exists y_n converging to y in Y such that

$$\limsup_{n \to \infty} f(x_n, y_n) \leq f(x, y)$$

but $f_n(x_n, y_n) \leq f(x_n, y_n)$ for any $n \in N$ then

$$\limsup_{n\to\infty} f_n(x_n, y_n) \le \limsup_{n\to\infty} f(x_n, y_n) \le f(x, y)$$

Lemma 4.3. Let f_n be an increasing sequence of functionals defined on $X \times Y$ and valued in R such that $f = \sup_n f_n$ satisfy (4.4) then the sequence f_n and f satisfy (L_2).

Proof. In the same line of proof of Lemma 4.2.

5. Approximation Method Using External Penality Functions

From now on we consider the following constrained two level problem

$$(S_c) \quad \begin{cases} \operatorname{Inf}_{x \in X} \ \operatorname{Sup}_{y \in M_2(x)} \ell_1(x, y) \\ g_1(x) \le 0 \\ \text{where } M_2(x) \text{ is theset of solutions to the parametrized lower problem:} \\ \quad P(x) \begin{cases} \operatorname{Inf} \ \ell_2(x, y) \\ y \in Y \\ g_2(x, y) \le 0 \end{cases} \end{cases}$$

with ℓ_i and g_i, $i = 1, 2$, real valued functionals defined on $X \times Y$.
For the sake of simplicity we only consider one constraint but we could take into account a finite number of constraints.
Now, we want to approximate problem (S_c) by an unconstrained one. For that, let us first define:

(5.1) $\quad K_2(x) = \{ y \in Y : g_2(x, y) \le 0 \}$

(5.2) $\quad K_1 = \{ x \in X : g_1(x) \le 0 \}$

(5.3) $\quad f_2(x, y) = \ell_2(x, y)$ if $y \in K_2(x)$ and $x \in X$
$\qquad = +\infty$ otherwise

(5.4) $\quad f_2(x, y) = \ell_1(x, y)$ if $x \in K_1$ and $y \in Y$
$\qquad = +\infty$ otherwise.

By using the functionals f_1 and f_2 the problem (S_c) can be rewritten, taking the form (S) of section 2. Let $(r_n)_{n \in N}$ be a sequence of real positive numbers strictly decreasing to 0 with $n \to +\infty$. Let us consider a function q_2 defined on $]0, +\infty[\times X \times Y$ with values in R^+ and satisfying the following conditions.

(5.5) for any $r \in]0, +\infty[$ the function $(x, y) \rightarrow q_2(r, x, y)$ is continuous and non negative

(5.6) $q_2(r, x, y) = 0$ for any $r > 0$ and for any (x, y) such that $y \in K_2(x)$

(5.7) for any (x, y) such that $y \notin K_2(x)$, $x \in X$, the function $r \rightarrow q_2(r, x, y)$ is strictly monotone and increasing to $+\infty$ if $r \rightarrow 0$, uniformly on any compact of $X \times Y$ which do not intercept the graph of the multifunction K_2.

An elementary exemple of such a function q_2 is the following:
$q_2(r, x, y) = 1/r \max(0, g_2(x, y))$. Moreover let q_1 be a function defined on $]0, +\infty[\times X$ with values in R^+ and satisfying the following conditions:

(5.8) for any $r \in]0, +\infty[$ the function $x \rightarrow q_1(r, x)$ is continuous and non negative

(5.9) $q_1(r, x) = 0$ for any $r > 0$ and any $x \in K_1$

(5.10) for any $x \notin K_1$ the function $x \rightarrow q_1(r, x)$ is strictly monotone and increasing to $+\infty$ if $r \rightarrow 0$ uniformly on any compact of X which do not intercept the set K_1.

Finally, let us define

(5.11)
$$f_{2,n}(x, y) = \begin{cases} \ell_2(x, y) + q_2(r_n, x, y) & \text{if } x \in X, y \in Y \\ = +\infty & \text{otherwise} \end{cases}$$

(5.12)
$$f_{1,n}(x, y) = \begin{cases} \ell_1(x, y) + q_1(r, x) & \text{if } x \in X, y \in Y \\ = +\infty & \text{otherwise} \end{cases}$$

then we can define the unconstrained approximate two level problem

$$(S_n) \quad \begin{cases} \inf \ \sup \ f_{1,n}(x, y) \\ x \in X \quad y \in M_{2,n}(x) \\ \text{where } M_{2,n}(x) \text{ is the set of solutions of } P_n(x) : \inf\limits_{y \in Y} f_{2,n}(x, y) \end{cases}$$

It is obvious that $(f_{1,n})_{n \in N}$ is an increasing sequence so, in order to apply the results of section 2 or 3, it is sufficient to verify the assumptions of proposition 4.1.

Lemma 5.1. $f_i = \sup\limits_n f_{i,n} \quad i = 1, 2$

Proof.

1) If $y \in K_2(x)$ and $x \in X$, from (5.6) $q(r_n, x, y) = 0$ then $f_{2,n}(x, y) = \ell_2(x, y)$ for any $n \in N$ and $(\sup_n f_{2,n})(x, y) = \ell_2(x, y) = f_2(x, y)$

If $y \notin K_2(x)$, $x \in X$ $\lim_{n \to \infty} q_2(r_n, x, y) = +\infty$ and $\sup_n f_{2,n}(x, y) = +\infty = f_2(x, y)$

2) If $f_{1,n}(x, y) = \ell_1(x, y)$ then $\sup_n f_{1,n}(x, y) = \ell_1(x, y) = f_1(x, y)$

If $x \notin K_1$, $y \in Y$ $\lim_{n \to \infty} q_1(r_n, x) = +\infty$ and $\sup_n f_{2,n}(x, y) = +\infty = f_1(x, y)$

Proposition 5.2. If moreover the functionals ℓ_i and q_i, $i = 1, 2$ are sequentially continuous, the multifunction K_2 is sequentially lower semicontinuous on X then the functionals $(f_{1,n})_{n \in N}$ and f_i, $i = 1, 2$ defined by (5.3), (5.4), (5.11) and (5.12) satisfy all the assumptions of proposition 4.1.

Proof.

1) Assumption (4.2) is obviously satisfied.

2) f_2 satisfy (4.3). In effect let $(x, y) \in X \times Y$ and x_n converging to x.

3) Let $y \in K_2(x)$ and $x \in X$. From sequentially lower semi continuity of the multifunction K_2 we have $K_2(x) \subset \liminf_{n \to \infty} K_2(x_n)$ for any $x \in X$ and any x_n converging to x. Then there exists $n_0 \in N$ and $y_n \in Y$ such that for $n \geq 0$, $y_n \in K_2(x_n)$ and y_n converges to y. Hence $f_2(x_n, y_n) = \ell_2(x_n, y_n)$. But $\ell_2($ is upper semi continuous then

$\limsup_{n \to \infty} f_2(x_n, y_n) = \limsup_{n \to \infty} \ell_2(x_n, y_n) \leq \ell_2(x, y) = f_2(x, y)$

If $y \notin K_2(x)$ $f_2(x, y) = +\infty$ so for any y_n converging to y:

$\limsup_{n \to \infty} f_2(x_n, y_n) \leq f_2(x, y)$

3) f_1 satisfy (4.4) as it is easy to prove by similar arguments to the ones used for (4.3).

Remark 5.1. So, if moreover we supposed K_1 closed and convex and $y \to f_2(x, y)$ strictly quasi convex and sequentially lower semicontinuous for any $x \in X$ on Y sequentially compact we can applied to this approximation method the results given in proposition 3.4 and 3.5.

Remark 5.2. The considered situation is quite different from the one described in [11]. We have eventually infinite dimensional spaces, even not necessarily satisfying the first axiom of countability, no uniqueness of the solution to the lower level and penalty functions acting simultaneously on the lower and on the upper level. But in [11] the upper level constraints could depend not only of the strategy of the leader but also of the strategy of the follower.

Acknowledgements. One of the authors (J. M.) would like to thank very much Professor Francis Clarke and the CRM for hospitality.

References

[1] **Attouch, H.**, Variational Convergence for Functions and Operators, Pitman, Boston, 1984.

[2] **Avriel M.**, Nonlinear programming: Analysis and Methods, Prentice Hall Inc. Englefood Cliffs, New Jersey, 1976.

[3] **Bagchi A.**, Stackelberg Differential Games in Economic Models, Lecture Notes in Control and Information Sciences, vol. 64, Springer-Verlag, Berlin, 1984.

[4] **Banks B., Klatte J., Kummer B. and Tammer K.**, Nonlinear Parametric Optimization, Birkhauser, Basel, 1983.

[5] **Bard J. F.**, Convex two-Level Optimization, Math. Programming, vol. 40, 1988, pp.15-27.

[6] **Basar T. and Olsder G. J.**, Dymanic Noncooperative Game Theory, Academic Press, New York, 1982.

[7] **Chen C. I. and Cruz J.**, Stackelberg Solution for two-Person Games with biase Information Patterns, IEEE Transactions on Automatic Control, vol. AC-17, 1972, pp.791-797.

[8] **De Giorgi E. and Franzoni T.**, Su un Tipo di Convergenza Variazionale, Rendiconti del Seminario Matematico di Brescia, vol. 3, 1979, pp.63-101.

[9] **Dolecki S.**, Convergence of Minima in Convergence Spaces, Optimization, vol. 17, 1986, pp.553-572.

[10] **Kuratowski C.**, Topology, Academic Press, New York, 1966.

[11] **Loridan P. and Morgan J.**, Approximation Results for a two-Level Optimization Problem and Application to Penalty Methods, Pubblicazioni del Dipartimento di Matematica e Applicazioni, dell'Università di Napoli, Preprint n. 52, 1985.

[12] **Loridan P. and Morgan J.**, A Theoretical Approximation Scheme for Stackelberg Problems. J.O.T.A., to appear in April 1989.

[13] **Loridan P. and Morgan J.**, Approximate Solutions for two-Level Optimization Problems, International Series of Numerical Mathematics, vol. 84, Birkhauser Verlag, Basel, 1988, pp. 181-196.

[14] **Loridan P. and Morgan J.**, New Results on Approximate Solutions in two-Level Optimization, Preprint n. 8, 1988, Pubblicazioni del Dipartimento di Matematica e Applicazioni dell'Università di Napoli.

[15] **Molodtsov D. A. and Fedorov V. V.**, Approximation of two-Person Games with Information Exchange, USSR Computational Mathematics and Mathematical Physics, vol. 13, 1973, pp. 123-142.

[16] **Moscariello G.**, Γ Convergenza negli Spazi Sequenziali, Rend. Accad. Scienze Fis. Mat. Società Nazionale, Scienze Lettere Arti in Napoli, 43, 1976, pp. 333-350.

[17] **Shimizu K. and Aiyoshi E.**, A New Computational Method for Stackelberg and Min-Max Problems by Use of a Penalty Method, IEE Transactions on Automatic Control, vol. AC-26, 1981, pp. 460-466.

[18] **Simaan M. and Cruz J.**, On the Stackelberg Strategy in Nonzero-Sum Games, J.O.T.A., vol. 11, 1973, pp.533-555.

[19] **Simaan M. and Cruz J.**, Additional Aspects of the Stackelberg Strategy in Nonzero sum Games, J.O.T.A., vol. 11, 1973, pp. 613-626.

[20] **Von Stackelberg H.**, The Theory of Market Economy, Oxford University Press, Oxford, 1952.

[21] **Zolezzi T.**, On Convergence of Minima, Bollettino U.M.I., vol. 8, 1973, pp. 246-257.

[22] **Zolezzi T.**, On Stability Analysis in Mathematical Programming, Math. Programming Study, vol. 21, 1984, pp. 227-242.

PROBLEMS OF CONVEX ANALYSIS IN ECONOMIC DYNAMICAL MODELS

E. Pessa - B. Rizzi

When dealing with economic dynamical models, one of the most interesting problems of convex analysis is that of searching for the extremum of a suitable functional J of the form:

(*)
$$J = \int_{t_0}^{t_1} \varphi\,(x(t), u(t))\, dt$$

where $x(t) \in R^n$ is a state vector and $u(t) \in R^n$ a parameter vector, linked together by the dynamical equations:

(1)
$$x(t) = F\,(x(t), u(t))$$

This problem appears in political economy, where sometimes there is the possibility of acting upon the parameters which characterize a given economic system and one must decide which is the right time course of these actions in order to have the maximum effectiveness or to follow some optimality criteria.

Generally speaking, the problem of finding the extremum of (*) under the conditions (1) is a very difficult one; here we will take a particular case, the one of the well-known Goodwin's nonlinear economic dynamical model described by the system of ordinary differential equations (Gandolfo 1977, Goodwin et al. 1984, Pessa & Rizzi 1987):

(2)
$$\begin{cases} \dot{v} = Av - Bvw \\ \dot{w} = -mw + nvw \end{cases}$$

where $v = l/n$, $A = (1/c) - \beta$, $B = 1/ac$

w = real wage rate

l - labor employed

n - labor supply

q = consumption + capital goods

k = real capital stock

$n = n_0 \exp(\beta t)$

a - q/l - const. - labor productivity

c = k/q = const.

whereas $m, \beta > 0$ are suitable parameters.
If we must take into account realistic situations, we will see that only the parameter a can be varied because, acting upon the organization of the production, by introducing improvements in the used techinique, etc., it is possible to obtain some (little) variations of a, whereas the other parameters, in this model, can be viewed only as rigidly fixed; otherwise we would destroy the model.
Of course, if we vary a, we will vary also B so that we can use also this latter as a control parameter; we will suppose that B can undergo only small variations around an equilibrium value B_0, so that we can write:

$$B = B_0 + \mu \tag{3}$$

where:

$$|\mu| \leq M_0 \tag{4}$$

and M_0 is a suitable (small) constant.
One of the most interesting problems for (2) is the one in which one would go from a certain initial state $v(t_0), w(t_0)$ to a given final state $v(t_1), w(t_1)$ in the minimum time; then we have an optimal time control problem where J is given by:

$$J = t_1 - t_0 \tag{5}$$

Also this problem is difficult to solve when (2) is nonlinear, as is the case for Goodwin's model; in order to overcome this difficulty, we will suppose that the variables v, w are very near to the equilibrium point of (2) given by:

$$v_0 = m/n \quad , \quad w_0 = A/B_0 \tag{6}$$

By introducing the new variables:

$$v = v_0 + \xi \quad , \quad w = w_0 + \eta \tag{7}$$

and neglecting the powers higher than the first in ξ, η, μ, we will obtain the linearized approximation:

$$\dot{\xi} = -K_1\eta - K_2\mu , \quad \dot{\eta} = K_3\xi \tag{8}$$

where:

$$K_1 = (mB_0/n), \; K_2 = (mA/nB_0), \; K_3 = (nA/B_0) \tag{9}$$

We can now solve the otimal time control problem for (8) by using the Pontrjagin maximum principle (Pontrjagin et al. 1974, Alexeev et al. 1982); to this end we will introduce the auxiliary functions $\psi_1(t)$, $\psi_2(t)$, which satisfy the system:

(10)
$$\frac{d\psi_i}{dt} = -\sum_{\alpha=1}^{2} \frac{\partial F^\alpha}{\partial x_i} \psi_\alpha \qquad (i=1,2) \quad x_1 = \xi,\, x_2 = \eta.$$

We will now construct the so-called Pontrjagin function:

(11)
$$H = \sum_{\alpha=1}^{2} \psi_\alpha F^\alpha$$

which, in our case, is:

(12)
$$H = -\psi_1(K_1\eta + K_2) + K_3\psi_2\xi$$

Then the equations (1), (10) can be written under the hamiltonian form:

(13)
$$\frac{dx_i}{dt} = \frac{\partial H}{\partial \psi_i}, \quad \frac{d\psi_i}{dt} = -\frac{\partial H}{\partial x_i}$$

For the system (8), we have that $\psi_1(t)$, $\psi_2(t)$ satisfy the equations:

(14)
$$\frac{d\psi_1}{dt} = -K_3\psi_2, \quad \frac{d\psi_2}{dt} = K_1\psi_1.$$

whence:

(15)
$$\psi_1 = A_0 \sin(\omega_0 t - \varphi_0), \quad \omega_0 = \sqrt{K_1K_3}, \qquad A_0,\, \varphi_0 = \text{const.}$$

Then the Pontrjagin function (12) meets its maximum for:

(16)
$$\mu(t) = M_0 \operatorname{sgn}(-\psi_1) = M_0 \operatorname{sgn}(\sin(\omega_0 t + \pi - \varphi_0))$$

According to the Pontrjagin maximum principle this is the optimal control law.
Now, if we put:

(17) $$\xi = \psi_2, \quad \eta = -\psi_1 + \rho_0$$

where:

(18) $$\rho_0 = \begin{cases} (K_2/K_1)M_0 & \text{if} \quad \mu(t) = -M_0 \\ -(K_2/K_1)M_0 & \text{if} \quad \mu(t) = M_0 \end{cases}$$

we will see that (8) becomes identical to (14), so that (17), once known the solutions of (14), give, at the same time, the solutions of (8). By substituting (17) into (12) we will obtain:

(19) $$M = K_1\psi_1^2 + K_2\psi_2^2 \geq 0$$

so that the condition of fundamental Pontrjagin's theorem is satisfied. By solving directly (14) it is easy to see that M = const. Then, we can conclude that the trajectory associated to (16) is optimal.

When $\mu = 0$, i.e. without control parameter, the system (8) has only the equilibrium point (0,0), which is a center. Because $\mu(t)$ has only the values M_0, $-M_0$, we can consider the auxiliary systems:

(20.a) $$\begin{cases} \dot{\xi} = -K_1\eta - K_2M_0 \\ \dot{\eta} = K_3\xi \end{cases}$$

(20.b) $$\begin{cases} \dot{\xi} = -K_1\eta + K_2M_0 \\ \dot{\eta} = K_3\xi \end{cases}$$

The equilibrium point of (20.a) is $P_1 : (0, -(K_2/K_1)M_0)$ and the one of (20.b) is $P_2 : (0, (K_2/K_1)M_0)$; they are both centers. It is, then, easy to see, by observing the phase plane, that, given every pair $x_i(t_0)$, $x_i(t_1)$ $(i=1, 2)$, it is always possible to have an admissible control, piecewise continuous in t, with constant values M_0 and $-M_0$, which let the system go from $x_i(t_0)$ to $x_i(t_1)$. Now, because it is well-known that if there is at least one admissible control process which let the system go from $x_i(t_0)$ to $x_i(t_1)$, then there is one optimal time control process for these latter points and this process is unique, we will obtain immediately the existence theorem:

Theorem: For the system (8) there is a unique optimal time control process which solves the optimal time problem relative to $x_i(t_0)$, $x_i(t_1)$.

We will remark that our solution (16) can be obtained by translating of α_0 the function $\mu(t) = M_0 \operatorname{sgn}(\operatorname{sen}(\omega_0 t))$, where:

(21) $$\alpha_0 = (\varphi_0 - \pi)/\omega_0$$

The values M_0, $-M_0$ are assumed for intervals whose lenght is π/ω_0.

The explicit form of the phase trajectories is easy to find; it is, however, strongly

dependent from the relationship between $x_i(t_o)$, $x_i(t_1)$, P_1, P_2.
These reasonings can be repeated in many other cases; let us consider, for example, the generalization of Goodwin's model proposed by Jarsulic (Jarsulic 1986), where the second of (2) is substituted by:

$$\dot{w} = -mw + nvw - fvw^2 + \epsilon w^2 \tag{22}$$

where f, ϵ are suitable small parameters. By denoting with v_1, w_1 the components of $x_i(t_1)$, we can introduce the variables:

$$v = v_1 + \xi \quad , \quad w = w_1 + \eta \tag{23}$$

If we suppose that v_1, w_1 be very near to an equilibrium point of the generalized Goodwin model and if we choose B and f (supposed very small) as control parameters, we can do an approximate linearization of our system, by obtaining:

$$\begin{aligned} \dot{\xi} &= A\xi - B\, v_1 w_1 \\ \dot{\eta} &= -nw_1 \xi + (2\epsilon w_1 - m - nv_1)\eta - fv_1 w_1^2 \end{aligned} \tag{24}$$

This system belongs to the class:

$$\frac{dx_i}{dt} \; L_{ik}\, x_k + M_{il}\, u_l \qquad (i, k = 1, \dots, n) \qquad (l = 1, \dots, r) \tag{25}$$

In our case the optimal time control problem is, owing to (23), relative to the origin of the phase plane.
Besides the eigenvalues λ of the matrix L_{ik} associated to (24) are real and distinct and have the values:

$$\lambda_1 = A \quad , \quad \lambda_2 = 2\epsilon\, w_1 - m - nv_1 \tag{26}$$

so that, owing to well-known theorems, in order to have an optimal control process, it suffices to have:

$$A < 0 \quad , \quad 2\epsilon\, w_1 - m - nv_1 < 0 \tag{27}$$

i.e.:

$$1/c < \beta \quad , \quad 2\epsilon\, w_1 < m + nv_1 \tag{28}$$

If (28) are satisfied, this optimal time control process is given by a piecewise continuous function, piecewise constant, which has no more than one point of discontinuity of the first kind.

References

/1/ **Alexeev V., Tikhomirov V., Fomine S.**, Commande Optimale, Mir, Moscou (1982).

/2/ **Gandolfo G.**, Metodi di Dinamica Economica, ISEDI, Milano (1977).

/3/ **Goodwin R.M., Kruger M., Vercelli A.**, Nonlinear Models of Fluctuating Growth, Springer, Berlin-Heidelberg-New York (1984).

/4/ **Jarsulic M.**, Growth Cycles in A Classical-Keynesian Model, in W.Semmler (ed.), Competition, Instability, And Nonlinear Cycles, Springer, Berlin-Heidelberger-New York, (1986) pp. 252-268.

/5/ **Pessa E., Rizzi B.**, Modelli dinamici nonlineari e cicli economici, XI Convegno Naz. A.M.A.S.E.S., Aosta 1987.

/6/ **Pontrjagin L., Boltianski V., Gamkrelidze R., Mitchenko E.**, Théorie mathématique des processus optimaux, Mir. Moscou (1974).

RECENT BOUNDS IN CODING USING PROGRAMMING TECHNIQUES

B.K. Dass

The three major problems of coding theory can be listed as follows:

1. How good are the best codes?
2. How can we design good codes?
3. How can we decode such codes?

We shall concentrate on the first problem only in this lecture.
One of the earliest problems which arose in coding theory was that of finding perfect codes. If we view a code of length n over the finite field F_q as a subset $\{ x_1, x_2, \dots, x_M \}$ of the vector space $V_n(F_q)$, the code is said to be perfect (or closed packed) if for some integer e, the Hamming spheres of radius e around the M codewords completely fill $V_n(F_q)$ without overlap. By 1950 several kinds of perfect codes were known:

n	q	e	Codes
2e+1	2 (arbitrary prime power)	(arbitrary)	Repetition codes
$(q^m-1)/(q-1)$	-do-	1	Hamming codes
23	2	3	Binary Golay code
11	3	2	Ternary Golay code

Perfect codes (perhaps with the exception of the repetition codes) are beautiful combinatorial objects with remarkable error-correcting powers, and researchers naturally yearned to extend the above table. However, after a long series of extremely difficult investigations ending in the early 1970's the truth came out: there are no perfect codes except those with the parameters listed above.
Disappointed by the nonexistence of other perfect codes, researchers widened their search to codes that are almost as good: nearly perfect and uniformly packed codes. Over finite fields such codes are now all classified for $e \geq 4$, for $e \leq 3$ only partial results are known.

The existence of the (23,12) perfect Golay code tells us that the maximum number of vectors from $V_{23}(F_2)$ which can be chosen so that each pair of vectors is separated by a Hamming distance ≥ 7 is exactly $4096 = 2^{12}$. It is natural to ask what this maximum number is when 23 is replaced by n and 7 is replaced by d, and so we define

A(n, d) = the largest integer M such that there exist M codewords $\{x_1, \ldots, x_M\}$ from $V_n(F_2)$ such that

(1) $$d_H(x_i, x_j) \geq d \text{ if } i \neq j.$$

Alternatively,

A(n, d) = the largest number of codewords possible in a code of lenght n and minimum distance d.

The study of the numbers A(n, d) is redarded by many as the central problem of coding theory. Although A(n, d) is rarely known exactly unless n and d are relatively small, or $2d \geq n$, a great deal of first-rate research has gone into the problem of finding good upper and lower bounds for A(n, d).
For relatively small values of n and d, essentially the only known way of obtaining lower bounds of the form $A(n, d) \geq M$ is to explicitly exhibit M vectors from $V_n(F_2)$ satisfying the conditions cited in (1).
On the other hand, in order to obtain an upper bound of the form $A(n, d) < M$, it is necessary to show that every subset $\{x_1, \ldots, x_M\}$ of $V_n(F_2)$ contains at least one pair (i, j) such that $d_H(x_i, x_j) < d$. There are a great many possible ways of doing this, but currently the most powerful technique appears to be the linear programming approach, which we now describe.
If $C = \{x_1, \ldots, x_M\}$ is a binary code of lenght n, that is, a subset of $V_n(F_2)$, for each $i = 0, 1, \ldots, n$ and $x \in C$, let $A_i(x)$ be the number of coderwords in C at distance i from x. The distante distribution of C is defined to be the (n + 1)-tuple $(A_0, A_1, \ldots, A_n)$ of nonnegative real numbers, where

$$A_i = (1/M) \sum_{x \in C} A_i(x)$$

Certain elementary properties of the A_i are immediate:

(2) $$A_0 + A_1 + \ldots + A_n = M,$$

(3) $$A_0 = 1.$$

And if the code has minimum distance d, it also follows that

(4) $$A_1 = A_2 = \dots = A_{d-1} = 0.$$

The key to the linear programming approach is the Delsarte-MacWilliams (DM) inequalities: If $P_j(i)$ denotes the coefficient of z^j in the polynomial $(1-z)^i(1+z)^{n-i}$, then the A_i's must satisfy

(5) $$\sum_{i=0}^{n} A_i P_j(i) \geq 0 \, , \, j = 0, 1, \dots, n.$$

These facts suggest the following idea. Let $A_{LP}(n, d)$ denote the maximum possible value of the function $f(A_0, \dots, A_n) = A_0 + A_1 + \dots + A_n$ of $n+1$ nonnegative real numbers, subject to the linear constraints (3)-(5). Then clearly $A(n, d) \leq A_{LP}(n, d)$; this is called the linear programming bound.

At present, the best known upper bounds on $A(n, d)$ for almost every pair (n, d) come from the linear programming bound or a modification of it. For example, the bound on $A(13,6)$ obtained by combining the DM inequalities (5) with the ad hoc inequality $A_{10} + 4A_{12} \leq 4$ (which is easy to obtain combinatorially) is $A(13,6) \leq 32$. For several years before the discovery of the LP approach, $A(13,6)$ was the simplest unknown value of $A(n, d)$, and researchers had struggled with it, using heroically complicated combinatorial arguments, but $32 \leq A(13,6) \leq 35$ was the best they could do. Of course the old lower bound-which arises because ther is a known code, the Nadler code, with $n = 13$, $d = 6$, $M = 32$-combined with the new upper bound, settles the matter: $A(13,6) = 32$.
Perhaps, the best known version of the $A(n, d)$ problem is the asymptotic problem: If $(d_n)_{n=1}^{\infty}$ is a sequence of integers with $d_n/n \to \delta$, $0 \leq \delta \leq 1$, how does $A(n, d_n)$ behave as $n \to \infty$? More formally, since the rate of a code of length n with M codewords is $1/n \log_2 M$, we define

(6) $$\overline{R}(\delta) = \sup \ \overline{\lim_{n\to\infty}} \ 1/n \log_2 A(n, d_n)$$

(7) $$\underline{R}(\delta) = \inf \ \underline{\lim_{n\to\infty}} \ 1/n \log_2 A(n, d_n)$$

where the "sup" in (6) and the "inf" in (7) are both taken over all sequences $(d_n)_{n=1}^{\infty}$ satisfying $d_n/n \to \delta$. It is annoying that both an upper and a lower value of $R(\delta)$ must be defined; everybody believes (but nobody has proved) that $\overline{R}(\delta) = \underline{R}(\delta)$ for all δ.
For simplicity in what follows, however, we shall refer only to $R(\delta)$, it being understood that an upper bound on $R(\delta)$ is an upper bound on $\overline{R}(\delta)$, and lower bound is a lower bound on $\underline{R}(\delta)$.

It is known, and relatively easy to prove, that $R(0) = 1$ and $R(\delta) = 0$ for $1/2 \leq \delta \leq 1$, but $R(\delta)$ is unknown for $0 < \delta < 1/2$. The best known lower bound for $R(\delta)$ in this range is

(8) $$R(\delta) \geq 1 - H_2(\delta)$$

a result proved by Gilbert, [$H_2(p) = -p \log_2 p - (1-p) \log_2(1-p)$].

No one has succeeded in raising Gilbert's lower bound, but there has been a whole series of steadily decreasing upper bounds on $R(\delta)$. The current record-holder, obtained by McEliece, Rodemich, Rumsey and Welch, using the linear programming approach, is

(9) $$R(\delta) \leq \min_{0 \leq \mu \leq 1-2\delta} \{ 1 + g(u^2) - g(u^2 + 2\delta u + 2\delta) \},$$

where $g(x) = H_2[(1 - \sqrt{1-x})/2]$. It is widely conjectured that $R(\delta) = 1 - H_2(\delta)$, but the only real supporting evidence is the fact that over the years the upper bound has been decreasing while the lower bound has remained fixed.

References

/1/ **McEliece R.J.**, The Theory of Information and Coding, Addison-Wesley, 1977, pp.253-256.

LOGICAL ASPECTS CONCERNING SHEPHARD'S AXIOMS OF PRODUCTION THEORY

W. Eichhorn – U. Leopold

Abstract

R.W. Shephard's [1974] eight axioms for production correspondences are consistent but dependent. We show that certain subsets of seven of the eight axioms are independent. This remains true in the case when the stronger versions of (some of) the axioms as formulated in Shepard [1970] are considered.

1. Introduction

Ronald W. Shepard's [1970], [1974] system of axioms for production correspondences is classical by now. The more astonishing it is that up to now nearly nobody has thrown light on the logical structure of this system.

In what follows we are interested in the system of Shephard's eight most general axioms. For six of them Fare [1988] and Eichhorn [1988] proved that they are logically independent, i.e., that none of them is a logical consequence of the remaining ones.

We show among others, that Shephard's eight axioms are dependent, but that certain subsets of seven of the eight axioms are independent.

In the next section, after recalling the notion of a (production) correspondence, we list Shephard's axioms. We point out already here that axioms 7 and 8 have to do with convexity and quasiconcavity of sets of points in the $\mathbb{R}^k_+$ ($\mathbb{R}_+$ the nonnegative real numbers, k a natural number) and of correspondences, respectively.

In Section 3 the consistency of the axioms is simply shown by examples of correspondences that satisfy all of them simultaneously. Examples at the end of Section 3 and in Section 4 will give rise to the proofs of our theorems concerning dependency and independence of Shephard's axioms.

2. The Axioms

Let n and m be two natural numbers. Mappings K of the kind

$$K : \mathbb{R}^n_+ \to \mathcal{P}(\mathbb{R}^m_+)$$

that assign to each vector $x = (x_1, \dots, x_n) \in IR^n_+$ an element of the power set $\mathcal{P}(IR^m_+)$, that is the set of the subsets of IR^m_+, play an important role in the theory of production. Such mappings, often called correspondences, are excellent tools for describing the connections between inputs and outputs when several goods are produced.
For example, a production correspondence assigns to each input vector $x = (x_1, \dots, x_n) \in IR^n_+$ the set of all output vectors $u = (u_1, \dots, u_m) \in IR^m_+$ of which the production unit under consideration can produce (at least) one in unit time by use of x. Like Shephard [1970], [1974] we briefly express this as follows.

$$x \to P(x) = \{ u | \, u \text{ is obtainable by use of } x \text{ in unit time} \}$$

From now on we call a correspondence

$$P : IR^n_+ \to \mathcal{P}(IR^m_+)$$

a production correspondence or a production correspondence in the narrower sense if and only if it satisfies the six or eight axioms A.1 - A.6 or A.1 - A.8 of Shephard [1974, p. 1-14], resp., which we list in what follows.

A.1 Impossibility of paradise, possibility of inaction
$P(0) = \{ 0 \}$, $0 \in P(x)$ for all $x \in IR^n_+$.
In other words, there is no free lunch, i.e., inputs are required to produce outputs. Inaction is possible with every input.

A.2 Impossibility of unbounded output from bounded input
$P(x) = \{ u | \, u \in IR^m_+$ is obtainable by use of $x \in IR^n_+ \}$ is bounded for each fixed input vector $x \in IR^n_+$.

A.3 Weak disposability of inputs
$P(\lambda x) \supseteq P(x)$ for all $\lambda \in [1, \infty)$ and all $x \in IR^n_+$.
In other words, an input vector yielding the output vector u may be scaled upward to yield at least u.

A.4 Weak disposability of outputs
If $u \in P(x)$ then $\mu u \in P(x)$ for all $\mu \in [0, 1]$ and all $x \in IR^n_+$.
In other words, the output vectors u obtainable from an input vector x can be scaled downward to the null output vector.

A.5 Attainability of outputs

If there exists, for the input vector $x \geq 0$[1], an output vector $u \geq 0$ satisfying $u \in P(x)$ then for any $\mu \in (0, \infty)$ there exists $\lambda \in (0, \infty)$ satisfying $\mu u \in P(\lambda x)$.

In other words, any desired proportional output vector variation can be obtained by suitable proportional variation of the input vector.

A.6 Upper semicontinuity

The correspondence P is upper semicontinuous on IR^n_+ .

A correspondence $K: IR^n \to \mathcal{P}(IR^m)$ is called upper semicontinuos at $x^0 \in IR^n$ if for each sequence (x^ν) satisfying $x^\nu \to x^0$ (for $\nu \to +\infty$) and (u^ν) satisfying both $u^\nu \in K x^\nu$ and $u^\nu \to u^0$ (for $\nu \to +\infty$) it follows that $u^0 \in K(x^0)$.

Axiom A.6 is merely a mathematical convenience. It is equivalent with closure of the graph of P:

$$\{ (x, u) \mid x \in IR^n_+ , u \in P(x) \}.$$

In this connection we emphasize that given any correspondence

$$Q: D \subseteq IR^n_+ \to \mathcal{P}(R^m_+),$$

which describes real-world production, there exists an upper semicontinuous correpondence whose restriction on D coincides with Q.

The following two axioms, or rather properties, seem to us more restrictive and/or less natural than axioms A.1- A.6.

A.7 Convexity of the output sets

For each $x \in IR^n_+$ the set P(x) is convex.

In other words, if the output vectors u^1 and u^2 are obtainable by use of the input vector x then every output vector represented by a point on the straight line between u^1 and u^2 can be obtained from x.

A.8 Quasiconcavity of P

$P: IR^n_+ \to \mathcal{P}(IR^m_+)$ is quasiconcave on IR^n_+ ,

i.e., for every fixed $\sigma \in [0, 1]$ and for every pair of vectors $x \in IR^n_+$ and $y \in IR^n_+$,

$$P(\sigma x + (1-\sigma) y) \supseteq P(x) \cap P(y).$$

---

[1] For $a = (a_1, \dots, a_n)$, $b = (b_1, \dots, b_n)$ we define

$(a_1, \dots, a_n) > (b_1, \dots, b_n)$ if and only if $a_1 > b_1, \dots, a_n > b_n$,

$(a_1, \dots, a_n) \geqq (b_1, \dots, b_n)$ if and only if $a_1 \geqq b_1, \dots, a_n \geqq b_n$,

$(a_1, \dots, a_n) \geq (b_1, \dots, b_n)$ if and only if $a \geqq b$, $a \neq b$.

We notice that this relation can be written $F(\sigma x+ (1-\sigma) y) \geq \min (F(x), F(y))$ in this case where P is a real-valued quasiconcave function F.
It is well known (see Shephard [1970, p. 182]) that axiom A.8 is synonymous with

A.8* Convexity of the inputs sets
For each $u \in IR^m_+$ the inverse correspondence $P^{-1} =: L$ given by
$L(u) := \{x \mid u \in IR^m_+$ is obtainable by use of $x \in IR^n_+ \}$ is convex.

3. The Consistency of the Axioms

The following examples of correspondences P satisfy axioms A.1 - A.8. Hence, the axioms are consistent and the existence of production correspondences is proven; see our definition in Section 2.

$$P(x) = \{ u \mid u_1+ ... + u_m \leq x_1+ ... + x_n \},$$

$P(x) = \{ u \mid u_1 \leq F_1(x), ..., u_m \leq F_m(x);\ F_j: IR^n_+ \rightarrow IR_+,\ F_j(0) = 0,\ F_j$ continuous, strictly increasing and quasiconcave; $F_j(\lambda x) = \lambda F_j(x)$ for all $\lambda \in IR_+$, $j=1, ..., m \}$

we point out here that in the second example quasiconcavityof the F_j is needed in order to meet axiom A.8, i.e., A.8*.
The 'level sets' $\{ x \mid F_j(x) \geq u_j\}$ are convex for all $u_j \in IR_+$ if and only if F_j is quasiconcave; see, e.g., Shephard [1974, p. 297]. Note that a continuous, strictly increasing and lineraly homogeneous function $F: IR^n_+ \rightarrow IR_+$, e.g.,

$$F(x) = | x | = (x^2_1 + ... + x^2_n)^{1/2},$$

is not necessarily quasiconcave.
Since $F(x) = | x |$ is not quasiconcave, the correspondence

$$P(x) = \{ u \mid u_1 \leq | x |, ..., u_m \leq | x | \}$$

does not satisfy axiom A.8, i.e., A.8*. Hence it fails to be a production correspondence in the narrower sense according to our definition in Section 2. But since it satisfies axioms A.1 - A.7, it is a production correspondence.
Our preceding considerations can also be formulated as follows. Axioms A.8 is not a logical consequence from axioms A.1 - A.7, i.e., axiom A.8 is independent of axioms A.1 - A.7.
The following example shows that axiom A.7 is independent of axioms A.1 - A.6 and A.8.

The production correspondence P given by

$$P(x) = \{ u \mid u_1 + \dots + u_m \le x_1 + \dots + x_n ,\ u_1 u_2 \dots u_m \le (x_1 + \dots + x_n)/2n \}$$

does not satisfy A.7, but obviuosly A.1 - A.6 and also A.8, since

$$P^{-1}(u) = \{ x \mid x_1 + \dots + x_n \ge u_1 + \dots + u_m ,\ \ge (x_1 + \dots + x_n) / 2n\, u_1 u_2 \dots u_m \}$$

is convex for each u.
We have proved

Theorem 1. Axioms A.1 - A.8 are consistent. Axiom A.8 is independent of axioms A.1 - A.7. Axiom A.7 is independent of axioms A.1 - A.6 and A.8.

4. Are the Axioms Independent?

We point out here that Shephard's axioms A.1 - A.8 are not (logically) independent. For instance, it is obvious that axioms A.1 and A.7 imply axiom A.4.
But what we can show is:

Theorem 2. Axioms A.1 - A.6 and A.8 are independent. In a theory based on these axioms property A.7 is not necessarily fulfilled.

Theorem 3. Axioms A.1 - A.3, A.5 - A.8 are independent. These axioms imply axiom A.4.

As we have already seen the second statements of these theorems are true. To prove the first assertion it remains to show that any one of the axioms A.i (i = 1, 2, 3, 4, 5, 6, 8) or, in case of Theorem 3, A.j (j = 1, 2, 3, 5, 6, 7, 8) does not follow from the axioms A.k ($k \neq j$) respectively.
We do this by presenting examples of correspondences K_i or K_j, o respectively, that satisfy axioms A.k (k = 1, ... , 8) with the exception of A.i or A.j, respectively. The, axiom A.i or A.j cannot be a logical consequence from axioms A.k. (Would it be such a consequence satisfying these axioms also satisfyied axiom A.i or A.j, respectively).
In Section 3 we have already shown two correspondences, where the first one does not fulfill axiom A.8, but axioms A.1 - A.7, and the second one does fulfill axiom A.7, but axioms A.1 - A.6, A.8.
What remains is to present examples $K_1, \dots, K_6$.

$$K_1(x) = \{ u \mid u_1 + \dots + u_m \le G_1(x) = x_1 + \dots + x_n + 1 \}$$

does not satisfy A.1, since $(1/m, \dots, 1/m) \in K_1(0)$,

$$K_2(x) = \{ u \mid u_1 + \dots + u_m \le G_2(x) = \begin{cases} x_1 + \dots + x_n \text{ for } x_1 + \dots + x_n < 1 \\ \infty \text{ otherwise} \end{cases} \}$$

does not satisfy A.2, because $K_2(x)$ is not bounded for x fulfilling $x_1 + \dots + x_n \ge 1$,

$$K_3(x) = \{ u \mid u_1 + \dots + u_m \le G_3(x) = (\Sigma x_j - 1)(\Sigma x_j - 2)^2 + 4 \}$$

does not satisfy A.3 for $x = (4/3n, \dots, 4/3n)$ and $\lambda = 3/2$,

$$K_5(x) = \{ u \mid u_1 + \dots + u_m \le G_5(x) = \begin{cases} x_1 + \dots + x_n \text{ for } x_1 + \dots + x_n \le 1 \\ \infty \text{ otherwise} \end{cases} \}$$

does not satisfy A.5, since there does not exist any vector x such that vectors u fulfilling $u_1 + \dots + u_m > 1$ are contained in $K_5(x)$.

It is easy to show that these correspondences K_i $(i = 1, 2, 3, 5)$ satisfy axioms A.k $(k \ne i;\ k = 1, \dots, 8)$.

$$K_6(x) = \begin{cases} \{ 0 \} & \text{for } x = 0, \\ \{ u \mid u_1 + \dots + u_m < x_1 + \dots + x_n \} & \text{otherwise} \end{cases}$$

satisfies axioms A.1 - A.5, A.7, A.8, but not A.6, since the graph of K_6 is not closed. This concludes the proof of Theorem 3.

In order to complete the proof of Theorem 2 we still have to present an example K_4.

$$K_4(x) = \begin{cases} \{ 0 \} & \text{for } x_1 + \dots + x_n < 1, \\ \{ u \mid u_1 + \dots + u_m = 2 \} \cup \{ 0 \} & \text{for } 1 \le x_1 + \dots + x_n < 2, \\ \{ u \mid u_1 + \dots + u_m \le x_1 + \dots + x_n \} & \text{for } 2 \le x_1 + \dots + x_n \end{cases}$$

does not satisfy A.4, since $(2/m, \dots, 2/m) \in K_4(1/n, \dots, 1/n)$, but $1/2(2/m, \dots, 2/m) \notin K_4(1/n, \dots, 1/n)$. All the other axioms with the exception of axiom A.7 are fulfilled.

A final remark should be made on free disposability (of inputs and outputs) which is also considered in Shephard [1970, p.185f], [1974, p.4]. Free disposability of inputs and outputs means stronger versions of axioms A.3 and A.4 as follows.

A.3* Free disposability of inputs

If $x' \ge x$, $P(x') \supseteq P(x)$.

A.4* Free disposability of outputs
$u \in P(x)$ implies $\{ v \mid 0 \leq v \leq u \} \subseteq P(x)$.

As can also be shown by our examples, Theorem 2 remains true whenever A.3, A.4 are replaced by A.3*, A.4*. The same can be said if A.3* stands for A.3 in Theorem 3.

References

/1/ **Eichhorn W.**, Unabfangigkeit der Shephardschen Axiome, in "Statistik, Informatik und Okonomie, edited by W.Janko. Springer-Verlag, Berlin-Heidelberg-New York, 1988, pp.49-54.

/2/ **Fare R.**, Fundamentals of Production Theory. Lecture Notes in Economics and Mathematical Systems, vol.311. Springer-Verlag, Berlin-Heidelberg-New York 1988.

/3/ **Shephard R.W.**, Theory of Cost and Production Functions. Princeton University Press, Princeton 1970.

/4/ **Shephard R.W.**, Indirect Production Functions. Verlag Anton Hain, Meisenheim 1974.

Contributing Authors

BECTOR C.R., Department of Actuarial and Management Sciences - The University of Manitoba - Winnipeg, Manitoba R 3T 2Z5 - Canada.

BENADADA Y., Department of Computer Science and Operations Research - University of Montreal - Canada.

CAMBINI A., Department of Statistics and Applied Mathematics - University of Pisa, Via Ridolfi 10, 56100 Pisa - Italy.

CASTAGNOLI E., Institute of Quantitative Methods - Bocconi University, Via Sarfatti 25 - 20136 Milano - Italy.

CHANDRA S., Department of Mathematics - Indian Institute of Technology, Hanz Khas, New Delhi - 110016 India.

CROUZEIX J.-P., Department of Applied Mathematics - University of Clermont II - France.

DASS B.K., F-23 Model Town 110009 Delhi - India.

EICHHORN W., Institute of Operations Research-University of Karlsruhe, D-7500 Karlsruhe Federal Republic of Germany.

ELSTER K.H., Department of Mathematics - Technische Hochschule Ilmenau - DDR-6300 Ilmenau.

FERLAND J.A., Department of Computer Science and Operations Research - University of Montreal - Canada.

HACKMAN S.T., School of Industrial and Systems Engineering-Georgia Institute of Technology , Atlanta - Georgia - USA.

LEOPOLD U., Department of Statistics - University of Graz - Federal Republic of Germany.

LORIDAN P., University of Bourgogne - Laboratoire d'Analyse Numerique, B.P.138 - 21004, Dijon Cedex - France.

MARTEIN L., Department of Statistics and Applied Mathematics - University of Pisa, Via Ridolfi 10, 56100 Pisa - Italy.

MARTINEZ-LEGAZ J.E., Department of Applied Mathematics and Analysis - University of Barcelona - Spain.

MAZZOLENI P., Institute of Mathematics-- University of Verona, Via dell'Artigliere 10, 37129 Verona - Italy.

MONTRUCCHIO L., Institute of Mathematics - Politecnico di Torino, Corso Duca degli Abruzzi 24, 10129 Torino - Italy.

MORGAN J., Department of Mathematics and Applications - University of Napoli, Via Mezzocannone 8, 80134 Napoli - Italy.

PASSY U., Faculty of Industrial Engineering and Management - Technion - Israel Institute of Technology - Haifa - Israel.

PATKAR V., Bombay Metropolitan Region Development Authority, Griha Nirman Bhavan, Bombay - 400051 India.

PECCATI L., Institute of Mathematics - Facoltà di Economia e Commercio, Via Assarotti 3, 10122 Torino - Italy.

PENOT J.P., U.A. CNRS n°1024, Faculty of Sciences - Avenue de l'Université, 6400 Pau, France.

PESSA E., Department of Mathematics "G. Castelnuovo", Piazzale Aldo Moro 2, 00185 Roma, Italy.

RIZZI B., Department of Mathematics and Statistics - University of Napoli, Via Mezzocannone 8, 80134 Napoli - Italy.

SCHAIBLE S., Graduate School of Management- University of California - Riverside, California 92521 - USA.

SIDERI E., Department of Mathematics - University of Genova, Via L.B. Alberti 4, 16132 Genova - Italy.

SINGH C., Department of Mathematics - St. Lawrence University - Canton, New York 13617 - USA.

SODINI C., Department of Statistics and Applied Mathematics - University of Pisa, Via Ridolfi 10, 56100 Pisa - Italy.

STANCU-MINASIAN I.M., Economic Cybernetics Department Academy of Economic Studies, Bucharest - 71131 - Romania.

TIGAN S., Territorial Computing Center - St. Republicii 107, 3400 Cluy-Napoca - Romania.

VLACH M., Department of Computer Science and Operations Research - Faculty of Mathematics and Phisics - Charles University, 118.00 Prague 1 - Czechoslovakia.

VOLLE M., Faculty of Sciences - 33 Rue L. Pasteur, 84000 Avignon - France.

WARD D., Department of Mathematics and Statistics - Miami University - Oxford, Ohio 45056-1641 - USA.

WOLF A., Department of Mathematics - Technische Hochschule Ilmenau - DDR-6300 Ilmenau.

Vol. 236: G. Gandolfo, P.C. Padoan, A Disequilibrium Model of Real and Financial Accumulation in an Open Economy. VI, 172 pages. 1984.

Vol. 237: Misspecification Analysis. Proceedings, 1983. Edited by T.K. Dijkstra. V, 129 pages. 1984.

Vol. 238: W. Domschke, A. Drexl, Location and Layout Planning. IV, 134 pages. 1985.

Vol. 239: Microeconomic Models of Housing Markets. Edited by K. Stahl. VII, 197 pages. 1985.

Vol. 240: Contributions to Operations Research. Proceedings, 1984. Edited by K. Neumann and D. Pallaschke. V, 190 pages. 1985.

Vol. 241: U. Wittmann, Das Konzept rationaler Preiserwartungen. XI, 310 Seiten. 1985.

Vol. 242: Decision Making with Multiple Objectives. Proceedings, 1984. Edited by Y.Y. Haimes and V. Chankong. XI, 571 pages. 1985.

Vol. 243: Integer Programming and Related Areas. A Classified Bibliography 1981–1984. Edited by R. von Randow. XX, 386 pages. 1985.

Vol. 244: Advances in Equilibrium Theory. Proceedings, 1984. Edited by C.D. Aliprantis, O. Burkinshaw and N.J. Rothman. II, 235 pages. 1985.

Vol. 245: J.E.M. Wilhelm, Arbitrage Theory. VII, 114 pages. 1985.

Vol. 246: P.W. Otter, Dynamic Feature Space Modelling, Filtering and Self-Tuning Control of Stochastic Systems. XIV, 177 pages. 1985.

Vol. 247: Optimization and Discrete Choice in Urban Systems. Proceedings, 1983. Edited by B.G. Hutchinson, P. Nijkamp and M. Batty. VI, 371 pages. 1985.

Vol. 248: Plural Rationality and Interactive Decision Processes. Proceedings, 1984. Edited by M. Grauer, M. Thompson and A.P. Wierzbicki. VI, 354 pages. 1985.

Vol. 249: Spatial Price Equilibrium: Advances in Theory, Computation and Application. Proceedings, 1984. Edited by P. T. Harker. VII, 277 pages. 1985.

Vol. 250: M. Roubens, Ph. Vincke, Preference Modelling. VIII, 94 pages. 1985.

Vol. 251: Input-Output Modeling. Proceedings, 1984. Edited by A. Smyshlyaev. VI, 261 pages. 1985.

Vol. 252: A. Birolini, On the Use of Stochastic Processes in Modeling Reliability Problems. VI, 105 pages. 1985.

Vol. 253: C. Withagen, Economic Theory and International Trade in Natural Exhaustible Resources. VI, 172 pages. 1985.

Vol. 254: S. Müller, Arbitrage Pricing of Contingent Claims. VIII, 151 pages. 1985.

Vol. 255: Nondifferentiable Optimization: Motivations and Applications. Proceedings, 1984. Edited by V.F. Demyanov and D. Pallaschke. VI, 350 pages. 1985.

Vol. 256: Convexity and Duality in Optimization. Proceedings, 1984. Edited by J. Ponstein. V, 142 pages. 1985.

Vol. 257: Dynamics of Macrosystems. Proceedings, 1984. Edited by J.-P. Aubin, D. Saari and K. Sigmund. VI, 280 pages. 1985.

Vol. 258: H. Funke, Eine allgemeine Theorie der Polypol- und Oligopolpreisbildung. III, 237 pages. 1985.

Vol. 259: Infinite Programming. Proceedings, 1984. Edited by E.J. Anderson and A.B. Philpott. XIV, 244 pages. 1985.

Vol. 260: H.-J. Kruse, Degeneracy Graphs and the Neighbourhood Problem. VIII, 128 pages. 1986.

Vol. 261: Th.R. Gulledge, Jr., N.K. Womer, The Economics of Made-to-Order Production. VI, 134 pages. 1986.

Vol. 262: H.U. Buhl, A Neo-Classical Theory of Distribution and Wealth. V, 146 pages. 1986.

Vol. 263: M. Schäfer, Resource Extraction and Market Structure. XI, 154 pages. 1986.

Vol. 264: Models of Economic Dynamics. Proceedings, 1983. Edited by H.F. Sonnenschein. VII, 212 pages. 1986.

Vol. 265: Dynamic Games and Applications in Economics. Edited by T. Başar. IX, 288 pages. 1986.

Vol. 266: Multi-Stage Production Planning and Inventory Control. Edited by S. Axsäter, Ch. Schneeweiss and E. Silver. V, 264 pages. 1986.

Vol. 267: R. Bemelmans, The Capacity Aspect of Inventories. IX, 165 pages. 1986.

Vol. 268: V. Firchau, Information Evaluation in Capital Markets. VII, 103 pages. 1986.

Vol. 269: A. Borglin, H. Keiding, Optimality in Infinite Horizon Economies. VI, 180 pages. 1986.

Vol. 270: Technological Change, Employment and Spatial Dynamics. Proceedings 1985. Edited by P. Nijkamp. VII, 466 pages. 1986.

Vol. 271: C. Hildreth, The Cowles Commission in Chicago, 1939–1955. V, 176 pages. 1986.

Vol. 272: G. Clemenz, Credit Markets with Asymmetric Information. VIII, 212 pages. 1986.

Vol. 273: Large-Scale Modelling and Interactive Decision Analysis. Proceedings, 1985. Edited by G. Fandel, M. Grauer, A. Kurzhanski and A.P. Wierzbicki. VII, 363 pages. 1986.

Vol. 274: W.K. Klein Haneveld, Duality in Stochastic Linear and Dynamic Programming. VII, 295 pages. 1986.

Vol. 275: Competition, Instability, and Nonlinear Cycles. Proceedings, 1985. Edited by W. Semmler. XII, 340 pages. 1986.

Vol. 276: M.R. Baye, D.A. Black, Consumer Behavior, Cost of Living Measures, and the Income Tax. VII, 119 pages. 1986.

Vol. 277: Studies in Austrian Capital Theory, Investment and Time. Edited by M. Faber. VI, 317 pages. 1986.

Vol. 278: W.E. Diewert, The Measurement of the Economic Benefits of Infrastructure Services. V, 202 pages. 1986.

Vol. 279: H.-J. Büttler, G. Frei and B. Schips, Estimation of Disequilibrium Models. VI, 114 pages. 1986.

Vol. 280: H.T. Lau, Combinatorial Heuristic Algorithms with FORTRAN. VII, 126 pages. 1986.

Vol. 281: Ch.-L. Hwang, M.-J. Lin, Group Decision Making under Multiple Criteria. XI, 400 pages. 1987.

Vol. 282: K. Schittkowski, More Test Examples for Nonlinear Programming Codes. V, 261 pages. 1987.

Vol. 283: G. Gabisch, H.-W. Lorenz, Business Cycle Theory. VII, 229 pages. 1987.

Vol. 284: H. Lütkepohl, Forecasting Aggregated Vector ARMA Processes. X, 323 pages. 1987.

Vol. 285: Toward Interactive and Intelligent Decision Support Systems. Volume 1. Proceedings, 1986. Edited by Y. Sawaragi, K. Inoue and H. Nakayama. XII, 445 pages. 1987.

Vol. 286: Toward Interactive and Intelligent Decision Support Systems. Volume 2. Proceedings, 1986. Edited by Y. Sawaragi, K. Inoue and H. Nakayama. XII, 450 pages. 1987.

Vol. 287: Dynamical Systems. Proceedings, 1985. Edited by A.B. Kurzhanski and K. Sigmund. VI, 215 pages. 1987.

Vol. 288: G.D. Rudebusch, The Estimation of Macroeconomic Disequilibrium Models with Regime Classification Information. VII, 128 pages. 1987.

Vol. 289: B.R. Meijboom, Planning in Decentralized Firms. X, 168 pages. 1987.

Vol. 290: D.A. Carlson, A. Haurie, Infinite Horizon Optimal Control. XI, 254 pages. 1987.

Vol. 291: N. Takahashi, Design of Adaptive Organizations. VI, 140 pages. 1987.

Vol. 292: I. Tchijov, L. Tomaszewicz (Eds.), Input-Output Modeling. Proceedings, 1985. VI, 195 pages. 1987.

Vol. 293: D. Batten, J. Casti, B. Johansson (Eds.), Economic Evolution and Structural Adjustment. Proceedings, 1985. VI, 382 pages. 1987.

Vol. 294: J. Jahn, W. Krabs (Eds.), Recent Advances and Historical Development of Vector Optimization. VII, 405 pages. 1987.

Vol. 295: H. Meister, The Purification Problem for Constrained Games with Incomplete Information. X, 127 pages. 1987.

Vol. 296: A. Börsch-Supan, Econometric Analysis of Discrete Choice. VIII, 211 pages. 1987.

Vol. 297: V. Fedorov, H. Läuter (Eds.), Model-Oriented Data Analysis. Proceedings, 1987. VI, 239 pages. 1988.

Vol. 298: S.H. Chew, Q. Zheng, Integral Global Optimization. VII, 179 pages. 1988.

Vol. 299: K. Marti, Descent Directions and Efficient Solutions in Discretely Distributed Stochastic Programs. XIV, 178 pages. 1988.

Vol. 300: U. Derigs, Programming in Networks and Graphs. XI, 315 pages. 1988.

Vol. 301: J. Kacprzyk, M. Roubens (Eds.), Non-Conventional Preference Relations in Decision Making. VII, 155 pages. 1988.

Vol. 302: H.A. Eiselt, G. Pederzoli (Eds.), Advances in Optimization and Control. Proceedings, 1986. VIII, 372 pages. 1988.

Vol. 303: F.X. Diebold, Empirical Modeling of Exchange Rate Dynamics. VII, 143 pages. 1988.

Vol. 304: A. Kurzhanski, K. Neumann, D. Pallaschke (Eds.), Optimization, Parallel Processing and Applications. Proceedings, 1987. VI, 292 pages. 1988.

Vol. 305: G.-J.C.Th. van Schijndel, Dynamic Firm and Investor Behaviour under Progressive Personal Taxation. X, 215 pages. 1988.

Vol. 306: Ch. Klein, A Static Microeconomic Model of Pure Competition. VIII, 139 pages. 1988.

Vol. 307: T.K. Dijkstra (Ed.), On Model Uncertainty and its Statistical Implications. VII, 138 pages. 1988.

Vol. 308: J.R. Daduna, A. Wren (Eds.), Computer-Aided Transit Scheduling. VIII, 339 pages. 1988.

Vol. 309: G. Ricci, K. Velupillai (Eds.), Growth Cycles and Multisectoral Economics: the Goodwin Tradition. III, 126 pages. 1988.

Vol. 310: J. Kacprzyk, M. Fedrizzi (Eds.), Combining Fuzzy Imprecision with Probabilistic Uncertainty in Decision Making. IX, 399 pages. 1988.

Vol. 311: R. Färe, Fundamentals of Production Theory. IX, 163 pages. 1988.

Vol. 312: J. Krishnakumar, Estimation of Simultaneous Equation Models with Error Components Structure. X, 357 pages. 1988.

Vol. 313: W. Jammernegg, Sequential Binary Investment Decisions. VI, 156 pages. 1988.

Vol. 314: R. Tietz, W. Albers, R. Selten (Eds.), Bounded Rational Behavior in Experimental Games and Markets. VI, 368 pages. 1988.

Vol. 315: I. Orishimo, G.J.D. Hewings, P. Nijkamp (Eds.), Information Technology: Social and Spatial Perspectives. Proceedings, 1986. VI, 268 pages. 1988.

Vol. 316: R.L. Basmann, D.J. Slottje, K. Hayes, J.D. Johnson, D.J. Molina, The Generalized Fechner-Thurstone Direct Utility Function and Some of its Uses. VIII, 159 pages. 1988.

Vol. 317: L. Bianco, A. La Bella (Eds.), Freight Transport Planning and Logistics. Proceedings, 1987. X, 568 pages. 1988.

Vol. 318: T. Doup, Simplicial Algorithms on the Simplotope. VIII, 262 pages. 1988.

Vol. 319: D.T. Luc, Theory of Vector Optimization. VIII, 173 pages. 1989.

Vol. 320: D. van der Wijst, Financial Structure in Small Busine VII, 181 pages. 1989.

Vol. 321: M. Di Matteo, R.M. Goodwin, A. Vercelli (Eds.), Te nological and Social Factors in Long Term Fluctuations. Proce ings. IX, 442 pages. 1989.

Vol. 322: T. Kollintzas (Ed.), The Rational Expectations Equilibri Inventory Model. XI, 269 pages. 1989.

Vol. 323: M.B.M. de Koster, Capacity Oriented Analysis and Des of Production Systems. XII, 245 pages. 1989.

Vol. 324: I.M. Bomze, B.M. Pötscher, Game Theoretical Foun tions of Evolutionary Stability. VI, 145 pages. 1989.

Vol. 325: P. Ferri, E. Greenberg, The Labor Market and Busine Cycle Theories. X, 183 pages. 1989.

Vol. 326: Ch. Sauer, Alternative Theories of Output, Unemployme and Inflation in Germany: 1960–1985. XIII, 206 pages. 1989.

Vol. 327: M. Tawada, Production Structure and International Trac V, 132 pages. 1989.

Vol. 328: W. Güth, B. Kalkofen, Unique Solutions for Strateg Games. VII, 200 pages. 1989.

Vol. 329: G. Tillmann, Equity, Incentives, and Taxation. VI, 1 pages. 1989.

Vol. 330: P.M. Kort, Optimal Dynamic Investment Policies of a Val Maximizing Firm. VII, 185 pages. 1989.

Vol. 331: A. Lewandowski, A.P. Wierzbicki (Eds.), Aspiration Bas Decision Support Systems. X, 400 pages. 1989.

Vol. 332: T.R. Gulledge, Jr., L.A. Litteral (Eds.), Cost Analys Applications of Economics and Operations Research. Proceeding VII, 422 pages. 1989.

Vol. 333: N. Dellaert, Production to Order. VII, 158 pages. 1989.

Vol. 334: H.-W. Lorenz, Nonlinear Dynamical Economics and Cha tic Motion. XI, 248 pages. 1989.

Vol. 335: A. G. Lockett, G. Islei (Eds.), Improving Decision Making in Org nisations. Proceedings. IX, 606 pages. 1989.

Vol. 336: T. Puu, Nonlinear Economic Dynamics. VII, 119 pages. 1989.

Vol. 337: A. Lewandowski, I. Stanchev (Eds.), Methodology and Softwa for Interactive Decision Support. VIII, 309 pages. 1989.

Vol. 338: J.K. Ho, R.P. Sundarraj, DECOMP: an Implementation Dantzig-Wolfe Decomposition for Linear Programming. VI, 206 page 1989.

Vol. 339: J. Terceiro Lomba, Estimation of Dynamic Econometric Mode with Errors in Variables. VIII, 116 pages. 1990.

Vol. 340: T. Vasko, R. Ayres, L. Fontvieille (Eds.), Life Cycles and Long Waves. XIV, 293 pages. 1990.

Vol. 341: G.R. Uhlich, Descriptive Theories of Bargaining. IX, 165 page 1990.

Vol. 342: K. Okuguchi, F. Szidarovszky, The Theory of Oligopoly with Mul Product Firms. V, 167 pages. 1990.

Vol. 343: C. Chiarella, The Elements of a Nonlinear Theory of Economi Dynamics. IX, 149 pages. 1990.

Vol. 344: K. Neumann, Stochastic Project Networks. XI, 237 pages. 1990.

Vol. 345: A. Cambini, E. Castagnoli, L. Martein, P. Mazzoleni, S. Schaibl (Eds.), Generalized Convexity and Fractional Programming with Economi Applications. Proceedings, 1988. VII, 361 pages. 1990.

GPSR Compliance

The European Union's (EU) General Product Safety Regulation (GPSR) is a set of rules that requires consumer products to be safe and our obligations to ensure this.

If you have any concerns about our products, you can contact us on ProductSafety@springernature.com

In case Publisher is established outside the EU, the EU authorized representative is:

Springer Nature Customer Service Center GmbH
Europaplatz 3
69115 Heidelberg, Germany

Batch number: 10712228

Printed by Printforce, the Netherlands